Supported by the National Fund for
Academic Publication in Science and Technology

The Boletes of China: *Tylopilus* s.l.

Yan-Chun LI · Zhu-Liang YANG

Yan-Chun LI · Zhu-Liang YANG
The Boletes of China: *Tylopilus* s.l.

Authors' address: Key Laboratory for Plant Diversity and Biogeography of East Asia, Chinese Academy of Sciences, and Yunnan Key Laboratory for Fungal Diversity and Green Development, Kunming Institute of Botany, Chinese Academy of Sciences, Lanhei Road #132, Heilongtan, Kunming 650201, Yunnan, China.
Email addresses: liyanch@mail.kib.ac.cn, fungi@mail.kib.ac.cn

Brief introduction to the book
China is a diversity hotspot for fungi of the Boletaceae family (boletes). Many boletes have been traditionally treated as members of the genus *Tylopilus* based on the color of the hymenophores or the spore prints. The aims of this book are to delimit and recognize the diversity both of the genera and species of *Tylopilus* s.l. in China. Our morphological and multi-locus phylogenetic studies reveal that the traditionally defined *Tylopilus* species cluster in 19 genera in Boletaceae, including two new genera, viz. *Abtylopilus* Yan C. Li & Zhu L. Yang and *Anthracoporus* Yan C. Li & Zhu L. Yang and 17 known genera, viz. *Austroboletus* (Corner) Wolfe, *Chiua* Yan C. Li & Zhu L. Yang, *Fistulinella* Henn., *Harrya* Halling, Nuhn & Osmundson, *Hymenoboletus* Yan C. Li & Zhu L. Yang, *Indoporus* A. Parihar, K. Das, Hembrom & Vizzini, *Leccinellum* Bresinsky & Manfr. Binder, *Mucilopilus* Wolfe, *Porphyrellus* E.-J. Gilbert, *Pseudoaustroboletus* Yan C. Li & Zhu L. Yang, *Retiboletus* Manfr. Binder & Bresinsky, *Royoungia* Castellano, Trappe & Malajczuk, *Sutorius* Halling, Nuhn & N.A. Fechner, *Tylocinum* Yan C. Li & Zhu L. Yang, *Tylopilus* P. Karst., *Veloporphyrellus* L.D. Gómez & Singer, and *Zangia* Yan C. Li & Zhu L. Yang. A total of 19 genera and 105 species including 34 new species and four new combinations are documented and illustrated in this monographic work. This is a reprint of the 2021 edition (Li and Yang 2021), with minor revisions, in accordance with the relevant contract between Springer and Science Press.

Map Content Approval Number: GS (2021) 6178
ISBN 978-7-03-072411-3
© Science Press 2022, Beijing

Preface

The boletes, in particular the fleshy-pored boletes, are a conspicuous element of the summer and fall mushroom flora of China. Many are very attractive because of their bright colors or curious ornamentation in the form of scales on the pileus or reticulum or other ornamentation on the stipe. In addition, many species stain blue immediately when the surface of the tubes or the context is injured, and this never fails to excite the interest of the collectors. Although boletes are found in forested areas throughout the world, nowhere is species-richer than China, which also has many endemic species. Over 300 species in the family Boletaceae have been reported from China over the past few decades.

A large number of the Chinese species of boletes have white, pinkish, purplish pink, reddish, reddish brown, or brownish hymenophores or spore prints. Most of these species were traditionally placed in the genus *Tylopilus*. These species are not only of great ecological and economic values but are also of scientific interest. They are very diverse in morphology, complex in structure, and occupy diverse ecological niches. Our present and previous molecular phylogenetic studies indicate that *Tylopilus* is polyphyletic and many new genera belonging to divergent clades are based on species formerly placed in this genus. Although the color of the hymenophores, or more exactly, the color of the spore prints, traditionally distinguished the major groups of boletes, it is not enough in the modern taxonomy of boletes. A combination of several morphological characteristics, including the color of spore print, the structure of pileipellis, the morphology of stipe, the ornamentation of basidiospore, the discoloration of the context, and the host plants, has been found to be a minimum for the taxonomy of boletes.

The objective of this book is to provide a monograph to those boletes from China that have pallid to pinkish, purplish pink, or red to reddish brown hymenophores and spore prints, i.e., *Tylopilus* s.l. In total, 105 species, including 34 new species, are documented here together with descriptions of macro- and microscopic characteristics and color photos. For the species new to science or new record for China, line drawings of the microscopic characteristics are also provided. Information on the habitats and distributions of each species, as well as keys to the genera and species are provided. In addition, the distribution and differences from similar species are provided, and voucher specimens are listed.

It is impossible to feature each *Tylopilus* s.l. species in a region as vast and diverse as China in a limited time span. There are likely to be many species that have not been scientifically described or remain to be discovered. We therefore offer this book as a first step in the identification of the currently known *Tylopilus* s.l. species.

The authors
27 April, 2022

Acknowledgments

We studied and cited many specimens that were collected in different parts of China and deposited in different herbaria by various mycologists during long-term field investigations. Without these collections, it would have been impossible to complete this book.

We acknowledge the following individuals who have helped them in various ways with this work. For sharing preserved specimens, mycological notes, and photographs, we are grateful to Prof. Tolgor Bau, Jilin Agriculture University; Profs. Zuo-Hong Chen and Ping Zhang, Hunan Normal University; Prof. Pei-Gui Liu, the late Prof. Mu Zang, Drs. Qing Cai, Bang Feng, Zai-Wei Ge, Ting Guo, Li-Hong Han, Yan-Jia Hao, Jiao Qin, Xiao-Fei Shi, Gang Wu, Kuan Zhao, Qi Zhao, Pan-Meng Wang, Xiang-Hua Wang, and Fu-Qing Yu, Messrs Cong Huang, Liu-Kun Jia, Chen Yan, Xin Xu and Jian-Wei Liu, Misses Xiao-Xia Ding, Mei-Xiang Li and Xin Meng, Kunming Institute of Botany, Chinese Academy of Sciences; Prof. Fang Li, Sun Yat-Sen University; Profs. Tai-Hui Li and Wang-Qiu Deng, Institute of Microbiology, Guangdong Academy of Sciences; Prof. Jun-Feng Liang, Research Institute of Tropical Forestry, Chinese Academy of Forestry; Dr. Chun-Ying Deng, Guizhou Academy of Sciences; Prof. Li-Ping Tang, Kunming Medical University; Dr. Tie-Zheng Wei, Institute of Microbiology, Chinese Academy of Sciences; Prof. Nian-Kai Zeng, Hainan Medical University; Prof. Xiu-Guo Zhang, Shandong Agricultural University; and Prof. Xue-Tai Zhu, Northwest Normal University. We thank Dr. Jane Marczewski for polishing the English of the manuscript.

This work was supported by the Strategic Priority Research Program of the Chinese Academy of Sciences (XDB31000000), the International Partnership Program of Chinese Academy of Sciences (151853KYSB20170026), the Funds for International Cooperation and Exchange of the National Natural Science Foundation of China (31210103919), the National Natural Science Foundation of China (Nos. 31370001, 31750001, 31872618 and 32070024), the Natural Science Foundation of Yunnan Province (2018FB027), the Youth Innovation Promotion Association of Chinese Academy of Sciences (2016348 and 2017436), the Ten Thousand Talents Program of Yunnan (YNWR-QNBJ-2018-125), and the Key Research Program of Frontier Sciences of Chinese Academy of Sciences (QYZDY-SSW-SMC029). The publication of this book was supported by the National Found for Academic Publication in Science and Technology (NFAPST). A part of the work was conducted at the Model Worker Innovation Studio of Yunnan Provincial Federation of Trade Unions.

Inevitably, some shortcomings or even mistakes will have found their way into this book. Any comments from the reader for the improvement and revision of the book would be highly appreciated.

Contents

Chapter 1 Introduction ··········1
1.1 Ecological and economic values of boletes ··········1
1.2 Historical research in the Boletaceae ··········2
1.3 Historical classification in *Tylopilus* s.l. ··········4

Chapter 2 Materials and methods ··········6
2.1 Materials ··········6
2.2 Methods ··········6

Chapter 3 Morphology and structures in *Tylopilus* s.l. ··········22

Chapter 4 Phylogenetic results and systematic treatments of *Tylopilus* s.l. ··········29

Chapter 5 Taxonomic part ··········33

Chapter 6 *Abtylopilus* Yan C. Li & Zhu L. Yang ··········36
6.1 *Abtylopilus alborubellus* Yan C. Li & Zhu L. Yang ··········37
6.2 *Abtylopilus scabrosus* Yan C. Li & Zhu L. Yang ··········40

Chapter 7 *Anthracoporus* Yan C. Li & Zhu L. Yang ··········43
7.1 *Anthracoporus cystidiatus* Yan C. Li & Zhu L. Yang ··········45
7.2 *Anthracoporus holophaeus* (Corner) Yan C. Li & Zhu L. Yang ··········48
7.3 *Anthracoporus nigropurpureus* (Hongo) Yan C. Li & Zhu L. Yang ··········51

Chapter 8 *Austroboletus* (Corner) Wolfe ··········54
8.1 *Austroboletus albidus* Yan C. Li & Zhu L. Yang ··········57
8.2 *Austroboletus albovirescens* Yan C. Li & Zhu L. Yang ··········60
8.3 *Austroboletus dictyotus* (Boedijn) Wolfe ··········64
8.4 *Austroboletus fusisporus* (Imazeki & Hongo) Wolfe ··········66

8.5 *Austroboletus olivaceobrunneus* Yan C. Li & Zhu L. Yang ········ 68
8.6 *Austroboletus olivaceoglutinosus* K. Das & Dentinger ········ 71
8.7 *Austroboletus subvirens* (Hongo) Wolfe ········ 73

Chapter 9 *Chiua* Yan C. Li & Zhu L. Yang ········ 76
9.1 *Chiua angusticystidiata* Yan C. Li & Zhu L. Yang ········ 78
9.2 *Chiua olivaceoreticulata* Yan C. Li & Zhu L. Yang ········ 80
9.3 *Chiua virens* (W.F. Chiu) Yan C. Li & Zhu L. Yang ········ 82
9.4 *Chiua viridula* Yan C. Li & Zhu L. Yang ········ 84

Chapter 10 *Fistulinella* Henn. ········ 86
10.1 *Fistulinella olivaceoalba* T.H.G. Pham, Yan C. Li & O.V. Morozova ········ 87
10.2 *Fistulinella salmonea* Yan C. Li & Zhu L. Yang ········ 91

Chapter 11 *Harrya* Halling, Nuhn & Osmundson ········ 94
11.1 *Harrya alpina* Yan C. Li & Zhu L. Yang ········ 96
11.2 *Harrya atrogrisea* Yan C. Li & Zhu L. Yang ········ 99
11.3 *Harrya chromipes* (Frost) Halling, Nuhn, Osmundson & Manfr. Binder ········ 101
11.4 *Harrya moniliformis* Yan C. Li & Zhu L. Yang ········ 104
11.5 *Harrya subalpina* Yan C. Li & Zhu L. Yang ········ 106

Chapter 12 *Hymenoboletus* Yan C. Li & Zhu L. Yang ········ 108
12.1 *Hymenoboletus filiformis* Yan C. Li & Zhu L. Yang ········ 110
12.2 *Hymenoboletus griseoviridis* Yan C. Li & Zhu L. Yang ········ 113
12.3 *Hymenoboletus jiangxiensis* Yan C. Li & Zhu L. Yang ········ 116
12.4 *Hymenoboletus luteopurpureus* Yan C. Li & Zhu L. Yang ········ 119

Chapter 13 *Indoporus* A. Parihar, K. Das, Hembrom & Vizzini ········ 121
13.1 *Indoporus squamulosus* Yan C. Li & Zhu L. Yang ········ 122

Chapter 14 *Leccinellum* Bresinsky & Manfr. Binder ········ 125
14.1 *Leccinellum castaneum* Yan C. Li & Zhu L. Yang ········ 127
14.2 *Leccinellum citrinum* Yan C. Li & Zhu L. Yang ········ 130
14.3 *Leccinellum cremeum* Zhu L. Yang & G. Wu ········ 133
14.4 *Leccinellum griseopileatum* Yan C. Li & Zhu L. Yang ········ 134
14.5 *Leccinellum onychinum* Fang Li, Kuan Zhao & Q.L. Deng ········ 137
14.6 *Leccinellum sinoaurantiacum* (M. Zang & R.H. Petersen) Yan C. Li & Zhu L. Yang ········ 139

Chapter 15 *Mucilopilus* Wolfe — 142

15.1 *Mucilopilus cinnamomeus* Yan C. Li & Zhu L. Yang — 143
15.2 *Mucilopilus paracastaneiceps* Yan C. Li & Zhu L. Yang — 147
15.3 *Mucilopilus ruber* Yan C. Li & Zhu L. Yang — 150

Chapter 16 *Porphyrellus* E.-J. Gilbert — 153

16.1 *Porphyrellus castaneus* Yan C. Li & Zhu L. Yang — 155
16.2 *Porphyrellus cyaneotinctus* (A.H. Sm. & Thiers) Singer — 157
16.3 *Porphyrellus griseus* Yan C. Li & Zhu L. Yang — 160
16.4 *Porphyrellus orientifumosipes* Yan C. Li & Zhu L. Yang — 163
16.5 *Porphyrellus porphyrosporus* (Fr. & Hök) E.–J. Gilbert — 165
16.6 *Porphyrellus pseudofumosipes* Yan C. Li & Zhu L. Yang — 167
16.7 *Porphyrellus scrobiculatus* Yan C. Li & Zhu L. Yang — 170

Chapter 17 *Pseudoaustroboletus* Yan C. Li & Zhu L. Yang — 173

17.1 *Pseudoaustroboletus valens* (Corner) Yan C. Li & Zhu L. Yang — 174

Chapter 18 *Retiboletus* Manfr. Binder & Bresinsky — 176

18.1 *Retiboletus ater* Yan C. Li & T. Bau — 178
18.2 *Retiboletus brunneolus* Yan C. Li & Zhu L. Yang — 180
18.3 *Retiboletus fuscus* (Hongo) N.K. Zeng & Zhu L. Yang — 183
18.4 *Retiboletus nigrogriseus* N.K. Zeng, S. Jiang & Zhi Q. Liang — 185
18.5 *Retiboletus pseudogriseus* N.K. Zeng & Zhu L. Yang — 187
18.6 *Retiboletus zhangfeii* N.K. Zeng & Zhu L. Yang — 189

Chapter 19 *Royoungia* Castellano, Trappe & Malajczuk — 191

19.1 *Royoungia coccineinana* (Corner) Yan C. Li & Zhu L. Yang — 193
19.2 *Royoungia grisea* Yan C. Li & Zhu L. Yang — 195
19.3 *Royoungia reticulata* Yan C. Li & Zhu L. Yang — 197
19.4 *Royoungia rubina* Yan C. Li & Zhu L. Yang — 199

Chapter 20 *Sutorius* Halling, Nuhn & N.A. Fechner — 201

20.1 *Sutorius alpinus* Yan C. Li & Zhu L. Yang — 203
20.2 *Sutorius eximius* (Peck) Halling, Nuhn & Osmundson — 206
20.3 *Sutorius microsporus* Yan C. Li & Zhu L. Yang — 209
20.4 *Sutorius obscuripellis* Vadthanarat, Raspé & Lumyong — 212
20.5 *Sutorius pseudotylopilus* Vadthanarat, Raspé & Lumyong — 215

20.6 *Sutorius subrufus* N.K. Zeng, H. Chai & S. Jiang ········ 218

Chapter 21 *Tylocinum* Yan C. Li & Zhu L. Yang ········ 220

21.1 *Tylocinum griseolum* Yan C. Li & Zhu L. Yang ········ 221

Chapter 22 *Tylopilus* P. Karst. ········ 223

22.1 *Tylopilus albopurpureus* Yan C. Li & Zhu L. Yang ········ 227
22.2 *Tylopilus alpinus* Yan C. Li & Zhu L. Yang ········ 230
22.3 *Tylopilus argillaceus* Hongo ········ 232
22.4 *Tylopilus atripurpureus* (Corner) E. Horak ········ 234
22.5 *Tylopilus atroviolaceobrunneus* Yan C. Li & Zhu L. Yang ········ 236
22.6 *Tylopilus aurantiacus* Yan C. Li & Zhu L. Yang ········ 238
22.7 *Tylopilus brunneirubens* (Corner) Watling & E. Turnbull ········ 241
22.8 *Tylopilus castanoides* Har. Takah. ········ 244
22.9 *Tylopilus felleus* (Bull.) P. Karst. ········ 246
22.10 *Tylopilus fuligineoviolaceus* Har. Takah. ········ 248
22.11 *Tylopilus fuscatus* (Corner) Yan C. Li & Zhu L. Yang ········ 250
22.12 *Tylopilus griseipurpureus* (Corner) E. Horak ········ 253
22.13 *Tylopilus griseiviridus* Yan C. Li & Zhu L. Yang ········ 255
22.14 *Tylopilus griseolus* Yan C. Li & Zhu L. Yang ········ 258
22.15 *Tylopilus himalayanus* D. Chakr., K. Das & Vizzini ········ 261
22.16 *Tylopilus jiangxiensis* Kuan Zhao & Yan C. Li ········ 264
22.17 *Tylopilus neofelleus* Hongo ········ 267
22.18 *Tylopilus obscureviolaceus* Har. Takah. ········ 269
22.19 *Tylopilus olivaceobrunneus* Yan C. Li & Zhu L. Yang ········ 272
22.20 *Tylopilus otsuensis* Hongo ········ 275
22.21 *Tylopilus phaeoruber* Yan C. Li & Zhu L. Yang ········ 277
22.22 *Tylopilus plumbeoviolaceoides* T.H. Li, B. Song & Y.H. Shen ········ 280
22.23 *Tylopilus pseudoalpinus* Yan C. Li & Zhu L. Yang ········ 282
22.24 *Tylopilus pseudoballoui* D. Chakr., K. Das & Vizzini ········ 285
22.25 *Tylopilus purpureorubens* Yan C. Li & Zhu L. Yang ········ 287
22.26 *Tylopilus rubrotinctus* Yan C. Li & Zhu L. Yang ········ 291
22.27 *Tylopilus rufobrunneus* Yan C. Li & Zhu L. Yang ········ 294
22.28 *Tylopilus subotsuensis* T.H.G. Pham, A.V. Alexandrova & O.V. Morozova ··· 298
22.29 *Tylopilus vinaceipallidus* (Corner) Watling & E. Turnbull ········ 301
22.30 *Tylopilus violaceobrunneus* Yan C. Li & Zhu L. Yang ········ 304
22.31 *Tylopilus violaceorubrus* Yan C. Li & Zhu L. Yang ········ 306

22.32 *Tylopilus virescens* (Har. Takah. & Taneyama) N.K. Zeng, H. Chai & Zhi Q. Liang ··· 310

Chapter 23 *Veloporphyrellus* L.D. Gómez & Singer ·············· 313

23.1 *Veloporphyrellus alpinus* Yan C. Li & Zhu L. Yang ································ 315
23.2 *Veloporphyrellus castaneus* Yan C. Li & Zhu L. Yang ···························· 317
23.3 *Veloporphyrellus gracilioides* Yan C. Li & Zhu L. Yang ························ 320
23.4 *Veloporphyrellus pseudovelatus* Yan C. Li & Zhu L. Yang ···················· 322
23.5 *Veloporphyrellus velatus* (Rostr.) Yan C. Li & Zhu L. Yang ··················· 325

Chapter 24 *Zangia* Yan C. Li & Zhu L. Yang ····························· 327

24.1 *Zangia chlorinosma* (Wolfe & Bougher) Yan C. Li & Zhu L. Yang ··············· 329
24.2 *Zangia citrina* Yan C. Li & Zhu L. Yang ··· 331
24.3 *Zangia erythrocephala* Yan C. Li & Zhu L. Yang ··································· 333
24.4 *Zangia olivacea* Yan C. Li & Zhu L. Yang ·· 335
24.5 *Zangia olivaceobrunnea* Yan C. Li & Zhu L. Yang ································ 337
24.6 *Zangia roseola* (W.F. Chiu) Yan C. Li & Zhu L. Yang ···························· 340

Chapter 25 Summary and conclusion ··· 343

References ··· 351

Index of scientific names ··· 361

Chapter 1

Introduction

The Boletaceae Chevall., which belongs to the Boletales (Agaricomycetes) in Basidiomycota, is a family that comprises species of fungi that produce a fleshy fruit body with tubular (or sometimes lamellar) hymenophore. In this book, we adopt the family concept as defined by Binder and Hibbett (2006), i.e., the fruit body is fleshy, the hymenophore is tubular or lamellate, the tube trama is often gelatinous, and the spore prints are mostly vinaceous to reddish brown. As the color and texture of this kind of fungi are supposedly similar to those of the livers of cattle, boletes are known as "Niuganjun" (cow liver) in China (Lan 1436). There are nearly 1000 species from about 80 genera described from this family worldwide, and all of them have great ecological value, and many are also economically important.

1.1 Ecological and economic values of boletes

The family Boletaceae is both ecologically and economically important. Most of the species are ectomycorrhizal symbionts of woody plants, with more than ten families of plants (including Betulaceae, Dipterocarpaceae, Fagaceae, Pinaceae, Salicaceae) known to form mycorrhizal relationships with boletes. These fungi play vital roles in maintaining forest ecosystems, vegetation restoration, and water and soil protection (van der Heijden *et al.* 1998; Kernaghan 2005; Rinaldi *et al.* 2008; Eastwood *et al.* 2011). The symbiotic ectomycorrhizal relationships are thought to contribute to the diversification of both fungi and their host plants (Wang and Qiu 2006), and it is thought to be this mycorrhizal association that strongly drives evolution and leads to the richness of genera and species, the diverse morphology, the complicated microscopic structure, the wide ecological distribution, and the presence of both convergent and parallel evolution in the Boletaceae (Singer 1986; Bruns *et al.* 1989; Binder and Bresinsky 2002a; Binder and Hibbett 2006; Nuhn *et al.* 2013; Wu *et al.* 2014, 2016a).

Many species in the Boletaceae are very important worldwide as delicacies, including the species of the *Boletus edulis* complex, which are famous across the world markets as porcini mushrooms (Dentinger *et al.* 2010), and are considered among the best of the edible fungi. During the Chinese mushroom season, it is usual to see people searching in forests for the "Huang Niugan" (*Tylopilus pseudoballoui* D. Chakr., K. Das & Vizzini) or "Bai Cong" [*Butyriboletus roseoflavus* (M. Zang and H.B. Li) D. Arora and J.L. Frank] (Yang *et al.* 2021). Some species are also used in herbal medicines, or have bioactive metabolites (Heleno *et al.* 2011), while other species are toxic, and perhaps even lethally so (Benjamin 1995; Matsuura *et al.* 2007). However, due to excessive long-term over-collection, habitat damage, and global warming, some species or populations are probably in danger of extinction. The study of boletes is therefore crucial for their conservation, utilization, and to prevent poisoning.

1.2 Historical research in the Boletaceae

The type genus of the Boletaceae is *Boletus* L. The name *Boletus* was first applied to the fleshy, stipitate, poroid fungi by Dillenius in 1719 (Dillenius 1719). Linnaeus (1753) included all poroid fungi (boletes and polypores) under *Boletus,* but Fries (1821) accepted Dillenius' generic concept and created infrageneric taxa within *Boletus* based on the presence or absence of a veil and color of the hymenium and spore print. Since the establishment of *Boletus*, a large number of taxonomic studies on the genus have been carried out worldwide, and a great deal of research has been published (e.g., Peck 1873, 1887; Snell and Dick 1941; Heinemann 1951; Hongo 1963, 1968, 1974a, 1974b, 1984a,1984b, 1985; Snell and Dick 1970; Watling 1970; Smith and Thiers 1971; Corner 1972; Hongo and Nagasawa 1976; Wolfe 1979a, 1979b; Singer 1986; Binder and Bresinsky 2002a, 2002b; Nagasawa 1997; Li and Watling 1999; Watling and Li 1999; Bessette et al. 2000; Li *et al.* 2002; den Bakker and Noordeloos 2005; Zang 2006; Desjardin *et al.* 2008, 2009; Nelson 2010; Horak 2011; Neves *et al.* 2012). However, most of these studies focused on morphology, subcellular structure, and ecology. Species in the family Boletaceae are diverse in morphology, complicated in structure, and have wide ecological distribution.

Some boletes are quite distinctive and easy to recognize, while others can be exceedingly difficult to identify due to either parallel or convergent evolution, even for experienced boletologists. Consequently, it is hard to understand the systematics and evolution of boletes effectively through a single morphological approach. This may be the reason why Boletaceae is regarded as a taxonomically difficult family. Even E.M. Fries, well-known as the "Linnaeus of fungal taxonomy" once said, "Nullum genus quam Boletorum magnis me molestavit (no genus has given me more trouble than that of the boleti)" (Bessette *et al.* 2000).

The application of molecular methods in the early 21st century revolutionized the study of taxonomy, systematics, and phylogeny in the Boletaceae. One of the most important works is that of Binder and Hibbett (2006), in which the authors divided the Boletales into six suborders, and which shows the Boletaceae is a genus- and specie-rich family of 38 genera falling in the suborder Boletineae. The authors also merged the families Boletellaceae Locq., Strobilomycetaceae E.-J. Gilbert, and Xerocomaceae Pegler & T.W.K. Young, which have ornamented basidiospores, and the families Octavianiaceae Locq. ex Pegler & T.W.K. Young and Chamonixiaceae Jülich, which have sequestrate forms, into the family Boletaceae, and demonstrated that the basidiospore ornamentation and sequestrate forms have evolved independently several times from the boletoid ancestors rather than evolved as single events (Binder and Hibbett 2006). These treatments give quite different results for the family Boletaceae to those from the morphology-based approach.

However, despite these global achievements in elucidating the taxonomy of the Boletaceae, several key scientific issues remain: the phylogenetic relationships among different genera in the Boletaceae are unresolved; the taxonomic positions of many species are unknown, and the new evolutionary clades represented by the known or new species need to be clarified. Accordingly, the use of an integrated approach, including morphology, ultrastructure, ecology, multi-locus phylogeny and genomics, should be employed to gain insights into the systematics, diversity, speciation, and co-evolution of Boletaceae. Based on this idea, Wu *et al.* (2014) conducted systematic and phylogenetic studies into Boletaceae based on four gene fragments (nrLSU, *tef1-α*, *rpb1*, and *rpb2*), along with studies of the macroscopic characteristics, microscopic structures, and ultrastructures, as well as related data from the literatures. In their work, the phylogenetic frame of Boletaceae was constructed, seven subfamilies including four new subfamilies were proposed, and 59 generic (or sub-generic) clades were revealed, of which 22 were new, and 11 known genera were monophyletic. Their study revealed that several key morphological traits of Boletaceae, such as fruit body discoloration and morphology, basidiospore ornamentation, and stuffed pores (the immature poroid hymenophore is covered with a layer of tangled white hyphae) have evolved independently several times. In a subsequent paper, based on molecular phylogenetic, morphological and ecological evidence, a number of important species from China belonging to different genera in the Boletaceae were documented and illustrated (Wu *et al.* 2016a).

The genus *Boletus* s.l. is not a monophyletic group, but represents a number of evolutionary lineages, and further division at generic level is necessary. A striking example is that only in one year, more than ten genera were split from *Boletus* (Arora and Frank 2014; Gelardi *et al.* 2014; Li *et al.* 2014a; Vizzini 2014a, b, c, d, e; Zeng *et al.* 2014; Zhao *et al.* 2014).

China is a center of both diversity and distribution for fungi. The earliest important academic record of Chinese boletes was probably provided by Chiu (1948, 1957), who made major contribution to the taxonomy of boletes, and reported 57 boletes, including 22 new species and 2 new varieties, from Yunnan province, southwestern China. Subsequently, Teng (1963) reported 2400 species of fungi in China including 36 boletes. Bi and Li (1990) and Bi *et al.* (1994, 1997) investigated the diversity of macrofungi, including boletes from Guangdong and Hainan provinces. Yeh and Chen (1980, 1981, 1982, 1983, 1985), Chen *et al.* (1997a, b, c, 1998a, b) and Chen and Yeh (2000) studied the diversity of boletes from Taiwan province and recorded 72 species alone from that island. Zang (1980, 1985, 1986, 1996, 1999) devoted a great deal of time to the study of Chinese boletes and published two monographs of the Boletaceae in China, the "*Flora Fungorum Sinicorum. Vol 22, Boletaceae (I)*" (Zang 2006) and "*Flora Fungorum Sinicorum. Vol 44, Boletaceae (II)*" (Zang 2013), in which 218 species were documented and illustrated, and nine species had been arranged in *Tylopilus* s.l. However, these two monographs are written in Chinese, which is a barrier for people from non-Chinese speaking countries. Moreover, most species are illustrated with black and white line drawings and described with textual descriptions of colors which often seem vague or imprecise to citizen, mycologists, or enthusiasts. In addition, as with other groups of fungi, the Boletaceae were studied in the European and American countries earlier than was the case in Asia, and consequently, many species from East Asia, particularly from China, were identified as species originally described from Europe and North America based on general morphological similarities. Our previous studies (Li *et al.* 2009; Cui *et al.* 2016) indicated that a few species described from Europe or North America do occur in China, especially in northeastern and northwestern China. However, most species found in China have evolved independently in East Asia. Thus, identification of some species, especially those reported from China but given the names of European or North American species, needs to be reconfirmed.

To date, nearly 1000 species of Boletaceae in about 80 genera have been reported worldwide (Henkel 1999; Jarosch 2001; Yang *et al.* 2006; Binder and Hibbett 2006; Halling *et al.* 2007; Desjardin *et al.* 2008, 2009; Kirk *et al* 2008; Orihara *et al.* 2010; Gelardi 2011; Horak 2011; Li *et al.* 2011; Halling *et al.* 2012a, b, 2015; Lebel *et al*. 2012; Neves *et al*. 2012; Husbands *et al*. 2013; Wu *et al*. 2014, 2016a, b; Henkel *et al*. 2017; Noordeloos *et al*. 2018; Kou and Ortiz-Santana 2020), and more than 300 species belonging to the Boletaceae have been reported from China alone in the past few decades (Li *et al*. 2011, 2014a, b; Zeng *et al*. 2013; Cui *et al*. 2016; Wu et al. 2016a, b), of which approximately one-third were reported as new species. We anticipate that numerous species new to science will be described, and some monophyletic groups representing generic lineages will be recognized within the Boletaceae, over the next few years.

1.3 Historical classification in *Tylopilus* s.l.

Following the establishment of the genus *Boletus*, mycologists began to recognize the generic distinctness of many groups within this genus, and many new genera were described to accommodate both known and newly described species with pallid to pinkish or purplish pink hymenophores or spore prints. Karsten (1881) recognized the generic distinctness of *Boletus felleus* Bull., which was diagnosed by rosy basidiospores and tubes adnate to the stipe, and gave this new monotypic genus the name *Tylopilus* P. Karst., with *T. felleus* (Bull.) P. Karst. as the type species. Hennings (1901) described a new genus *Fistulinella* Henn., with *F. staudtii* Henn. as the type. However, although the type species in *Fistulinella* was characterized by the pallid hymenophore, the other species subsequently described in this genus were characterized by pallid to pinkish or purplish pink hymenophores (Pegler and Young 1981; Fulgenzi *et al*. 2010; Horak 2011). *Porphyrellus* E.-J. Gilbert was proposed as a genus in 1931, with *Boletus porphyrosporus* Fr. & Hök designated as the type. *Porphyrellus* was distinguished by the white flesh without color change or staining blue on injury, the fleshy and tomentose stipe, the concolorous pores and tubes, the club-shaped and fusiform cystidia, and the smooth, red-brown basidiospores. Singer (1945) emended and expanded the generic limits of *Porphyrellus* based on the color of spore print and the size of spore, and included taxa with perforate-punctate basidiospores and reticulate to lacerate-reticulate stipes, which he named *Porphyrellus* sect. *Graciles* Singer. Smith and Thiers (1968, 1971) suggested that Singer (1945, 1962) had not demonstrated any significant differences between *Porphyrellus* and *Tylopilus* at the generic rank, and treated *Porphyrellus* Singer as a subgenus within *Tylopilus* based on the similarity of spore print colors, basidiospore ornamentation, and cystidia characteristics.

Corner (1972) adopted broad generic concepts within the boletes, and considered *Tylopilus* as a subgenus of *Boletus*. As spore print color barriers did not apparently exist between *Tylopilus* and *Porphyrellus*, Corner preferred to place the smooth-spored taxa of *Porphyrellus* into *Boletus* subg. *Tylopilus*, and the perforate-punctate-spored taxa in *P*. sect. *Graciles* into a new subgenus, *Boletus* subg. *Austroboletus* Corner. Subsequently, the following genera with pinkish, purplish pink, reddish, reddish brown or brownish hymenophores or spore prints were described, and mostly based on the species that had been traditionally placed in *Tylopilus*: *Austroboletus* (Corner) Wolfe, *Mucilopilus* Wolfe, *Veloporphyrellus* L.D. Gómez & Singer, *Royoungia* Castellano, Trappe & Malajczuk, *Zangia* Yan C. Li & Zhu L. Yang, *Harrya* Halling, Nuhn & Osmundson, *Sutorius* Halling, Nuhn & N.A. Fechner, *Pseudoaustroboletus* Yan C. Li & Zhu L. Yang, *Chiua* Yan C. Li & Zhu L. Yang, *Hymenoboletus* Yan C. Li & Zhu L. Yang, *Tylocinum* Yan C. Li & Zhu L. Yang and *Indoporus* A. Parihar, K. Das, Hembrom & Vizzini (Wolfe 1979a, b; Gómez and Singer

1984; Castellano *et al.* 1992; Binder and Bresinsky 2002b; Li *et al.* 2011; Halling *et al.* 2012a, b; Li *et al.* 2014b; Wu *et al.* 2016a, b; Parihar *et al.* 2018). From the study of these and other works regarding the Boletaceae, the present authors have come to the conclusion that the genus *Tylopilus* in a wider concept, as adopted by many authors (e.g., Smith and Thiers 1968, 1971; Horak 2011), is a heterogeneous mixture of several groups of species. These groups can be separated from each other based on morphological and multi-locus phylogenetic studies.

Species in *Tylopilus* s.l. are very diverse in morphology, complex in structure, and have wide ecological niches. China is one of the diversity hotspots for boletes, however, before our study, fewer than 30 species with pallid to purplish pink or reddish brown hymenophores or spore prints had been reported (Li and Song 2003; Fu *et al.* 2006; Li *et al.* 2011; Li *et al.* 2014a, b; Zang 2006, 2013), and our work identified a large number of species that needed scientific descriptions. In our previous study (Wu *et al.* 2016a), we documented and illustrated 38 species belonging to *Tylopilus* s.l. However, many species in *Tylopilus* s.l. still lack detailed illustrations and documentations.

Chapter 2
Materials and methods

2.1 Materials

China, which is on the East Asian shore of the western Pacific Ocean, has a land area of about 9.6 million square kilometers and is one of the largest countries in land area on the planet. The highly variable climate and diverse topography of southern (especially southwestern) China mean that its species richness is very great, and the mountains of southwest China are ranked as one of the world's 34 biodiversity hotspots (Myers *et al*. 2000). Thus, China has unexpectedly high species diversity in Boletaceae. Over the last two decades, our research group has been conducting a broad field survey of the species in *Tylopilus* s.l. across China, and 23 of a total of 34 provincial administrative areas rich in *Tylopilus* s.l. species have been investigated (Fig. 2.1). Whenever possible, specimens from other countries in Africa, Europe, New Zealand, North/Central America, and Southeast Asia were also studied. Voucher specimens from our field investigations were deposited in the fungal herbarium of the Herbarium KUN (Kunming Institute of Botany, Chinese Academy of Sciences, Yunnan, China). Additional collections on loan from the following herbaria: GDGM (Guangdong Institute of Microbiology, Guangdong Academy of Sciences, Guangdong, China), HGAS (Guizhou Academy of Sciences, Guizhou, China), HMAS (Chinese Academy of Sciences, Beijing, China), HMJAU (Jilin Agricultural University, Jilin, China), MHHNU (Mycological Herbarium of Hunan Normal University, Hunan, China), PDD (Manaaki Whenua - Landcare Research, Auckland, New Zealand), RITF (Research Institute of Tropical Forestry, Chinese Academy of Forestry, Guangdong, China), TNS (National Museum of Nature and Science, Tsukuba, Japan) were also studied. Herbarium acronyms follow Thiers (2018) with two exception: RITF and MHHNU, which are not listed in Index Herbariorum.

2.2 Methods

Field investigation: Each collection is accompanied by digital images made at the time, and detailed field records including the colors of different parts of the fruit body, and the discoloration of the fresh fruit body when injured. In addition, plants growing around the target species were also noted, and the most probably host plants were recorded. This ecological symbiotic information is very important in species identification. A sample from each collection was preserved in silica gel for molecular phylogenetic studies; and specimens were dried in a fungal dryer for further morphological studies in the laboratory.

Morphological studies: Macroscopic characteristics assessed in this work include the size, color, odor, and discoloration of the fruit body, were derived from field notes of the fresh fruit bodies, and from photographs. Color codes (code in parentheses, e.g., 30E7–8) used for the newly described

Chapter 2 Materials and methods   7

species, follow Kornerup and Wanscher (1981). Microscopic structures assessed include the structure of the pileipellis, the morphology of the basidia and cystidia, the morphology and ornamentation of the basidiospores, were examined under light microscopy after sectioning and mounting in 5%–10% KOH

Fig. 2.1 A map of China, the administrative areas from which the collections were made by our research group are marked with black dots.

solution. Melzer's reagent was also used to test the amyloidity and dextrinoidity of the basidiospores. All microscopic structures were drawn by freehand from rehydrated material by the first author. In the descriptions of basidiospores, the abbreviation n/m/p means n basidiospores measured from m basidiomata of p collections; the notation of the form (a) b–c (d) means the dimensions of the basidiospores; the range b–c stands for a minimum of 90% of the measured values, a or d given in parentheses stands for the extreme values; Q is used to mean "length/width ratio" of a basidiospore in side view; Q_m is the average Q of all basidiospores ± sample standard deviation. To observe basidiospore ornamentations, tiny hymenophoral fragments were taken from dried specimens, mounted on aluminum stubs with double-sided adhesive tape, coated with gold palladium, and then observed under a Hitachi S4800 scanning electron microscope (SEM). Abbreviations of generic names mentioned in this work are presented in Table 2.1.

Table 2.1   The generic names mentioned in this work and their abbreviations.

Generic name	Abbreviation	Generic name	Abbreviation
Abtylopilus	*Ab.*	*Mucilopilus*	*M.*
Anthracoporus	*An.*	*Porphyrellus*	*P.*
Austroboletus	*A.*	*Pseudoaustroboletus*	*Ps.*
Boletus	*B.*	*Retiboletus*	*R.*
Chiua	*C.*	*Royoungia*	*Ro.*
Fistulinella	*F.*	*Sutorius*	*S.*
Harrya	*Ha.*	*Tylocinum*	*Ty.*
Hymenoboletus	*Hy.*	*Tylopilus*	*T.*
Indoporus	*I.*	*Veloporphyrellus*	*V.*
Leccinellum	*L.*	*Zangia*	*Z.*

Phylogenetic studies and taxa delimitation: Molecular phylogenetic analyses of species in *Tylopilus* s.l. were conducted, in order to understand the diversity and intra- and intergeneric relationships of species in *Tylopilus* s.l. in China. Total DNA was extracted from silica-gel dried or herbaria materials using the CTAB method (Doyle and Doyle 1987). A total of four nuclear loci were amplified and sequenced, including the nuclear ribosomal large subunit (nrLSU), the translation elongation factor 1-α gene (*tef1-α*), the largest subunit of RNA polymerase II (*rpb1*), and the second-largest subunit of RNA polymerase II (*rpb2*). The primer pairs used for amplifying the above gene fragments followed those in Li *et al.* (2009), Wu *et al.* (2014), and references therein. The PCR reaction was conducted on an ABI 2720 Thermal Cycler (Applied Biosystems, Foster City, CA, USA) or an Eppendorf Master Cycler (Eppendorf, Netheler-Hinz, Hamburg, Germany) under the following conditions: denaturalization at 95°C for 3 min, then 30 cycles of denaturalization at 95°C for 1 min, annealing at 55°C for 1 min, and polymerization at 72°C for 1 min, followed by a final elongation at 72°C for 8min. The PCR products were purified with a Gel Extraction & PCR Purification Combo Kit (Spin-column) (Bioteke, Beijing, China) according to the manufacturer's instructions, and prepared for the sequencing reaction using the BigDye Terminator Cycle Sequencing Kit v. 3.1 (Applied BioSystems), and then sequenced on an ABI-3730-XL DNA Analyzer (Applied Biosystems, Foster City, CA, USA) using the same primers as in the PCR amplification. When PCR products could not be sequenced directly, they were cloned into a PMD18-T vector (Takara, Japan) and then amplified and sequenced using the standard primers M13F and M13R. Sequences newly generated in this work have been submitted to GenBank. Detailed information of voucher specimens, including GenBank accession numbers, is given in Table 2.2.

In the multi-locus phylogenetic analysis, sequences generated in our study and from recently reported genera in the Boletaceae, i.e., *Indoporus* A. Parihar, K. Das, Hembrom & Vizzini, *Ionosporus* Khmeln., *Turmalinea* Orihara & N. Maek., *Jimtrappea* T.W. Henkel, M.E. Sm. & Aime, *Costatisporus* T.W. Henkel & M.E. Sm., and *Castellanea* T.W. Henkel & M.E. Sm. (Smith *et al.* 2015; Orihara *et al.* 2016; Parihar *et al.* 2018; Khmelnitsky *et al.* 2019) were aligned manually to the super multi-locus matrix of Wu *et al.* (2016a) that includes 62 genera in the Boletaceae (roughly 80% of the described genera in the family). This dataset was used to investigate the relationships among all of our samples and related taxa in GenBank (data not shown). The dataset was then narrowed by running heuristic searches in Paup* 4.0b10, pruning redundant sequences until a comprehensive dataset comprising of 762 sequences representing 181 species remained. A Maximum Likelihood (ML) analysis was then conducted using RAxML v7.9.1 (Stamatakis 2006), implementing the GTR + GAMMA model, and executing a rapid bootstrapping (MLB) with 1000 replicates.

The newly described genera referenced to the guidelines for establishment of genera suggested by Vellinga *et al.* (2015). In species delimitation, the Genealogical Concordance Phylogenetic Species Recognition (GCPSR) criterion (Taylor *et al.* 2000) was adopted which has been proven to be useful in fungi and is currently the most wildly used identification method in the kingdom of fungi (Dettman *et al.* 2003; Giraud *et al.* 2008; Hibbett *et al.* 2011; Vialle *et al.* 2013; Han *et al.* 2018, 2020).

In this study, species were defined according to the morphological characteristics and two GCPSR-based criteria proposed by Dettman *et al.* (2003), i.e., the genealogical concordance criterion: the lineages were genealogically concordant if they were present in at least some of the gene trees, and the genealogical non-discordance criterion: the lineages were genealogically non-discordant if they were strongly supported (maximum parsimony value: MP ⩾ 70%, maximum likelihood value: ML ⩾ 70%) in a single gene and not contradicted at or above this level of support in any other single-locus tree. When assigning independent evolutionary lineages to phylogenetic species, the combined four-locus analysis was also considered. For any divergent terminal branch represented by only one specimen, the branch was considered as a phylogenetic species if it also showed morphological differences from its closely related sister groups.

Table 2.2 Specimens used in multi-locus phylogenetic analysis and their GenBank accession numbers. Sequences newly generated in this study are indicated in bold; "—" represents missing data; names with quotes mean the named sequences in GenBank were identified incorrectly.

Taxon	Voucher ID	Coll. No.	Location	nrLSU	tef1-a	rpb1	rpb2
Abtylopilus alborubellus	HKAS 99704 (Type)	Wu 1585	China	**MT154713**	—	—	—
Abtylopilus scabrosus	HKAS 59826	Zeng 487	China	KT990558	**MT110336**	**MT110379**	—
Abtylopilus scabrosus	HKAS 50211 (Type)	Li 457	China	KT990552	KT990752	KT990920	KT990389
Afroboletus luteolus	00_436	—	Africa	KF030238	KF030397	KF030392	—
Alessioporus ichnusanus	AMB 12756	—	Italy	KJ729504	KJ729513	—	—
Anthracoporus cystidiatus	HKAS 55375	Feng 264	China	KT990622	KT990816	KT990969	**MT110410**
Anthracoporus cystidiatus	MHHNU 7312 (Type)	Zhang 812	China	**MT154710**	—	**MT110377**	**MT110411**
Anthracoporus holophaeus	HKAS 59407	Li 1660	China	KT990708	KT990888	KT991030	KT990506
Anthracoporus nigropurpureus	HKAS 52685	Li 998	China	KT990627	KT990821	KT990973	KT990459
Anthracoporus nigropurpureus	HKAS 53370	Li 1025	China	KT990628	KT990822	KT990974	KT990460
Aureoboletus tenuis	HKAS 75104	Wu 789	China	KT990518	KT990722	KT990897	KT990359
Aureoboletus zangii	HKAS 74751	Wu 440	China	KT990521	KT990725	KT990899	KT990362
Austroboletus albidus	HKAS 82463	Feng 1321	China	**MT154755**	—	—	—
Austroboletus albidus	HKAS 107148 (Type)	Shi 238	China	**MT154756**	—	—	—
Austroboletus albovirescens	HKAS 59624	Li 1876	China	KF112485	KF112217	KF112570	KF112765
Austroboletus albovirescens	HKAS 74743	Wu 432	China	KT990527	KT990730	—	KT990367
Austroboletus albovirescens	HKAS 107171 (Type)	Cai 1508	China	**MW114850**	—	—	—
Austroboletus albovirescens	HKAS 107172	Ding 433	China	**MW114849**	—	—	—
Austroboletus cf. *novaezelandiae*	CD 567	—	Australia	KC552061	KC552102	—	—
Austroboletus dictyotus	HKAS 53450	Li 1088	China	KF112487	KF112215	KF112573	KF112768
Austroboletus eburneus	REH 9487	—	Australia	JX889668	JX889708	—	—
Austroboletus fusisporus	HKAS 53461	Li 1099	China	KF112486	KF112214	KF112572	KF112767

Continued

Taxon	Voucher ID	Coll. No.	Location	nrLSU	tef1-a	rpb1	rpb2
Austroboletus lacunosus	REH 9146	—	Australia	JX889669	JX889709	—	—
Austroboletus niveus	MEL 2053830	—	Australia	KC552058	KC552099	—	—
Austroboletus occidentalis	MEL 2300518	—	Australia	KC552059	KC552100	—	—
Austroboletus olivaceobrunneus	HKAS 107144	530828 MF 0042	China	MT154758	—	MT110400	MT110434
Austroboletus olivaceobrunneus	HKAS 92428 (Type)	Zhao 798	China	MT154757	MT110363	KF112569	KF112764
Austroboletus olivaceoglutinosus	HKAS 57756	Wu 224	China	KF112383	KF112212	—	—
Austroboletus olivaceoglutinosus	HKAS 106456	Xu 132	China	MT154753	MW165263	—	—
Austroboletus olivaceoglutinosus	HKAS 91167	Liu 359	China	MT154754	—	—	—
"*Austroboletus subvirens*"	KPM-NC-0017836	—	Japan	JN378518	JN378458	—	—
Austroboletus subvirens	HKAS 107142	530828 MF 0013	China	MT154759	—	—	MW165285
Austroboletus subvirens	HKAS 107149	Wu 2027	China	MT154760	MW165262	—	—
Boletus edulis	HMJAU 4637	—	China	KF112455	KF112202	KF112586	KF112704
Boletus reticuloceps	HKAS 51232	Liang 521	China	KT990537	KT990739	KT990906	KT990376
Borofutus dhakanus	HKAS 73789	Hosen 198	Bangal	JQ928616	JQ928576	JQ928586	JQ928597
Buchwaldoboletus lignicola	HKAS 76674	Shi 524	China	KF112350	KF112277	KF112642	KF112819
Butyriboletus regius	HKAS 84878	MB 000287	Germany	MT264910	—	—	—
Butyriboletus subappendiculatus	HKAS 84880	MB 000260	Germany	KT002618	KT002642	KT002630	—
Chalciporus piperatus	HKAS 84882	MB 001169	Germany	KT990562	KT990758	—	KT990397
Chalciporus rubinelloides	HKAS 58728	Cai 61	China	KT990564	KT990760	—	KT990399
Chiua angusticystidiata	HKAS 50282	Li 528	China	KT990554	KT990754	MT110408	KT990390
Chiua olivaceoreticulata	HKAS 59675 (Type)	Li 1927	China	KT990699	KT990884	—	—
Chiua olivaceoreticulata	HKAS 59706	Li 1960	China	KT990593	KT990787	KT990941	KT990428
Chiua virens	HKAS 76678	Shi 677	China	KF112438	KF112272	KF112582	KF112793

Taxon	Voucher ID	Coll. No.	Location	nrLSU	tef1-a	rpb1	rpb2
Chiua virens	HKAS 49445	Yang 4604A	China	**KT990557**	**KT990757**	**KT990924**	**KT990393**
Chiua virens	HKAS 50543	Yang 4746	China	KT990550	KT990751	KT990918	**MT110452**
Chiua viridula	MHHNU 7346	Zhang 846	China	KT990561	—	—	KT990394
Chiua viridula	HKAS 74928 (Type)	Wu 614	China	KF112483	KF112273	KF112583	KF112794
Fistulinella olivaceoalba	HKAS 59845	Zeng 416	China	**MT154768**	**MT110369**	**MT110405**	**MT110443**
Fistulinella olivaceoalba	HKAS 53432	Li 1087	China	**MT154769**	**MT110370**	—	—
Fistulinella olivaceoalba	HKAS 53367	Li 1022	China	KF112439	KF112304	KF112615	KF112790
Fistulinella prunicolor	REH 9502	—	Australia	JX889648	JX889690	—	—
Fistulinella salmonea	HKAS 80671 (Type)	Zhao 246	China	**MT154766**	—	**MT110404**	—
Fistulinella salmonea	HKAS 106323	Li 289	China	**MT154767**	**MT110368**	—	—
Fistulinella viscida	PDD 75695	—	New Zealand	**MW114854**	—	—	—
Gymnogaster boletoides	NY 01194009	REH 9455	Australia	KT990572	KT990768	KT990928	KT990406
Harrya alpina	HKAS 52820 (Type)	Feng 99	China	KF112437	KF112270	KF112580	KF112792
Harrya atriceps	REH 7403	—	Costa Rica	JX889662	JX889702	—	—
Harrya atrogrisea	HKAS 50542 (Type)	Yang 4745	China	KT990694	KT990880	KT991024	KT990499
Harrya chromipes	HKAS 50527	Yang 4730	China	KF112437	KF112270	KF112580	KF112792
Harrya chromipes	HKAS 59217	Ge 2134	China	HQ326931	HQ326864	**MT110406**	**MT110451**
Harrya moniliformis	HKAS 49627	Li 429	China	KT990695	KT990881	KT991025	KT990500
Harrya moniliformis	HKAS 51181	Li 744	China	KT990591	KT990785	KT990939	KT990426
Harrya subalpina	HKAS 50546	Yang 4749	China	**KT990692**	**KT990879**	**KT991022**	**MT110450**
Hortiboletus amygdalinus	HKAS 54166 (Type)	Yang 5070	China	KT990581	KT990777	KT990933	KT990416
Hortiboletus subpaludosus	HKAS 59608	Li 1860	China	KF112371	KF112185	KF112551	KF112696

Continued

Continued

Taxon	Voucher ID	Coll. No.	Location	nrLSU	tef1-a	rpb1	rpb2
Hourangia cheoi	HKAS 52269	Yang 4952	China	KF112385	KF112286	KF112628	KF112773
Hourangia nigropunctata	HKAS 76657	Shi 390	China	KF112388	KF112287	KF112629	KF112774
Hymenoboletus filiformis	HKAS 82811 (Type)	Hao 1020	China	MT154777	MT110371	—	MT110446
Hymenoboletus griseoviridis	MHHNU 30371	Chen 30371	China	MT154780	MT110374	—	MT110449
Hymenoboletus griseoviridis	HKAS 75687 (Type)	Liu 76	China	MT154781	MW165269	—	—
Hymenoboletus griseoviridis	HKAS 78816	Zhao 1557	China	MT154782	MW165270	—	MW165288
Hymenoboletus jiangxiensis	HKAS 76988	Wu 816	China	MT154778	MT110372	—	MT110447
Hymenoboletus jiangxiensis	HKAS 77009 (Type)	Wu 837	China	MT154779	MT110373	—	MT110448
Hymenoboletus luteopurpureus	HKAS 55828	Liang 816	China	KF112471	KF112271	KF112581	KF112795
Hymenoboletus luteopurpureus	HKAS 41694 (Type)	Yang 3574	China	MW114853	—	—	—
Imleria badia	HKAS 74714	Yang M3	China	KF112375	—	KF112609	—
Imleria subalpina	HKAS 74712	Feng 773	China	KF112373	KF112189	KF112607	KF112706
Indoporus shoreae	AP 6693 (Type)	—	India	MK123973	—	—	MK243367
Indoporus shoreae	AP 6697	—	India	MK123976	—	—	MK243368
Indoporus squamulosus	HKAS 76299 (Type)	Hao 641	China	MT154708	MT110334	MT110375	—
Indoporus squamulosus	HKAS 107153	Zeng 784	China	MT154709	MT110335	MT110376	MT110409
Indoporus squamulosus	HKAS 84835	Han 539	China	MT154707	—	—	—
Leccinellum aff. *crocipodium*	HKAS 76658	Shi 397	China	KF112447	KF112252	KF112595	KF112728
Leccinellum castaneum	HKAS 55179 (Type)	Zhao 8184	China	MT154744	—	—	—
Leccinellum castaneum	HKAS 57592	Wu 60	China	KF112446	—	KF112594	KF112726
Leccinellum citrinum	HKAS 53410 (Type)	Li 1065	China	KT990585	—	KT990937	KT990421
Leccinellum citrinum	HKAS 53427	Li 1082	China	KF112488	KF112253	KF112596	KF112727
Leccinellum cremeum	HKAS 90639 (Type)	T 22183	China	—	KT990781	KT990936	KT990420

14　The Boletes of China: *Tylopilus* s.l.

Taxon	Voucher ID	Coll. No.	Location	nrLSU	tef1-a	rpb1	rpb2
Leccinellum crocipodium	930809/1	—	France	AF139694	—	—	—
Leccinellum corsicum	Buf 4507	—	USA	KF030347	KF030435	KF030389	—
Leccinellum griseopileatum	HKAS 77060 (Type)	Wu 888	China	MT154747 & MT154748	MT110357 & MT110358	—	MT110429
Leccinellum onychinum	HKAS 92188 (Type)	LF 700	China	KU321699	—	—	—
Leccinellum sinoaurantiacum	HKAS 89413	Li 2770	China	MT154745	MT110356	MT110394	MT110428
Leccinellum sinoaurantiacum	HKAS 36065 (Type)	Zang 13486	China	MT154746	—	—	—
Leccinum monticola	HKAS 76669	Shi 448	China	KF112443	KF112249	KF112592	KF112723
Leccinum quercinum	HKAS 63502	Wu 271	China	KF112444	KF112250	KF112593	KF112724
Leccinum scabrum	HKAS 57266	Feng 537	China	KF112442	KF112248	KF112590	KF112722
Mucilopilus cf. *castaneiceps*	HKAS 71039	Yang 5554	Japan	KT990547	KT990748	KT990915	KT990385
Mucilopilus cinnamomeus	HKAS 50229	Li 475	China	KF112423	KF112216	KF112574	KF112769
Mucilopilus cinnamomeus	HKAS 59572 (Type)	Li 1824	China	MT154761	—	MT110401	MT110435
Mucilopilus paracastaneiceps	HKAS 50338 (Type)	Li 584	China	KT990555	KT990755	KT990922	KT990391
Mucilopilus paracastaneiceps	HKAS 75065	Wu 750	China	KF112382	KF112211	—	KF112735
Mucilopilus ruber	HKAS 84555 (Type)	Han 259	China	MT154762	MT110364	—	MT110436
Mucilopilus ruber	HKAS 84632	Han 336	China	MT154763	MT110365	—	MT110437
Nigroboletus roseonigrescens	GDGM 43238 (Type)	—	China	KT220588	KT220595	KT220591	KT220594
Nigroboletus roseonigrescens	ZT 13553	—	China	KT220589	KT220596	KT220592	KT220594
Parvixerocomus aokii	HKAS 59812	Zeng 3	China	KF112378	KF112266	KF112597	—
Parvixerocomus pseudoaokii	HKAS 52633	Li 946	China	KF112379	KF112267	KF112598	KF112736
Phylloporus imbricatus	HKAS 68642	Feng 861	China	KF112398	KF112299	KF112637	KF112786
Phylloporus luxiensis	HKAS 75077	Wu 762	China	KF112490	KF112298	KF112636	KF112785

Continued

Chapter 2 Materials and methods

Continued

| Taxon | Voucher ID | Coll. No. | Location | GenBank accession numbers ||||
				nrLSU	tef1-a	rpb1	rpb2
Porphyrellus castaneus	HKAS 63076	Tang 1256	China	KT990548	KT990749	KT990916	KT990386
Porphyrellus castaneus	HKAS 52554 (Type)	Li 869	China	KT990697	KT990883	KT991026	KT990502
Porphyrellus cyaneotinctus	HKAS 80183	Hao 903	China	**MT154718**	**MT110340**	—	—
Porphyrellus cyaneotinctus	HKAS 80192	Hao 912	China	**MT154719**	—	—	—
Porphyrellus griseus	HKAS 82849 (Type)	Hao 1058	China	**MT154716**	—	—	**MT110414**
Porphyrellus orientifumosipes	HKAS 53372 (Type)	Li 1027	China	KT990629	KT990823	KT990975	KT990461
Porphyrellus orientifumosipes	HKAS 84710	Han 414	China	**MT154717**	**MT110339**	—	**MT110415**
Porphyrellus porphyrosporus	HKAS 49182	Ge 687	China	KT990544	KT990746	KT990912	KT990383
Porphyrellus pseudofumosipes	HKAS 103784 (Type)	530828 MF 0008	China	MW114845	—	—	—
Porphyrellus scrobiculatus	HKAS 53366 (Type)	Li 1021	China	KF112480	KF112241	KF112610	KF112716
Pseudoaustroboletus valens	HKAS 52602	Li 915	China	KM274869	KM274877	**MT110395**	**MT110430**
Pseudoaustroboletus valens	HKAS 82644	Li 690	China	**MT154749**	**MT110359**	**MT110396**	**MT110431**
Pulchroboletus roseoalbidus	AMB 12757	—	Italy	KJ729499	KJ729512	—	—
Pulveroboletus mirus	HKAS 59530	Li 1783	China	KT990617	KT990811	KT990967	KT990452
Pulveroboletus ravenelii	64/96	—	USA	KF030306	—	—	—
Retiboletus brunneolus	HKAS 106427	Li 764	China	**MT154752**	**MT110362**	—	—
Retiboletus brunneolus	HKAS 52680 (Type)	Li 993	China	KF112424	KF112179	—	KF112690
Retiboletus fuscus	HKAS 79727	Cui 47	China	MT010614	MT010624	**MT110399**	**MT110433**
Retiboletus fuscus	HKAS 63590	Wu 358	China	KF112417	KF112178	KF112537	KF112691
Retiboletus griseus	202/97	—	USA	AF456834	KF030414	KF030373	—
Retiboletus kauffmanii	HKAS 63548	Wu 317	China	KF112416	KF112177	KF112536	KF112689
Retiboletus nigrogriseus	FHMU 2800 (Type)	Jiang 66	China	MH367476	MH367488	—	—
Retiboletus nigrogriseus	FHMU 2045	Zeng 3084	China	MH367475	MH367487	—	—

15

Continued

Taxon	Voucher ID	Coll. No.	Location	nrLSU	tef1-a	rpb1	rpb2
Retiboletus pseudogriseus	HKAS 106467	Zeng 647	China	**MT154751**	**MT110361**	**MT110398**	**MT110432**
Retiboletus sinensis	HKAS 59832	Zeng 569	China	KT990633	KT990827	KT990978	KT990464
Retiboletus zhangfeii	HKAS 53420	Li 1075	China	**MT154750**	**MT110360**	**MT110397**	—
Retiboletus zhangfeii	HKAS 53418	Li 1073	China	KT990630	KT990824	KT990976	KT990462
Retiboletus zhangfeii	HKAS 59699	Li 1951	China	JQ928627	JQ928582	JQ928592	JQ928603
Retiboletus zhangfeii	HKAS 83962 (Type)	Zeng 1369	China	KP739296	KP739306	—	—
Royoungia boletoides	REH 8774	—	Australia	JX889660	JX889701	—	—
Royoungia coccineinana	HKAS 68927	Li 2165	China	KT990508	—	KT990889	KT990347
Royoungia coccineinana	HKAS 107311	Jia 645	China	**MW114848**	**MW165271**	—	**MW165287**
Royoungia grisea	HKAS 90183 (Type)	Shi 426	China	KT990546	—	KT990914	—
Royoungia palumana	REH 9445	—	Australia	JX889652	JX889693	—	—
Royoungia reticulata	HKAS 90197	Yang 4604B	China	KT990556	KT990756	KT990923	KT990392
Royoungia reticulata	HKAS 59704	Li 1957	China	KT990594	—	KT990942	**MT110445**
Royoungia rubina	HKAS 53379 (Type)	Li 1034	China	**MT154776**	KF112274	—	KF112796
Spongiforma thailandica	DED 7873	—	Thailand	EU685108	KF030436	KF030387	—
Strobilomyces echinocephalus	HKAS 59420	Li 1673	China	KF112463	KF112256	KF112600	KF112810
Strobilomyces floccopus	AFTOL-ID 716	MB 03-102	—	AY684155	AY883428	AY858964	AY858963
Strobilomyces seminudus	HKAS 59461	Li 1714	China	KF112479	KF112260	KF112606	KF112815
Suillus aff. *granulatus*	HKAS 57622	Wu 90	China	KF112429	KF112280	KF112645	KF112726
Suillus aff. *luteus*	HKAS 57748	Wu 216	China	KF112430	KF112281	KF112646	KF112824
Sutorius alpinus	HKAS 50420	Li 666	China	KT990549	KT990750	KT990917	KT990387
Sutorius alpinus	HKAS 59657	Li 1909	China	KT990707	KT990887	KT991029	KT990505
Sutorius alpinus	HKAS 52672 (Type)	Li 985	China	KF112399	KF112207	KF112584	KF112802

Continued

Taxon	Voucher ID	Coll. No.	Location	GenBank accession numbers			
				nrLSU	tef1-a	rpb1	rpb2
Sutorius australiensis	REH 9280	—	Australia	JQ327031	JQ327031	—	—
Sutorius eximius	HKAS 91261	Li 233	China	**MT154770**	—	—	**MT110444**
Sutorius eximius	REH 9400	—	USA	JQ327004	JQ327029	—	—
Sutorius microsporus	—	S.D. Yang 010	China	MH879697	MH879727	—	—
Sutorius microsporus	HKAS 56291	Li 1451	China	KF112400	KF112208	KF112585	KF112803
Sutorius microsporus	HKAS 68720 (Type)	Feng 939	China	**MT154773**	—	—	—
Sutorius obscuripellis	HKAS 106335	Li 432	China	**MT154771**	—	—	—
Sutorius obscuripellis	FHMU 2258	Zeng 3297	China	MH879701	MH879731	—	—
Sutorius obscuripellis	HKAS 107150	Wu 2070	China	**MT154772**	**MW165273**	—	—
Sutorius pseudotylopilus	HKAS 74813	Wu 499	China	**MT154774**	—	—	—
Sutorius pseudotylopilus	HKAS 77110	Wu 938	China	**MT154775**	—	—	—
Sutorius subrufus	FHMU 2004 (Type)	Zeng 3043	China	MH879698	MH879728	—	MH879745
Sutorius subrufus	FHMU 2258	Zeng 3297	China	MH879701	MH879731	—	—
Tengioboletus glutinosus	HKAS 53452	Li 1090	China	KT990655	KT990844	KT990994	KT990480
Tylocinum griseolum	HKAS 50209	Li 455	China	KT990551	—	KT990919	KT990388
Tylocinum griseolum	HKAS 50281 (Type)	Li 527	China	KF112451	KF112284	—	KF112730
Tylopilus alboater	TH 6941	—	USA	AY612832	—	—	—
Tylopilus albopurpureus	HKAS 84693	Han 397	China	**MT154739**	**MT110352**	**MT110390**	**MT110425**
Tylopilus albopurpureus	HKAS 88998 (Type)	Wu 1261	China	**MT154740**	**MT110353**	**MT110391**	**MT110426**
Tylopilus alpinus	HKAS 90204	Tian 385	China	KT990643	—	**MT110389**	—
Tylopilus alpinus	HKAS 55438	Feng 327	China	KF112404	KF112191	KF112538	KF112687
Tylopilus argillaceus	HKAS 90201	Li 814	China	KT990588	KT990783	—	**MT110419**
Tylopilus argillaceus	HKAS 90186	Li 558	China	KT990589	KT990784	MW165280	KT990424

18 The Boletes of China: *Tylopilus* s.l.

Continued

Taxon	Voucher ID	Coll. No.	Location	nrLSU	tef1-a	rpb1	rpb2
Tylopilus atripurpureus	HKAS 50208	Li 454	China	KF112472	KF112283	KF112620	KF112799
Tylopilus atronicotianus	SnWV	—	USA	KF030293	—	—	—
Tylopilus atroviolaceobrunneus	HKAS 84351 (Type)	Ge 3514	China	KT990625	KT990819	MT110384	MT110421
Tylopilus atroviolaceobrunneus	HKAS 107143	530828 MF 0035	China	MT154728	—	—	—
Tylopilus aurantiacus	Osmundson 1198	—	Thailand	EU430740	—	—	—
Tylopilus aurantiacus	HKAS 59700 (Type)	Li 1952	China	KF112458	KF112223	KF112619	KF112740
Tylopilus badiceps	MB03-052	—	USA	KF030336	—	—	—
"*Tylopilus balloui*"	Halling 8526	—	Belize	EU430736	—	EU434336	—
"*Tylopilus balloui*"	Halling 8521	—	Belize	EU430735	—	EU434339	—
Tylopilus balloui	Halling 8292	—	USA	EU430734	—	—	—
Tylopilus balloui	Osmundson 1030	—	USA	EU430737	—	EU434338	—
"*Tylopilus balloui*"	TH 6385	—	Guyana	AY612823	—	—	—
"*Tylopilus balloui*"	TH 8409	—	Guyana	HQ161873	—	HQ161842	—
Tylopilus brunneirubens	HKAS 59664	Li 1916	China	KT990709	MT110349	MT110387	MW165290
Tylopilus brunneirubens	HKAS 87073	Liu 426	China	MT154737	MT110350	—	MT110423
Tylopilus brunneirubens	HKAS 88992	Wu 1255	China	MT154738	MT110351	MT110388	MT110424
Tylopilus brunneirubens	HKAS 52609	Li 922	China	KT990626	KT990820	KT990972	—
Tylopilus castanoides	HKAS 92325	Liu 780	China	MT154723	MW165267	—	—
Tylopilus felleus	HKAS 54926	Rexer 8989	China	KF112411	HQ326866	KF112575	KF112737
Tylopilus formosus	PDD 72637	—	New Zealand	HM060319	—	—	—
Tylopilus fuligineoviolaceus	HKAS 92013	Cai 1393	China	MT154721	MT110341	MT110382	MT110416
Tylopilus fuscatus	HKAS 59838	Zeng 118	China	MW114857	—	—	—
Tylopilus griseiolivaceus	JBSD 127430	—	—	MN115801	—	—	—

Continued

Taxon	Voucher ID	Coll. No.	Location	nrLSU	tef1-a	rpb1	rpb2
Tylopilus griseiviridus	HKAS 99999 (Type)	Wu 1879	China	MW114856	—	—	—
Tylopilus griseipurpureus	HKAS 90200	Zeng 493	China	KT990624	KT990818	KT990971	KT990458
Tylopilus griseolus	HKAS 106432	Li 813	China	MT154734	MT110347	MT110386	—
Tylopilus griseolus	HKAS 106429	Li 782	China	MT154735	MT110348	MW165281	—
Tylopilus griseolus	HKAS 99967 (Type)	Wu 1847	China	MT154736	MW165266	—	MW165284
Tylopilus himalayanus	HKAS 107146	Hao 1445	China	MT154741	—	—	—
Tylopilus himalayanus	HKAS 91278	Li 250	China	MT154742	MT110354	MT110392	MT110427
Tylopilus himalayanus	HKAS 93425	Wang 419	China	MT154743	MT110355	MT110393	—
Tylopilus himalayanus	DC 17-25 (Type)	—	India	MG799328	—	—	—
Tylopilus jiangxiensis	HKAS 107152 (Type)	Wu 2177	China	MT154731	—	—	—
Tylopilus jiangxiensis	HKAS 105250	Zhao 1221	China	MN304779	MN304797	MN304785	MN304791
Tylopilus jiangxiensis	HKAS 105251	Zhao 1222	China	MN304780	MN304798	MN304786	MN304792
Tylopilus leucomycelinus	JBSD 127420	—	Dominican Republic	MN115803	—	—	MN095210
Tylopilus neofelleus	HMAS 84730 (Type)	—	China	KM975494	—	—	—
Tylopilus obscureviolaceus	HKAS 80115	Hao 835	China	MT154730	MT110345	MT110385	MT110422
Tylopilus obscureviolaceus	HKAS 90202	Zeng 444	China	KT990623	KT990817	KT990970	KT990457
Tylopilus olivaceobrunneus	HKAS 107145 (Type)	Hao 1433	China	MT154722	—	—	—
Tylopilus oradivensis	NY	Halling 8187	Costa Rica	EU430732	—	EU434342	—
Tylopilus otsuensis	HKAS 50240	Li 458	China	KT990553	KT990753	KT990921	MT110417
Tylopilus phaeoruber	HKAS 74925 (Type)	Wu 611	China	KF112473	KF112222	KF112577	KF112739
Tylopilus plumbeoviolaceoides	GDGM 21040	—	China	MT154720	—	—	—
Tylopilus plumbeoviolaceoides	GDGM 20311 (Type)	—	China	KM975498	—	—	—
Tylopilus plumbeoviolaceus	NYBG 0009	—	USA	KY432825	—	—	—

Taxon	Voucher ID	Coll. No.	Location	nrLSU	tef1-a	rpb1	rpb2
Tylopilus plumbeoviolaceus	190/83	–	USA	AF457405	–	–	–
Tylopilus pseudoalpinus	HKAS 84602 (Type)	Han 306	China	MW114855	MW165268	–	–
Tylopilus pseudoballoui	DC 17-30 (Type)	–	India	MG799327	–	–	–
Tylopilus pseudoballoui	HKAS 51151	Li 714	China	KT990590	MW165265	–	KT990425
Tylopilus purpureorubens	HKAS 80106	Hao 826	China	MT154726	MT110343	–	–
Tylopilus purpureorubens	HKAS 80605 (Type)	Wu 1230	China	MT154727	MT110344	–	MT110420
Tylopilus rubrobrunneus	BD 329	–	USA	HQ161876	–	HQ161845	–
Tylopilus rubrotinctus	HKAS 106346	Li 557	China	MW114852	–	–	–
Tylopilus rubrotinctus	HKAS 106435	Li 843	China	MW114851	–	–	MW165291
Tylopilus rubrotinctus	HKAS 78271	Li 1101	China	MT154732	MT110346	–	–
Tylopilus rubrotinctus	HKAS 80684 (Type)	Zhao 259	China	MT154733	MW165264	–	MW165283
Tylopilus rufobrunneus	HKAS 106331 (Type)	Li 389	China	MT154714	MT110337	MT110380	MT110413
Tylopilus rufobrunneus	HKAS 106466	Zeng 370	China	MT154715	MT110338	MT110381	–
Tylopilus rufonigricans	TH 6376	–	Guyana	AY612835	–	–	–
Tylopilus subotsuensis	HKAS107180	Zeng818	China	MW114847 & MW114846	–	–	–
Tylopilus subotsuensis	HKAS 106464	Zeng 119	China	MT154724 & MT154725	MT110342	MT110383	MT110418
Tylopilus variobrunneus	snHor 02	–	–	KF030316	–	–	–
Tylopilus vinaceipallidus	HKAS 50210	Li 456	China	KF112431	KF112221	KF112576	KF112738
Tylopilus violaceobrunneus	HKAS 89443 (Type)	Li 2800	China	KT990702	KT990886	KT991028	KT990504
Tylopilus violaceorubrus	HKAS 107154 (Type)	Wu 1956	China	MT154729	–	–	MW165286
Tylopilus virescens	FHMU 1004	Zeng 1464	China	MG365898	MG365904	–	–

Continued

Continued

Taxon	Voucher ID	Coll. No.	Location	nrLSU	tef1-a	rpb1	rpb2
Tylopilus virescens	FHMU 2812	Zeng 1360	China	MG365895	—	—	—
Veloporphyrellus alpinus	HKAS 68301	Zhu 125	China	JX984538	JX984550	—	**MT110438**
Veloporphyrellus alpinus	HKAS 57490 (Type)	Feng 761	China	KF112380	KF112209	KF112555	KF112733
Veloporphyrellus castaneus	MHHNU 30640	Chen 30640	China	**MT154765**	**MT110367**	—	**MT110442**
Veloporphyrellus castaneus	HKAS 107147 (Type)	Shi 368	China	**MT154764**	**MT110366**	—	**MT110441**
Veloporphyrellus gracilioides	HKAS 53590 (Type)	Ge 1504	China	KF112381	KF112210	KF112556	KF112734
Veloporphyrellus gracilis	112/96	—	USA	DQ534624	KF030425	KF030358	—
Veloporphyrellus pseudovelatus	HKAS 52258 (Type)	Yang 4941	China	JX984540	JX984551	**MT110402**	**MT110439**
Veloporphyrellus pseudovelatus	HKAS 59444	Li 1697	China	JX984542	JX984553	**MT110403**	**MT110440**
Veloporphyrellus velatus	HKAS 63668	Zeng 763	China	JX984546	JX984554	—	—
Xanthoconium sinense	HKAS 80118	Hao 838	China	KT990666	KT990855	KT991004	KT990490
Xanthoconium stramineum	3518	—	USA	KF030353	KF030428	KF030386	—
Xerocomellus chrysenteron	Xch 1	—	Germany	AF050647	KF030415	KF030365	—
Xerocomellus zelleri	REH 8724	—	USA	KF030271	KF030416	KF030366	—
Xerocomus fulvipes	HKAS 76666	Shi 438	China	KF112390	KF112292	KF112631	KF112789
Xerocomus subtomentosus	Xs 1	—	Germany	AF139716	JQ327035	KF030391	—
Zangia chlorinosma	HKAS 48695	Yang 4531	China	HQ326939	—	**MW165274**	—
Zangia citrina	HKAS 52684 (Type)	Li 997	China	HQ326941	HQ326872	**MW165278**	—
Zangia erythrocephala	HKAS 52843	Feng 122	China	HQ326943	**MW165272**	**MW165276**	—
Zangia erythrocephala	HKAS 52844 (Type)	Feng 123	China	HQ326944	HQ326873	**MW165277**	—
Zangia olivacea	HKAS 45445 (Type)	Yang 3960	China	HQ326945	HQ326873	**MW165275**	—
Zangia olivaceobrunnea	HKAS 52272	Yang 4955	China	HQ326948	HQ326876	**MW165279**	**MW165289**
Zangia roseola	HKAS 51137	Li 700	China	KF112414	KF112269	KF112579	KF112791

Chapter 3
Morphology and structures in *Tylopilus* s.l.

A typical mushroom in *Tylopilus* s.l. has a fruit body (basidioma) consisting of a pileus, hymenophore, and stipe (cap, tubes, and stalk, respectively, in nontechnical terms). The basidioma (pl. basidiomata or basidiomes) produces the basidiospores, and since the basidiospores are produced on a cell called basidium (pl. basidia), the structure on the fruit body which produces the basidia is called the hymenophore. Basidioma, pileus, hymenophore and stipe are the terms that will be used hereinafter.

Basidioma: The size of the basidioma in *Tylopilus* s.l. varies from 1.5 cm to 22 cm in diameter and mostly fall between 5 cm and 15 cm. The terms referring to the size of the basidiomata have been defined with reference to the standard proposed by Bas (1969), i.e., very small (pileus ⩽ 3 cm wide), small (pileus 3–5 cm wide), medium-sized (pileus 5–9 cm wide), large (pileus 9–15 cm wide), and very large (pileus ⩾ 15 cm wide).

Pileus: The pileal surface can have a wide range of colors in *Tylopilus* s.l. and sometimes the color can change due to age, exposure to the sun, or other environmental influences. The pileal surface may be glabrous, tomentose, fibrillose, or glutinous. The context of pileus is usually white or yellow and without discoloration when touched, but in some species the color changes when the context is touched or injured.

Hymenophore: Species in *Tylopilus* s.l. have a hymenophore made up of vertical tubes on the undersurface of the pileus, within which the reproductive basidiospores are developed; the color is pallid to pinkish when young and pink to purplish pink or reddish brown to vinaceous when mature.

Stipe: The surface of the stipe may be glabrous, tomentose, fibrillose, glutinous, or reticulate. The base of the stipe is always covered with a cottony mycelium. The context of the stipe is usually the same color as the pileus context, but sometimes, especially towards the base, the color may be different. The color and discoloration of the basal mycelium and context are often diagnostic characteristics.

Pileipellis: The pileipellis is the uppermost layer of hyphae in the pileus. It covers the pileal trama, the fleshy tissue of the pileus. Different groups of species may have different types of pileipellis (Fig. 3.1), and the type of pileipellis is therefore an important characteristic in the identification of species in *Tylopilus*. We have defined the terms of pileipellis types as follows:

Trichoderm (Fig. 3.1 a): a pileipellis made up of narrow (not more than 7 μm wide) vertically arranged hyphae;

Ixotrichoderm (Fig. 3.1 b): a trichoderm made up of gelatinizing hyphae;

Intricate trichoderm (Fig. 3.1 c): a trichoderm made up of interwoven hyphae;

Intricate ixotrichoderm (Fig. 3.1 d): an intricate trichoderm made up of gelatinizing hyphae;

Palisadoderm (Fig. 3.1 e): a pileipellis made up of broad (up to or more than 8 μm wide) hyphae;

Ixopalisadoderm (Fig. 3.1 f): a palisadoderm, made up of gelatinizing hyphae;

Epithelium (Fig. 3.1 g): a pileipellis made up of erect moniliform hyphae;

Hyphoepithelium (Fig. 3.1 h): an epithelium with an upper layer of filamentous hyphae;

Ixohyphoepithelium (Fig. 3.1 i): a hyphoepithelium made up of gelatinizing hyphae;

Chapter 3 Morphology and structures in *Tylopilus* s.l. 23

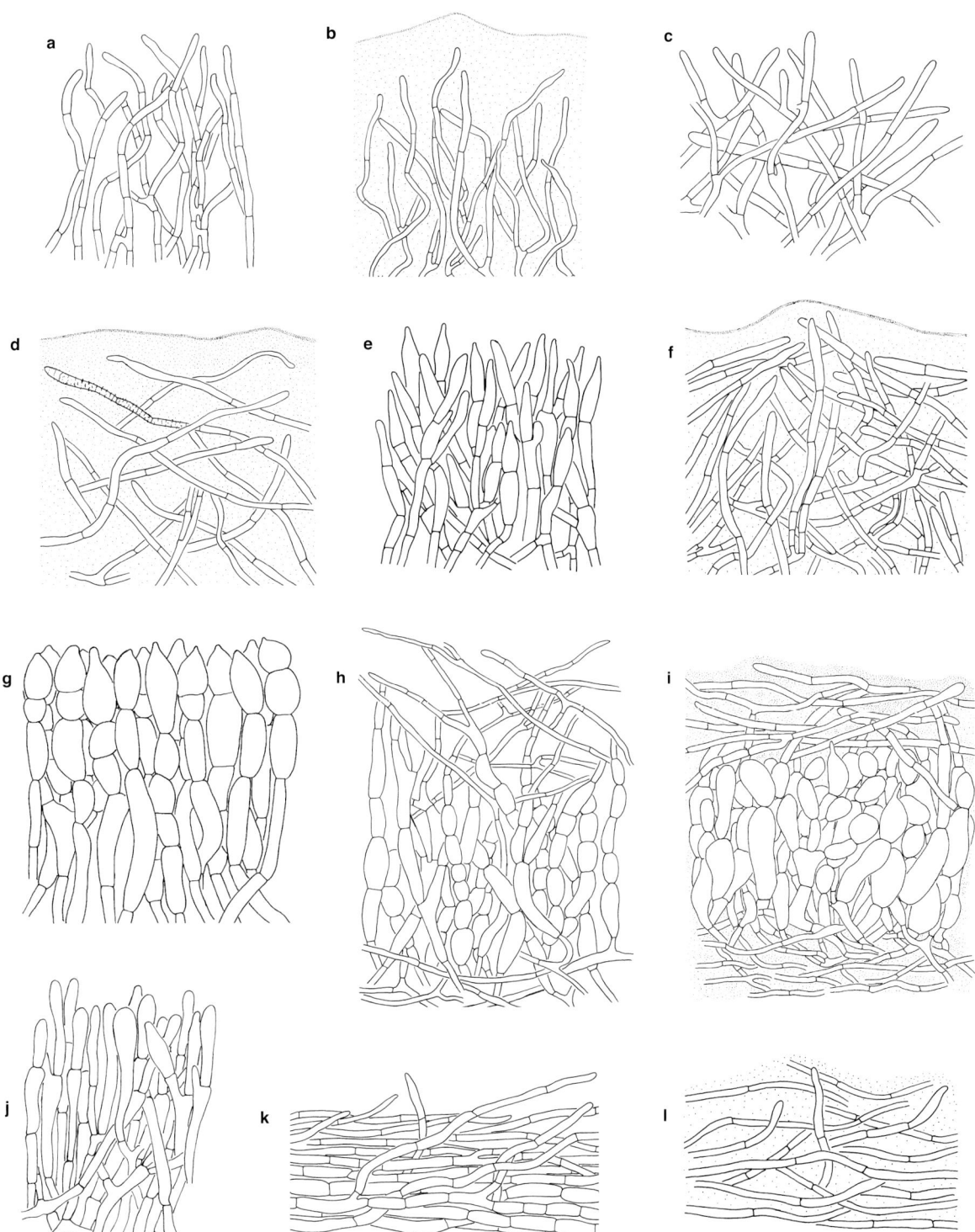

Fig. 3.1 The types of pileipellis in *Tylopilus* s.l.
a. Trichoderm; b. Ixotrichoderm; c. Intricate trichoderm; d. Intricate ixotrichoderm; e. Palisadoderm; f. Ixopalisadoderm; g. Epithelium; h. Hyphoepithelium; i. Ixohyphoepithelium; j. Hymeniderm; k. Cutis; l. Ixocutis.

Hymeniderm (Fig. 3.1 j): a pileipellis made up of a layer of cells and looks like a hymenium;

Cutis (Fig. 3.1 k): a pileipellis consisting of repent non-gelatinizing filamentous hyphae;

Ixocutis (Fig. 3.1 l): a cutis made up of gelatinizing hyphae, resulting in a slime layer on the pileus.

Hymenium: The hymenium is a tissue layer on the tube surface, where the cells develop into basidia and cystidia. The subhymenium consists of the supportive hyphae from which the cells of the hymenium grow, and beneath which is the tube trama (Fig. 3.2).

Basidia: Basidia are the basidiospore-bearing cells and produced in the hymenium. Most species of *Tylopilus* s.l. have clavate basidia, most often 4-spored, sometimes 2-spored. Most species have basidia that are 25–50 μm long, but a few species have basidia that are relatively shorter and less than 25 μm long.

Basidiospores: The sizes and shapes of the basidiospores play an important part in distinguishing species. The shape of the basidiospores in *Tylopilus* ranges from subglobose, ellipsoid or elongated to cylindrical or fusiform (Fig. 3.3). As the variation in shape is mainly due to the variation of the length/

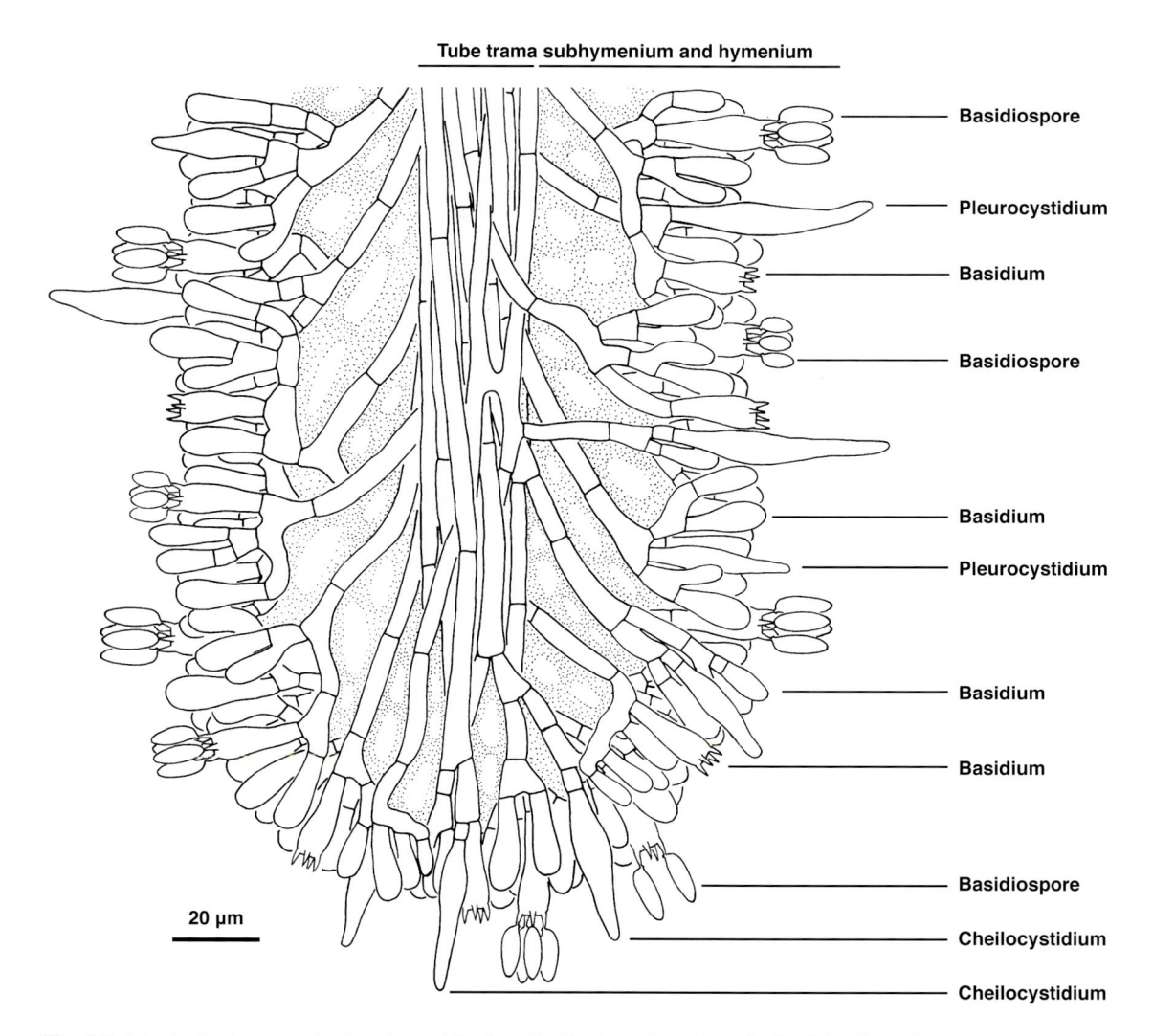

Fig. 3.2 A typical microscopic structure of the longitudinal section of a tube in *Tylopilus* s.l.

Chapter 3 Morphology and structures in *Tylopilus* s.l. 25

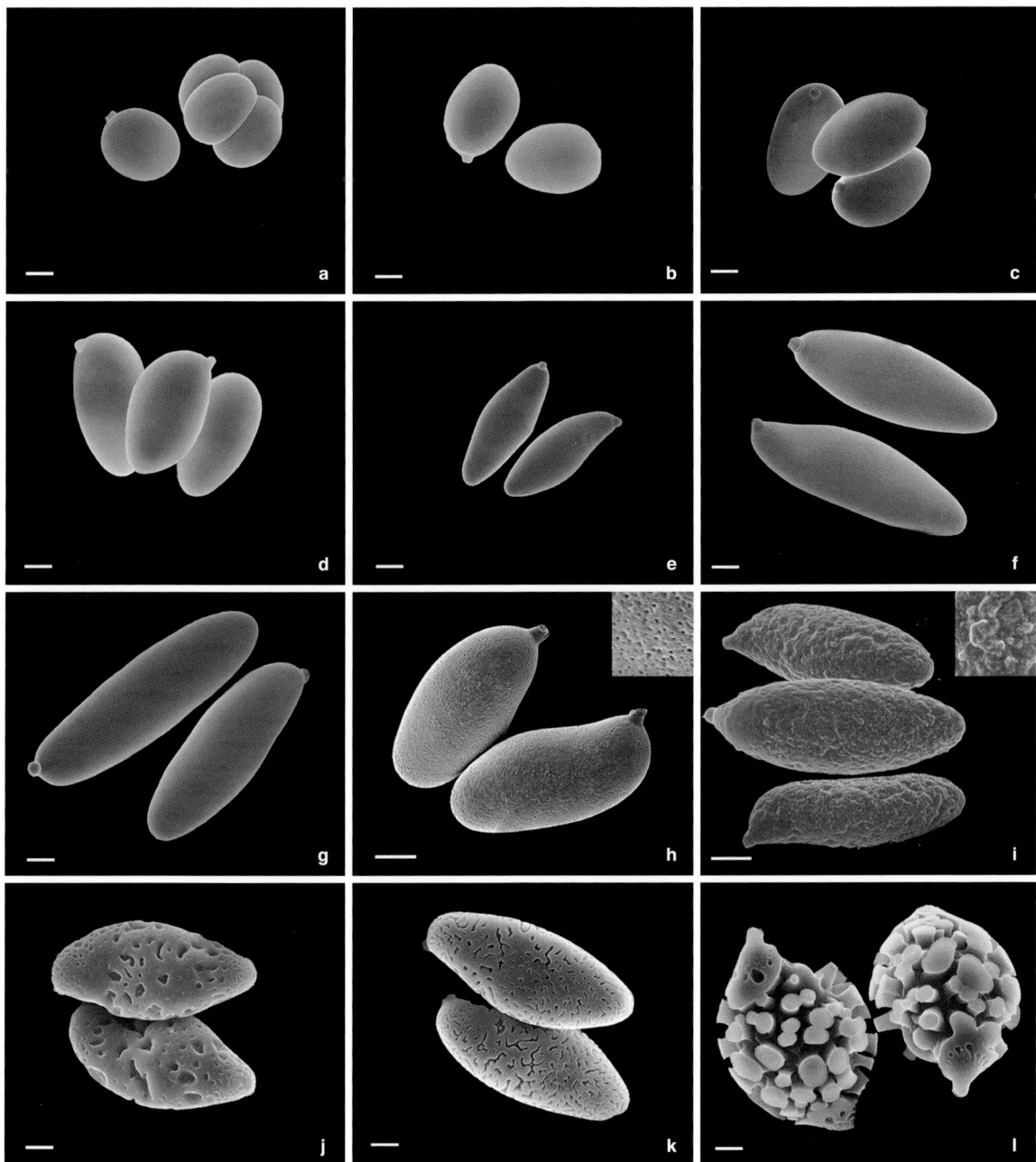

Fig. 3.3 Shapes and ornamentations of basidiospores in *Tylopilus* s.l. under SEM, a–g. shapes of basidiospores (surface smooth), h–l. ornamentations of basidiospores.
a. Subglobose to broadly ellipsoid: *Tylopilus pseudoballoui* D. Chakr. et al. (KUN-HKAS 51151); b. Ellipsoid: *Tylopilus otsuensis* Hongo (KUN-HKAS 50212); c, d. Elongated: c. *Anthracoporus nigropurpureus* (Hongo) Yan C. Li & Zhu L. Yang (KUN-HKAS 52685), d. *Mucilopilus paracastaneiceps* Yan C. Li & Zhu L. Yang (KUN-HKAS 50338); e, f. Fusiform: e. *Indoporus squamulosus* Yan C. Li & Zhu L. Yang (KUN-HKAS 76299), f. *Leccinellum castaneum* Yan C. Li & Zhu L. Yang (KUN-HKAS 55179); g. Cylindrical: *Sutorius alpinus* Yan C. Li & Zhu L. Yang (KUN-HKAS 50415); h. Finely pitted: *Mucilopilus cinnamomeus* Yan C. Li & Zhu L. Yang (KUN-HKAS 59572); i. Warty: *Veloporphyrellus castaneus* Yan C. Li & Zhu L. Yang (MHHNU 30640); j. Pitted: *Austroboletus albovirescens* Yan C. Li & Zhu L. Yang (KUN-HKAS 85661); k. Reticulate: *Austroboletus olivaceoglutinosus* K. Das & Dentinger (KUN-HKAS 57756); l. Cylindrical protuberance: *Austroboletus fusisporus* (Imazeki & Hongo) Wolfe (KUN-HKAS 53461). Scale bars = 2 μm.

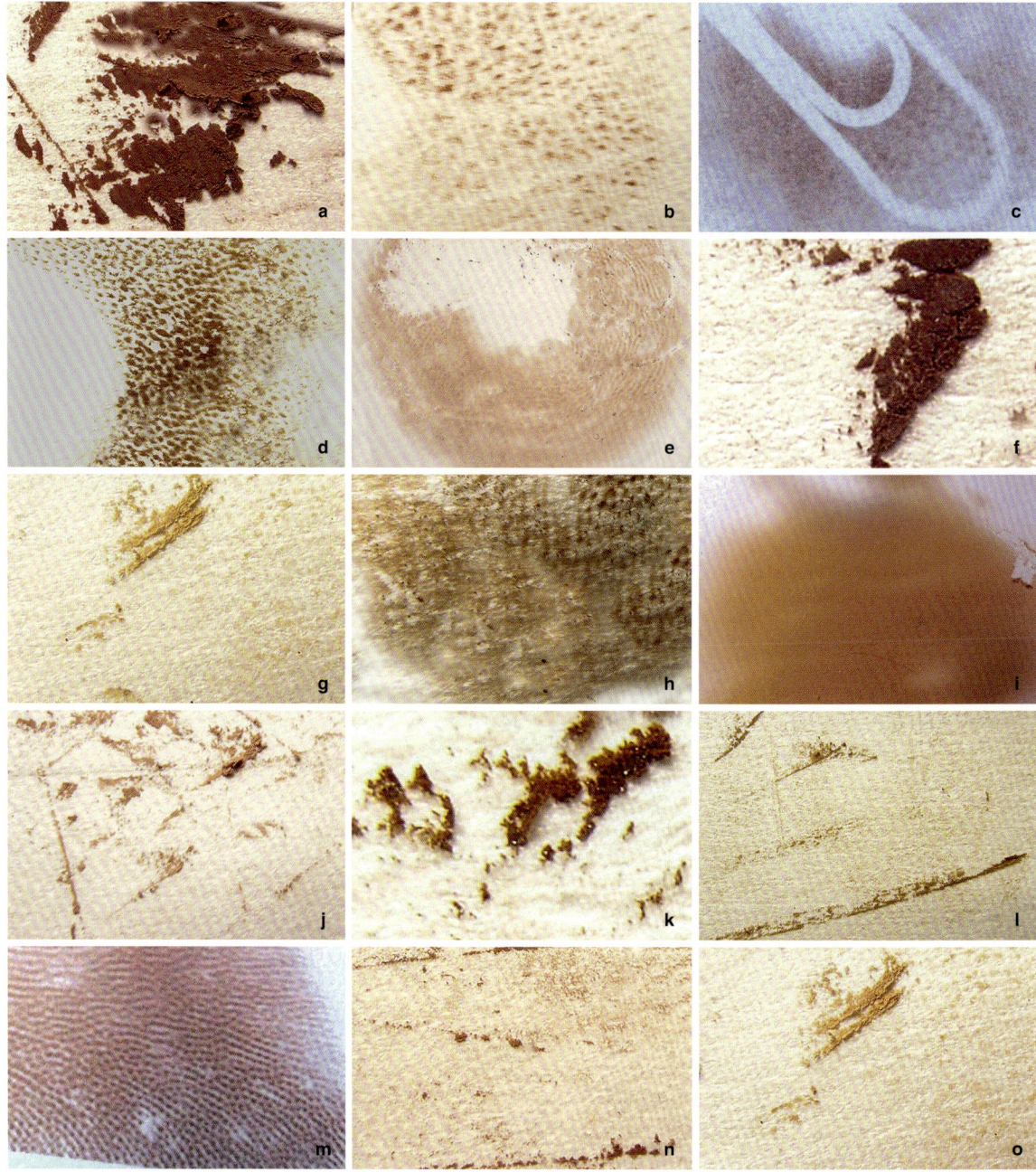

Fig. 3.4 Spore prints in *Tylopilus* s.l.
a. *Abtylopilus scabrosus* Yan C. Li & Zhu L. Yang (KUN-HKAS 50211); b. *Anthracoporus nigropurpureus* (Hongo) Yan C. Li & Zhu L. Yang (KUN-HKAS 52685); c. *Austroboletus fusisporus* (Imazeki & Hongo) Wolfe (KUN-HKAS 50238); d. *Chiua virens* (W.F. Chiu) Yan C. Li & Zhu L. Yang (KUN-HKAS 50364); e. *Harrya moniliformis* Yan C. Li & Zhu L. Yang (KUN-HKAS 49627); f. *Leccinellum castaneum* Yan C. Li & Zhu L. Yang (KUN-HKAS 55179); g. *Mucilopilus paracastaneiceps* Yan C. Li & Zhu L. Yang (KUN-HKAS 50338); h. *Porphyrellus castaneus* Yan C. Li & Zhu L. Yang (KUN-HKAS 50491); i. *Pseudoaustroboletus valens* (Corner) Yan C. Li & Zhu L. Yang (KUN-HKAS 52603); j. *Royoungia reticulata* Yan C. Li & Zhu L. Yang (KUN-HKAS 59704); k. *Sutorius alpinus* Yan C. Li & Zhu L. Yang (KUN-HKAS 50415); l. *Tylocinum griseolum* Yan C. Li & Zhu L. Yang (KUN-HKAS 50209); m. *Tylopilus neofelleus* Hongo (KUN-HKAS 52600); n. *Veloporphyrellus pseudovelatus* Yan C. Li & Zhu L. Yang (KUN-HKAS 52673); o. *Zangia erythrocephala* Yan C. Li & Zhu L. Yang (KUN-HKAS 56273).

width ratio, we have defined the terms of the shapes of basidiospores as follows:

Subglobose to broadly ellipsoid (Fig. 3.3 a): length/width ratio = 1.05–1.3;

Ellipsoid (Fig. 3.3 b): length/width ratio = 1.3–1.6;

Elongated (Fig. 3.3 c, d): length/width ratio = 1.6–2;

Fusiform (Fig. 3.3 e, f): the shape of this type is fusiform to subfusiform in side view with suprahilar slightly depressed at apex. This is the most common basidiospore shape in *Tylopilus* s.l. and even in the Boletaceae, and is often referred to as "boletoid basidiospores";

Cylindrical (Fig. 3.3 g): length/width ratio = 2–3.

The walls of the basidiospores in *Tylopilus* s.l. are always smooth. Exceptions are the ornamented basidiospores in *Mucilopilus* (Fig. 3.3 h), *Veloporphyrellus* (Fig. 3.3 i), and *Austroboletus* (Fig. 3.3 j–l). The color of the spore print is usually pinkish to purplish pink or reddish brown to brownish (Fig. 3.4 a–l).

Cystidia: The cystidia are relatively large cells found on the surface of the tube, often among clusters of basidia. They may occur on the edge of a tube (cheilocystidia), on the face of tube (pleurocystidia), or on the surface of stipe (caulocystidia). Since cystidia have highly varied and distinct shapes that are often unique to a particular species or genus, they are useful microscopic characteristics in the identification of *Tylopilus* s.l. species. The shapes of cystidia range from clavate to lanceolate, subfusiform, subfusoid-mucronate, ventricose, ventricose-mucronate, or ventricose-mucronate with a long pedicel, and finger-like (Fig. 3.5 a–i).

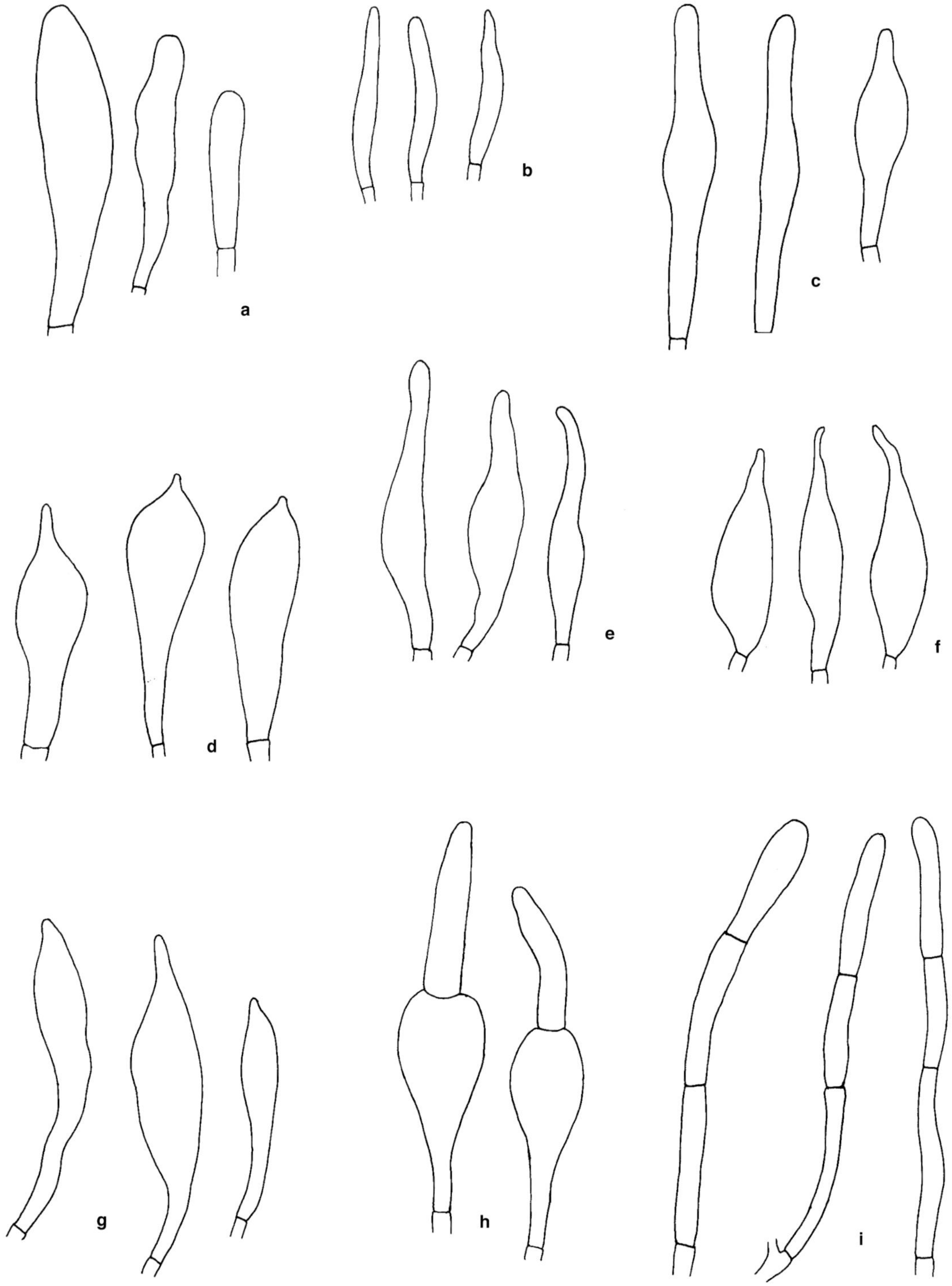

Fig. 3.5 Types of cystidia in *Tylopilus* s.l.
a. Clavate; b. Lanceolate; c. Fusiform to subfusiform; d. Subfusoid-mucronate; e. Ventricose; f. Ventricose-mucronate; g. Ventricose-mucronate with a long pedicel; h. Finger-like with lower cell inflated; i. Finger-like without inflated lower cell.

Chapter 4

Phylogenetic results and systematic treatments of *Tylopilus* s.l.

In the narrowed multi-locus matrix, there were 775 sequences comprising four genes (273 for nrLSU, 196 for *tef1-α*, 152 for *rpb1*, and 154 for *rpb2*) obtained from 274 samples. Of these, 238 sequences (88 for nrLSU, 55 for *tef1-α*, 42 for *rpb1*, and 53 for *rpb2*) from 88 samples were newly generated in the present study. An aligned matrix of 3286 bp was retrieved. The Maximum Likelihood (ML) analysis was conducted and the ML tree with bootstrap support (MLB) is displayed in Fig. 4.1.

Our phylogenetic analysis (Fig. 4.1) reveals that *Tylopilus* s.l. is polyphyletic and clusters into 19 genus-level clades corresponding to 19 genera, of which 17 genera (*Austroboletus*, *Chiua*, *Fistulinella*, *Harrya*, *Hymenoboletus*, *Indoporus*, *Leccinellum* Bresinsky & Manfr. Binder, *Mucilopilus*, *Porphyrellus*, *Pseudoaustroboletus*, *Retiboletus* Manfr. Binder & Bresinsky, *Royoungia*, *Sutorius*, *Tylocinum* Yan C. Li & Zhu L. Yang, *Tylopilus*, *Veloporphyrellus*, and *Zangia*) have been reported, and two new genera (*Abtylopilus* Yan C. Li & Zhu L. Yang and *Anthracoporus* Yan C. Li & Zhu L. Yang) were proposed in this book. All of the 19 genera can be classified into five of the currently recognized seven subfamilies (or groups) in Boletaceae, i.e., Austroboletoideae G. Wu & Zhu L. Yang, Boletoideae Singer, Leccinoideae G. Wu & Zhu L. Yang, Zangioideae G. Wu, *et al.*, and the *Pulveroboletus* group. Among the 19 genera, two genera, namely *Leccinellum* and *Retiboletus* not only have pallid to ferrugineous or pinkish to grayish pink hymenophores but also have yellow to yellow-brown hymenophores. In this book, only species with pallid to pale ferrugineus or pinkish hymenophores in these two genera were documented and illustrated. While all species in the remaining 18 genera have pallid to pinkish or rose-cored hymenophores, which is an important characteristic to identify species traditionally placed in *Tylopilus*, were documented and illustrated in this book.

In total, 105 species (including 34 new species, and four new combinations) can be delimited using the GCPSR approach together with morphological and ecological evidence from China (Fig. 4.1).

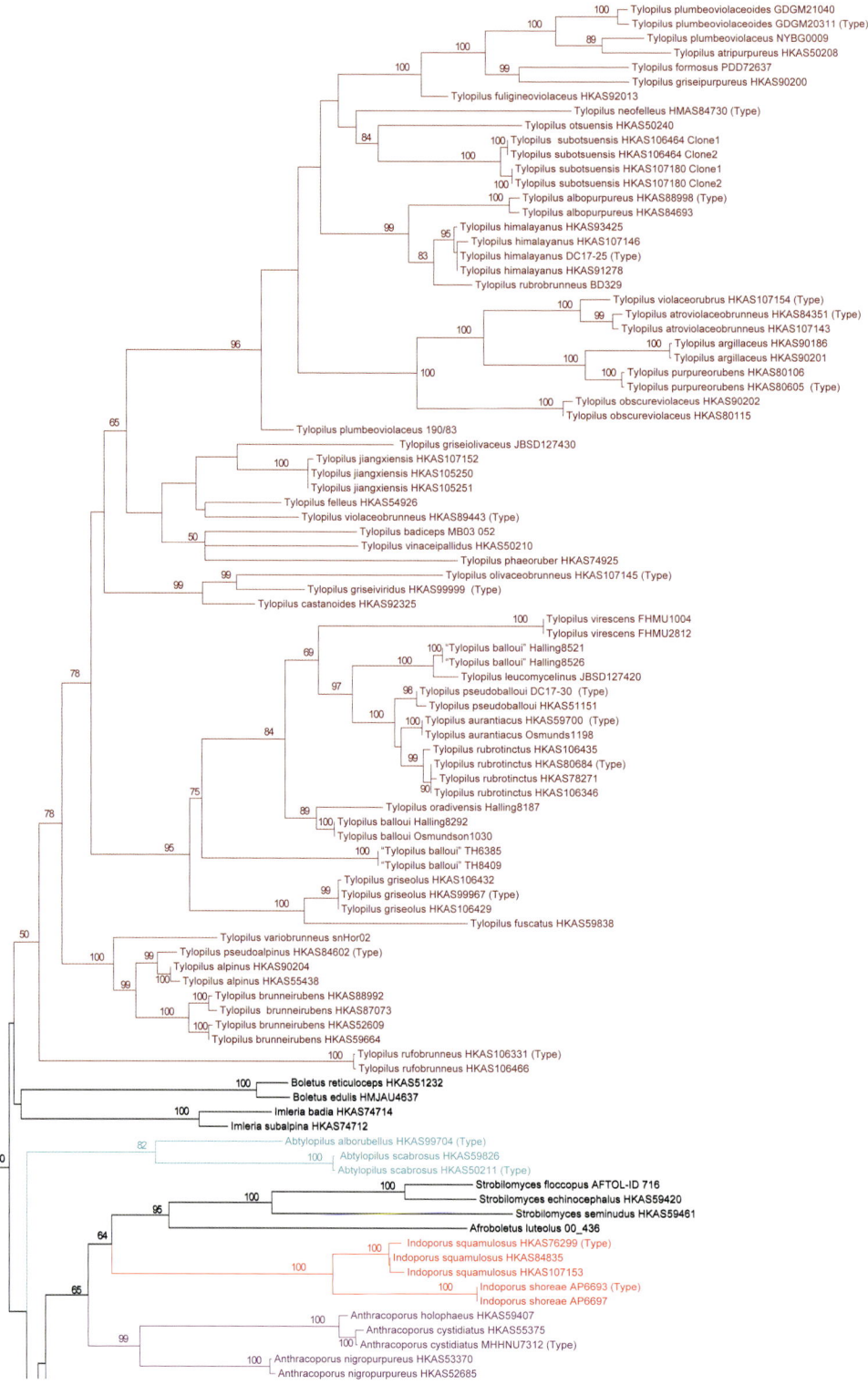

Fig. 4.1 Phylogenetic relationships among species in *Tylopilus* s.l. (shown in different colored fonts on the tree) inferred from a multigene (nrLSU, *tef1-a*, *rpb1* and *rpb2*) dataset using the Maximum Likelihood method. The bootstrap frequencies (≥ 50%) are shown at the branches. The seven subfamilies of Boletaceae are indicated with solid arrows on the left.

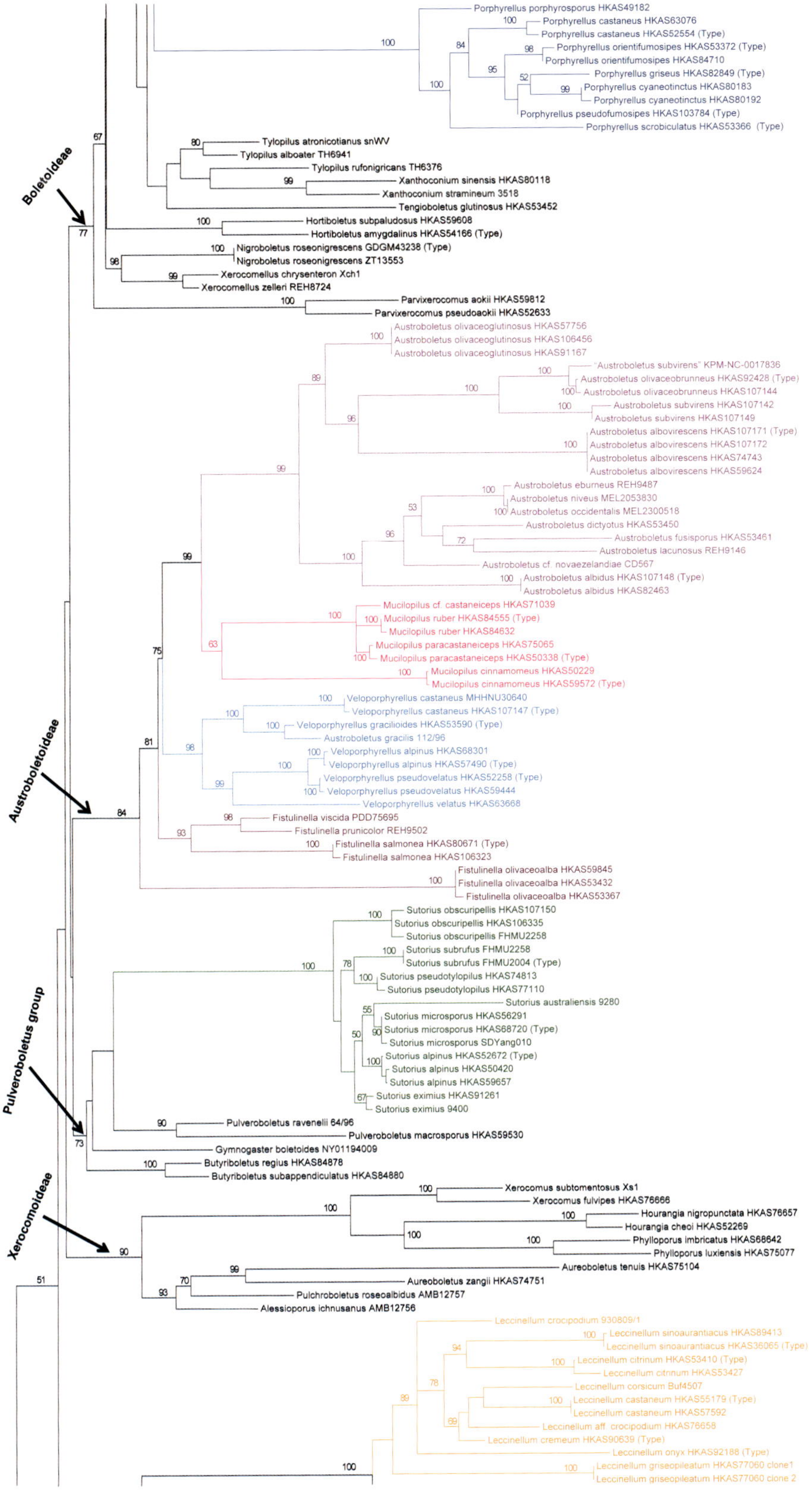

Chapter 4 Phylogenetic results and systematic treatments of *Tylopilus* s.l.   31

Continued Fig. 4.1

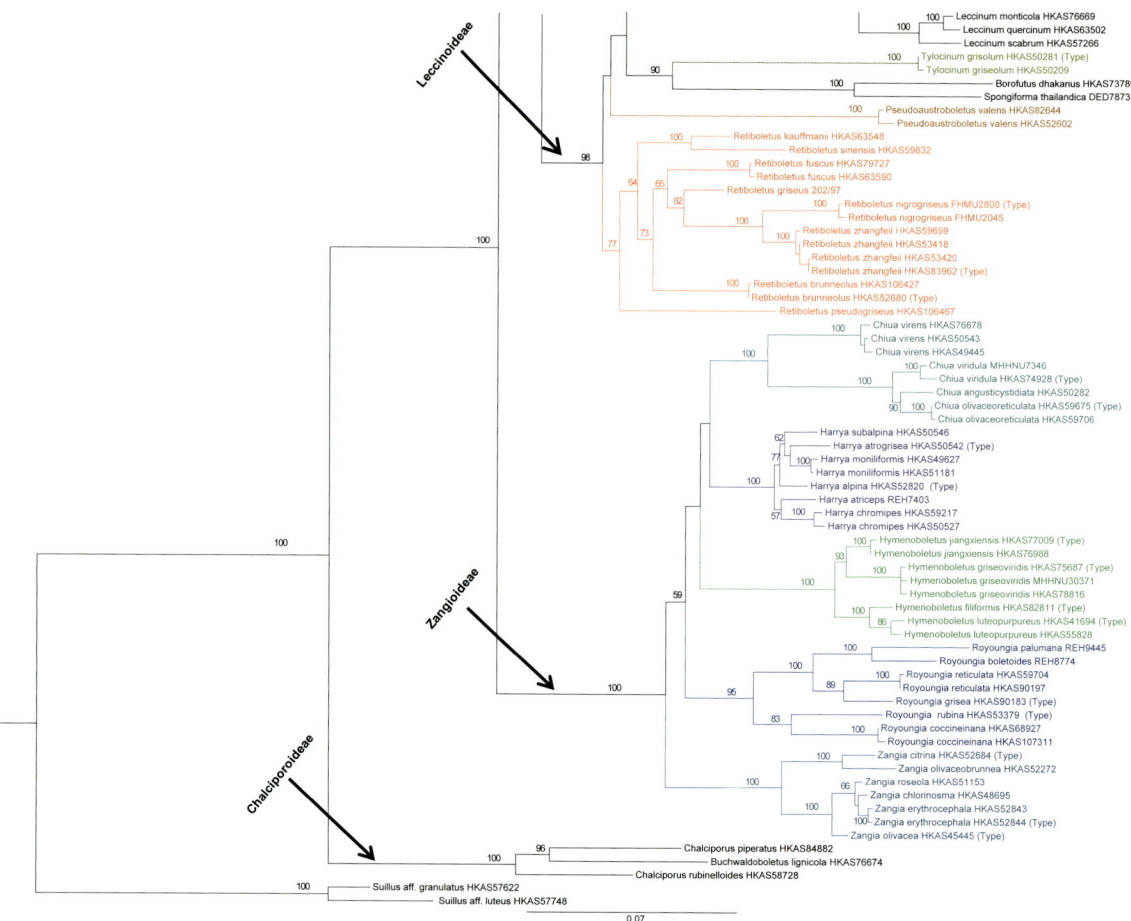

Continued Fig. 4.1

Chapter 5

Taxonomic part

To aid identification, a key to the 19 genera in China is provided below. For those genera with more than three species in China, a key to the species is also provided.

Key to the genera of *Tylopilus* s.l. in China

1. Younger basidiomata with distinct membranous veil or appendiculate pileal margin; cheilocystidia often finger-like with 1–2 septa ·· 2
1. Younger basidiomata with neither membranous veil nor appendiculate pileal margin; cheilocystidia rarely finger-like ·· 3
2. Basidiospores distinctly ornamented under light microscopy; stipe with distinct reticulum ········ ·· *Austroboletus*
2. Basidiospores smooth under light microscopy; stipe without reticulum ············ *Veloporphyrellus*
3. Context staining red first, then black when injured; pileus with umber or dark tinge ················ 4
3. Context unchanged or changed in color but rarely never staining red at first and then black when injured; pileus with bright or umber to dark tinge ·· 6
4. Hymenophore blackish or white to cream when young and pinkish when mature, pores of the tube narrow (up to 2 mm wide) ·· 5
4. Hymenophore gray to grayish white when young, grayish pink when mature, pores of the tube wide (2–4 mm wide) ·· *Indoporus*
5. Basidiomata mostly dark brown to blackish when mature; pileal surface rugose or covered with coarsely fibrillose squamules, always rimose when mature; hymenophore blackish when young and grayish pink when mature ·· *Anthracoporus*
5. Basidiomata without black tinge when mature; pileal surface nearly glabrous, not rimose; hymenophore white to cream when young, and white to pinkish when mature ············ *Abtylopilus*
6. Surface of stipe covered with yellow, red to pinkish red or purplish pink scabrous squamules; stipe base bright yellow to chrome-yellow ·· 7
6. Surface of stipe glabrous or covered with differently colored squamules or reticulum; stipe base neither bright yellow nor chrome-yellow ·· 11
7. Context of pileus white to cream, while context of stipe white to cream or pale yellow upwards and yellow to chrome-yellow downwards, without color change or staining blue asymmetrically when injured; stipe surface covered with rose to pink or reddish scabrous squamules ············ 8

7. Context of pileus and stipe yellow to bright yellow, without discoloration when injured; stipe surface covered with yellow or pink to reddish scabrous squamules ·································· *Chiua*
8. Pileal surface nearly glabrous or fibrillose; context of stipe without discoloration when injured; pileipellis not an ixohyphoepithelium ··· 9
8. Pileal surface often rugose; context always asymmetrically bluish when injured; pileipellis an ixohyphoepithelium, gelatinous ··· *Zangia*
9. Pileal surface rarely with rubescent tinge when touched or damaged by animals; stipe surface yellowish to yellow or red to purplish red in the middle part and yellow to chrome-yellow at base, always covered with rose to pink scabrous squamules ·· 10
9. Pileal surface always with distinct rubescent tinge when touched or damaged by animals; stipe surface white to cream but yellow to chrome-yellow at base, always densely covered with rose to pink scabrous squamules ··· *Harrya*
10. Stipe yellowish to yellow at apex, purplish red in the middle part and yellow to chrome-yellow at base; surface covered with purplish pink scabrous squamules; pileus olive-green or brownish to brownish red ··· *Hymenoboletus*
10. Stipe yellowish to yellow without purplish tinge, lower part yellow to chrome-yellow, without purplish tinge; surface covered with pink to reddish scabrous squamules; pileus yellowish green to grayish green, or reddish to red ·· *Royoungia*
11. Surface of stipe with distinct reticulum; pileal surface dry; context in the base of stipe staining yellow tinge when injured ·· 12
11. Surface of stipe without distinct reticulum or with reticulum but restrict to the upper part; pileal surface dry or gelatinous; context in the base of stipe rarely staining yellow tinge when injured ···· ·· 13
12. Surface of stipe covered with membranous ridged reticulum; context white, without color change, but occasionally with yellowish discoloration at the base of stipe when bruised ············· ·· *Pseudoaustroboletus*
12. Surface of stipe always with reticulum; context of pileus pallid or greenish, but always yellow at the base of stipe ··· *Retiboletus*
13. Pileal surface strongly gelatinous, or viscid when wet; pileal margin always extended ··········· 14
13. Pileal surface dry, non-gelatinous; pileal margin not extended ··· 15
14. Pileus and stipe strongly gelatinous; context of stipe white without discoloration when injured; spore print brown to reddish brown; pileipellis an ixocutis or an ixotrichoderm ········ *Fistulinella*
14. Pileus viscid when wet, but stipe unviscid; context in the base of stipe staining yellow when injured; spore print pink to grayish pink; pileipellis an ixotrichoderm ······················ *Mucilopilus*
15. Stipe surface nearly glabrous or reticulate in some species, without dotted or scabrous squamules ·· 16
15. Stipe surface with dotted or scabrous squamules which are darkened when mature or aged,

without reticulum ·· 17
16. Basidioma bright colored; context mostly with bitter taste, without color change or staining reddish but rarely blue when injured ··· *Tylopilus*
16. Basidioma umber or dark colored; context without bitter taste, always with a bluish discoloration when bruised ··· *Porphyrellus*
17. Surface of stipe densely covered with dotted squamules but never red or purplish brown; hymenophore without dark purple, purplish red or purplish brown tinge; context without color change or staining reddish when injured ·· 18
17. Surface of stipe covered with red or purplish brown dotted squamules; hymenophore dark purple, purplish red or purplish brown; context without color change or staining reddish when injured ··· *Sutorius*
18. Surface of stipe covered with small gray, dark gray to black granular squamules; hymenophore white to dingy white; pores of the tube wide (1–2 μm wide); context staining red when injured ····
·· *Tylocinum*
18. Surface of stipe covered with yellowish red or gray, dark gray to black rough and verrucose squamules; hymenophore white to dingy white or yellowish to orange-red; pores of the tube narrow (up to 1 μm wide); context unchanged in color when injured ························· *Leccinellum*

Chapter 6

Abtylopilus Yan C. Li & Zhu L. Yang

Abtylopilus Yan C. Li & Zhu L. Yang, The Boletes of China: *Tylopilus* s.l. 39 (2021)

MycoBank: MB 834691

Etymology: "*Abtylopilus*" refers to its similarity to *Tylopilus*.

Type species: *Abtylopilus scabrosus* Yan C. Li & Zhu L. Yang.

Diagnosis: This genus differs from other genera in the Boletaceae in the nearly glabrous pileus, the white to cream or grayish and then grayish pink hymenophore, the fine hymenophore pores (0.3–1 mm wide), the initially red and then black discoloration of the context when injured, and the palisadoderm pileipellis composed of broad vertically arranged hyphae.

Basidiomata stipitate-pileate with tubular hymenophore. Pileus hemispherical to applanate; surface nearly glabrous, dry, margin slightly extended; context solid, white, grayish white to cream, becoming red then black when injured. Hymenophore adnate or depressed around apex of stipe; surface cream to whitish when young, whitish to grayish pink when mature; tubes concolorous with hymenophoral surface; pores angular, fine (0.3–1 mm wide); staining at first red and then black when injured. Stipe central, often enlarged downwards, concolorous with pileal surface, becoming first red then black when injured; context solid to spongy, white to cream, becoming first red then black when bruised; basal mycelium cream. Spore print brownish red to brown-red (Fig. 3.4 a). Basidiospores subfusiform to cylindrical; surface smooth. Hymenial cystidia fusiform to subfusoid-ventricose. Pileipellis a palisadoderm composed of broad vertically arranged hyphae. Clamp connections absent in all tissues.

Commentary: Species of *Abtylopilus* are easily misidentified as *Tylopilus* based on the color of the hymenophore. However, this genus can be distinguished from genera in the Boletaceae by the combination of the glabrous pileus, the white to cream or grayish and then grayish pink hymenophore, the fine hymenophore pores, the initially red and then black discoloration, and the palisadoderm pileipellis. Phylogenetically, the genus forms a distinct clade and nests in the subfamily Boletoideae. However, its phylogenetically related genera are yet unknown (Fig. 4.1). The initially red and then black discoloration in *Abtylopilus* is similar to that of *Strobilomyces*, but species in *Strobilomyces* have distinctly ornamented basidiospores. Currently, two species, both newly described here, are known to occur in China.

6.1 *Abtylopilus alborubellus* Yan C. Li & Zhu L. Yang, The Boletes of China: *Tylopilus* s.l. 39 (2021)

MycoBank: MB 834692

Etymology: The epithet "*alborubellus*" refers to the color of the basidiomata.

Type: CHINA, HAINAN PROVINCE: Changjiang County, Bawangling National Nature Reserve, alt. 800 m, 26 June 2016, G. Wu 1585 (KUN-HKAS 99704, GenBank Acc. No.: MT154713 for nrLSU).

Basidioma small. Pileus 3–5 cm in diam., subhemispherical, pale gray (9C1) or gray (11D1) to brownish gray (9D1–2), often with grayish red tinge, becoming initially red to dark red (11C7–8) or violet-brown

Fig. 6.1 a *Abtylopilus alborubellus* Yan C. Li & Zhu L. Yang. Photos by G. Wu (KUN-HKAS 99704, type).

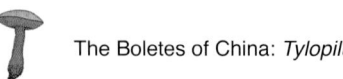

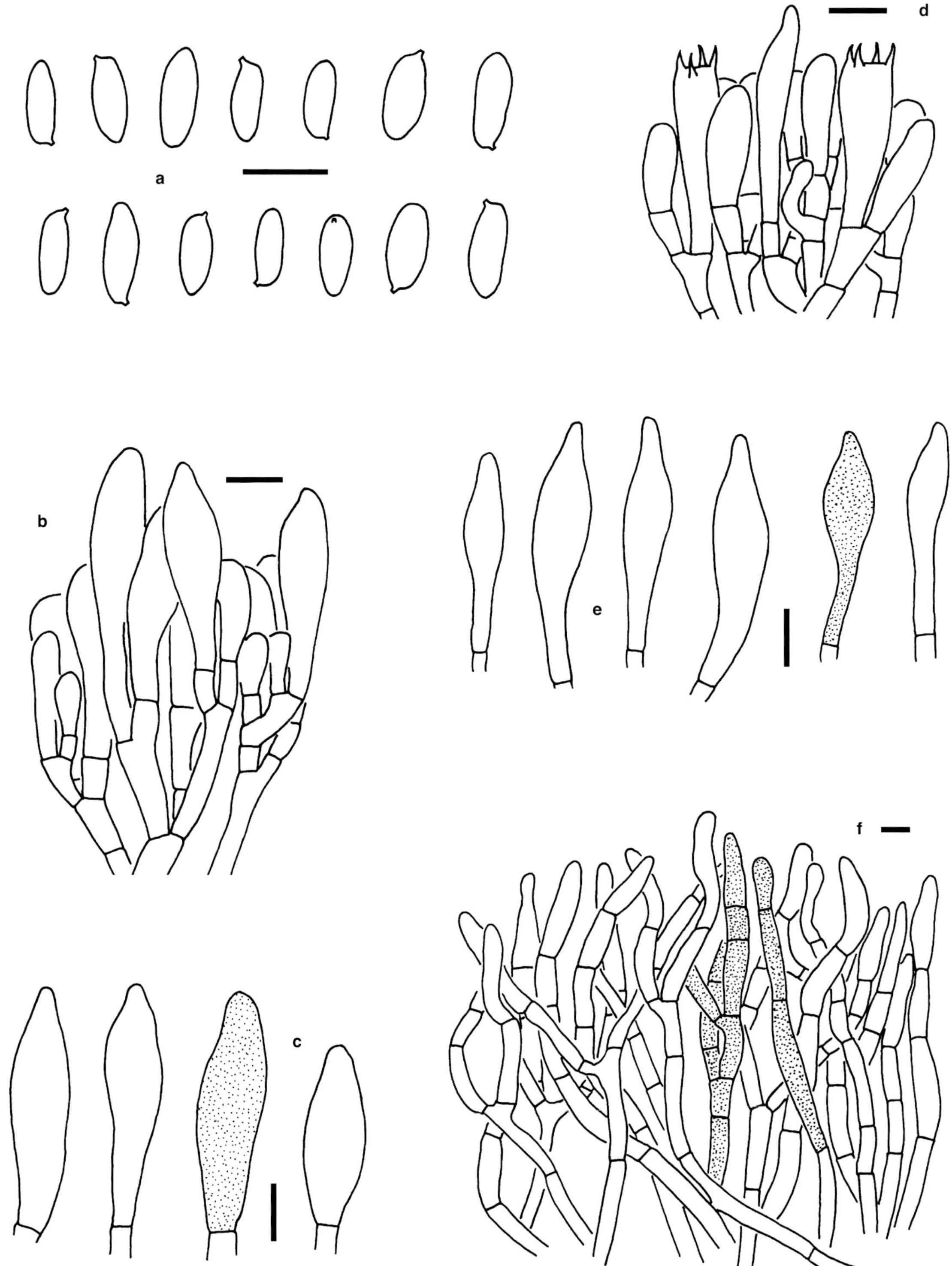

Fig. 6.1 b *Abtylopilus alborubellus* Yan C. Li & Zhu L. Yang (KUN-HKAS 99704, type).
a. Basidiospores; b. Basidia and cheilocystidia; c. Cheilocystidia; d. Basidia and pleurocystidium; e. Pleurocystidia; f. Pileipellis. Scale bars = 10 μm.

(10F7–8) to dark violet-brown (11F6–7), and then black when injured, slightly darker in the center; surface with white to pale gray (9C1) fine farinose squamules, dry, margin often extended; context solid, white to grayish, staining initially red and then black when hurt. Hymenophore adnate; surface whitish to pale gray (13B1) or gray (11C1), staining at first red and then black when bruised; pores subangular to roundish, 0.3–1 mm wide; tubes 4–8 mm long, concolorous or a little paler than hymenophoral surface, staining at first red and then black when bruised. Spore print brownish red (8D5–6). Stipe clavate, 5–7 × 1–1.5 cm, white to grayish at apex, grayish brown (11D1–2) in the lower part, often covered with white to grayish (9C1) farinose squamules, staining at first red and then black when touched; context solid, white, staining at first red and then black when hurt; basal mycelium white, becoming at first red and then black when hurt. Taste and odor unknown.

Basidia 25–35 × 7–10 μm, clavate to narrowly clavate, hyaline in KOH, 4-spored, sometimes 2-spored, hyaline in KOH. Basidiospores [40/1/1] 9–12 × 3–5 μm (Q = 2.25–3.33, Q_m = 2.53 ± 0.24), subcylindrical or subfusiform and inequilateral in side view with slight suprahilar depression, oblong to fusiform in ventral view, smooth, yellowish to brownish yellow in KOH, yellow to yellow-brown in Melzer's reagent. Hymenophoral trama boletoid composed of 2–11 μm wide filamentous hyphae, hyaline to yellowish in KOH, yellowish to yellow in Melzer's reagent. Cheilocystidia 32–51 × 9–12 μm, broadly subfusiform to fusoid-ventricose, brown to pinkish brown in KOH, thin-walled. Pleurocystidia 35–45 × 6–10 μm, similar to cheilocystidia, but much narrower, thin-walled. Pileipellis a palisadoderm composed of 4–9 μm wide vertically arranged hyphae, yellowish brown to brownish in KOH, and yellow-brown to dark brown in Melzer's reagent; terminal cells 18–40 × 5–8.5 μm, subfusiform to cystidioid. Pileal trama composed of 5–12 μm wide interwoven hyphae, yellowish brown to pale brownish in KOH, and yellowish brown to brownish in Melzer's reagent. Clamp connections absent in all tissues.

Habitat: Scattered on soil in tropical forests dominated by plants of the family Fagaceae.

Known distribution: Currently known from southern China.

Commentary: *Abtylopilus alborubellus* is characterized by the pale gray or gray to brownish gray pileus often with grayish red tinge and staining initially red to dark red or violet-brown and then black when injured, the white to grayish context staining at first red and then black when injured, the whitish to pale gray or grayish hymenophore staining at first red and then black when bruised, the fine hymenophoral pores 0.3–1 mm wide, the grayish brown stipe covered with concolorous farinose squamules, the smooth basidiospores, the palisadoderm pileipellis composed of 4–9 μm wide vertically arranged hyphae, the distribution in tropical forests, and the association with plants of the family Fagaceae. It is similar to *Ab. scabrosus* Yan C. Li & Zhu L. Yang in the discoloration of the context and the structure of pileipellis. Phylogenetically, they cluster together with high support (Fig. 4.1). However, *Ab. scabrosus* is characterized by its usually stubby stipe covered with dark scabrous squamules, and relatively short basidiospores measuring 8–10.5 × 3.5–4.5 μm.

6.2 *Abtylopilus scabrosus* Yan C. Li & Zhu L. Yang, The Boletes of China: *Tylopilus* s.l. 43 (2021)

MycoBank: MB 834693

Etymology: The epithet "*scabrosus*" refers to the scabrous stipe.

Type: CHINA, YUNNAN PROVINCE: Jinghong County, Dadugang Town, alt. 1350 m, 7 July 2006, Y.C. Li 457 (KUN-HKAS 50211, GenBank Acc. No.: KT990552 for nrLSU, KT990752 for *tef1-α*, KT990920 for *rpb1*, KT990389 for *rpb2*).

Basidioma small to large. Pileus 3–15 cm in diam., hemispherical to convex, grayish red (8C4–5) to brownish red (11D5–6) or ruby (12D3–5), slightly darker in the center; surface with tomentose squamules, dry, becoming at first red and then black when injured; context solid, white to cream, becoming at first red and then black when injured. Hymenophore adnate or sometimes depressed around apex of stipe; surface gray (14B1) to grayish pink (14C2), becoming at first red and then black when injured; tubes concolorous with hymenophoral surface; pores angular to roundish, 0.3–0.5 mm wide; tubes 6–30 mm long, concolorous or a little paler than hymenophoral surface, staining at first red and then black when injured. Spore print brownish red (8D5–6). Stipe 6.5–12 × 2.5–3.8 cm, stubby, subcylindrical, always enlarged downwards, white to dingy white, becoming at first red and then black when injured; surface always densely covered with dark scabrous squamules; context spongy, white to cream, becoming at first red and then black when bruised; basal mycelium cream to dingy white, becoming at first red and then black when injured. Taste and odor mild.

Basidia 28–55 × 16–17 μm, clavate, 4-spored, rarely 2-spored, hyaline in KOH. Basidiospores [160/8/5] 8–10.5 × 3.5–4.5 μm (Q = 2.29–2.71, Q_m = 2.58 ± 0.15), subcylindrical in side view with slight suprahilar depression, oblong to cylindrical in ventral view, smooth, hyaline to yellowish in KOH, yellow to brownish yellow in Melzer's reagent. Hymenophoral trama boletoid composed of 3.5–11 μm wide filamentous hyphae, hyaline to yellowish in KOH, yellowish to yellow in Melzer's reagent. Cheilo- and pleurocystidia 40–50 × 10–13 μm, fusiform to subfusoid-ventricose, yellowish to pale brownish in KOH, thin-walled. Pileipellis a palisadoderm composed of 5–11 μm wide filamentous hyphae, yellowish brown to brownish in KOH and

Fig. 6.2 a *Abtylopilus scabrosus* Yan C. Li & Zhu L. Yang. Photos by Y.C. Li (KUN-HKAS 50211, type).

yellow-brown to dark brown in Melzer's reagent; terminal cells 19–52 × 5–10.5 μm, clavate to subcylindrical. Pileal trama composed of 4–18 μm wide interwoven hyphae, yellowish brown to pale brownish in KOH and yellowish brown to brownish in Melzer's reagent. Clamp connections absent in all tissues.

Fig. 6.2 b *Abtylopilus scabrosus* Yan C. Li & Zhu L. Yang. Photos by F. Li (KUN-HKAS 90185).

42 The Boletes of China: *Tylopilus* s.l.

Habitat: Scattered on soil in tropical forests dominated by plants of the family Fagaceae.

Known distribution: Currently known from southern and southwestern China.

Additional specimens examined: CHINA, HAINAN PROVINCE: Ledong County, Jianfengling, alt. 850 m, 6 August 2009, N.K. Zeng 487 and N.K. Zeng 499 (KUN-HKAS 59826 and KUN-HKAS 90198, respectively). GUANGDONG PROVINCE: Fengkai County, Heishiding Nature Reserve, alt. 250 m, 23 May 2012, F. Li 350 and F. Li 372 (KUN-HKAS 106326 and KUN-HKAS 90185, respectively).

Commentary: *Abtylopilus scabrosus* is characterized by the grayish red to brownish red or ruby pileus, the gray to grayish pink hymenophore, the darkened scabrous stipe, the initially red and then black discoloration when touched and the palisadoderm pileipellis composed of 5–11 μm wide vertically arranged hyphae. It is similar to *Ab. alborubellus* in the discoloration when injured, the structure of pileipellis, and the distribution in tropical forests. However, *Ab. alborubellus* differs in its pale gray to brownish gray pileus, pale gray or grayish hymenophore, white to grayish or grayish brown stipe covered with white to grayish farinose squamules, and slightly longer basidiospores than *Ab. scabrosus*.

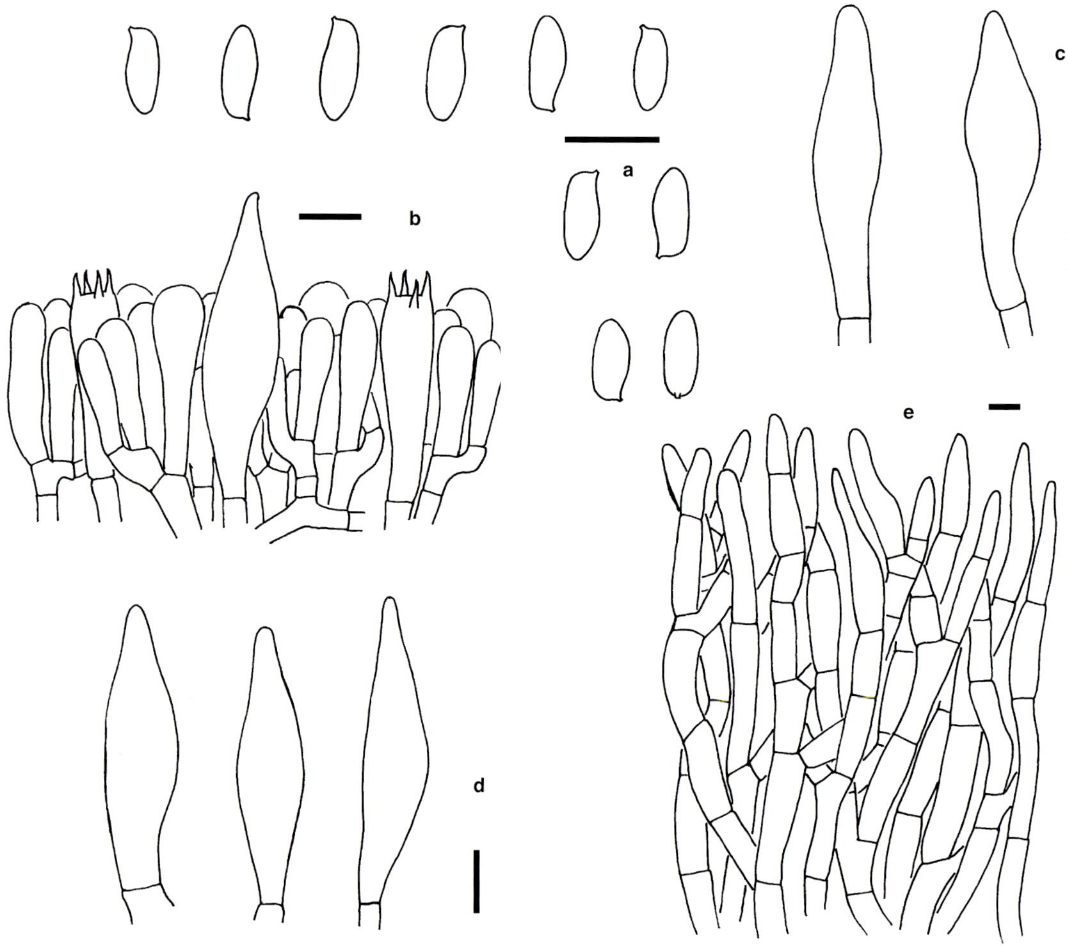

Fig. 6.2 c *Abtylopilus scabrosus* Yan C. Li & Zhu L. Yang (KUN-HKAS 50211, type).
a. Basidiospores; b. Basidia and cheilocystidium; c. Pleurocystidia; d. Cheilocystidia; e. Pileipellis. Scale bars = 10 μm.

Chapter 7
Anthracoporus Yan C. Li & Zhu L. Yang

Anthracoporus Yan C. Li & Zhu L. Yang, The Boletes of China: *Tylopilus* s.l. 49 (2021)

MycoBank: MB 834694

Etymology: "*Anthracoporus*" refers to the initially blackish hymenophore of the basidioma.

Type species: *Anthracoporus holophaeus* (Corner) Yan C. Li & Zhu L. Yang ≡ *Boletus holophaeus* Corner, Trans Br mycol Soc 59: 180 (1972).

Diagnosis: This genus differs from other genera in the Boletaceae in its tomentose or rugose pileus, black to grayish black hymenophore when young and then becoming grayish pink when mature, fine hymenophore pores (0.3–2 mm wide), initially red and then black discoloration of the context when injured, and trichoderm or epithelium pileipellis.

Basidiomata stipitate-pileate with tubular hymenophore. Pileus hemispherical to applanate; surface tomentose, usually rugose, dry; context solid, white to grayish white, becoming at first red and then black when injured. Hymenophore adnate or depressed around apex of stipe; surface black to grayish black when young, grayish white to grayish pink when mature, staining at first red and then black when injured; pores angular to roundish, fine; tubes concolorous with hymenophoral surface, staining at first red and then black when injured. Spore print grayish red. Stipe central, glabrous or reticulate, concolorous with pileal surface, becoming at first red and then black when injured; context white to cream, becoming at first red and then black when bruised; basal mycelium cream, becoming at first red and then black when bruised. Basidiospores smooth, elongated to cylindrical or subfusiform. Hymenophoral cystidia fusiform to subfusoid-ventricose. Pileipellis often a palisadoderm or an epithelium. Clamp connections absent in all tissues.

Commentary: The red and then black discoloration, the grayish pink hymenophore and the smooth basidiospores in *Anthracoporus* are similar to those of *Abtylopilus*. However, *Anthracoporus* differs from *Abtylopilus* in its darker basidiomata, tomentose or rugose pileus, and black to grayish black hymenophores in younger basidiomata. Currently, three species including two new combinations and one new species occur in China.

Key to the species of *Anthracoporus* in China

1. Stipe surface not reticulate, or, if reticulate only on the upper part; pileipellis an epithelium composed of inflated concatenated cells, or a palisadoderm composed of vertically arranged broad hyphae ·· 2
1. Stipe surface distinctly reticulate; pileipellis a trichoderm composed of interwoven filamentous hyphae ··· *An. nigropurpureus*
2. Pileus umbrous purple to umbrous red or blackish brown to blackish red; stipe with indistinct reticulum on upper half; basidiospores 10.5–13.5 × 4–5 µm; pileipellis a palisadoderm composed of 5.5–9.5 µm wide vertically arranged hyphae ···················· *An. holophaeus*
2. Pileus grayish red to brownish red or ruby red; stipe nearly glabrous; basidiospores short 9–10.5 × 4–5 µm; pileipellis an epithelium made up of 10–21 µm wide moniliform hyphae ·····················
·· *An. cystidiatus*

7.1 *Anthracoporus cystidiatus* Yan C. Li & Zhu L. Yang, The Boletes of China: *Tylopilus* s.l. 50 (2021)

MycoBank: MB 834695

Etymology: The epithet "*cystidiatus*" refers to the structure of the terminal cells in pileipellis.

Type: CHINA, HUNAN PROVINCE: Dong'an County, Shunhuangshan National Nature Reserve, alt. 900 m, 14 July 2010, P. Zhang 812 (MHHNU 7312, GenBank Acc. No.: MT154710 for nrLSU, MT110377 for *rpb1*, MT110411 for *rpb2*).

Basidioma small. Pileus 3–5 cm in diam., convex to applanate, grayish red (8C4–5) to brownish red (11D5–6) or ruby red (12D3–5), slightly darker in the center; surface with tomentose squamules, dry, staining at first red and then black when touched; context solid, white, staining at first red and then black when bruised. Hymenophore adnate or sometimes depressed around apex of stipe; surface black (4F3) to grayish black (4E2) when young and grayish pink (12A3–4) to blackish purple (12A4–5), staining at first red and then black when bruised; pores subangular to roundish, 0.3–2 mm wide; tubes short, 4–7 mm long, concolorous or a little paler than hymenophoral surface, staining at first red and then black when bruised. Spore print grayish red (7B3). Stipe 3–5 × 0.5–0.8 cm, clavate, black, nearly glabrous, sometimes covered

Fig. 7.1 a *Anthracoporus cystidiatus* Yan C. Li & Zhu L. Yang. Photo by P. Zhang (MHHNU7312, type).

46   The Boletes of China: *Tylopilus* s.l.

Fig. 7.1 b *Anthracoporus cystidiatus* Yan C. Li & Zhu L. Yang. Photo by Z.H. Chen (MHHNU 30418).

with fibrillose squamules, staining at first red and then black when touched; context solid white, staining at first red and then black when hurt; basal mycelium white. Taste and odor unknown.

Basidia 20–30 × 7–10 μm, clavate to narrowly clavate, 4-spored, sometimes 2-spored, hyaline in KOH. Basidiospores [160/8/3] (8) 9–10.5 (12) × 4–5 μm [Q = 2–2.33 (2.4), Q_m = 2.15 ± 0.11], subcylindrical and inequilateral in side view with slight suprahilar depression, elongated to cylindrical in ventral view, smooth, yellowish to brownish yellow in KOH, yellow to yellow-brown in Melzer's reagent. Hymenophoral trama boletoid composed of 3–11 μm wide filamentous hyphae, hyaline to yellowish in KOH, yellowish to yellow in Melzer's reagent. Cheilocystidia 40–57 × 10–16 μm, broadly subfusiform to fusoid-ventricose, brown to dark brown in KOH, thin-walled. Pleurocystidia 60–72 × 8–10 μm, morphologically similar to cheilocystidia, but much longer and narrower, thin-walled. Pileipellis an epithelium composed of 8–21 μm wide inflated concatenated cells, yellowish brown to brownish in KOH and yellow-brown to dark brown in Melzer's reagent; terminal cells 25–40 × 10–21 μm, subfusiform to cystidioid. Pileal trama composed of 7–11 μm wide interwoven hyphae, yellowish brown to pale brownish in KOH and yellow-brown to brown in Melzer's reagent. Clamp connections absent in all tissues.

Habitat: Scattered on soil in subtropical forests dominated by plants of the family Fagaceae.

Known distribution: Currently known from central and southwestern China.

Additional specimens examined: CHINA, HUNAN PROVINCE: Dong'an County, Shunhuangshan

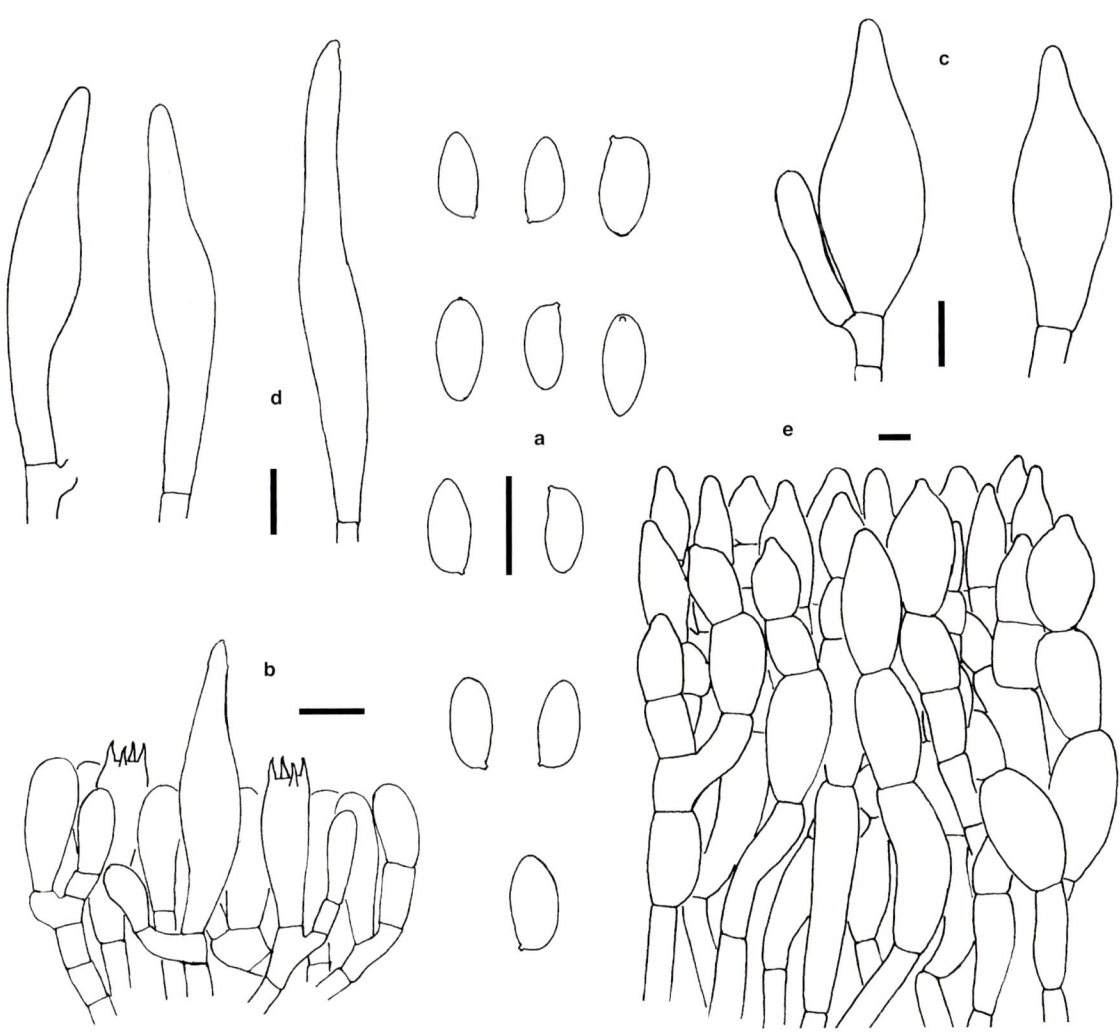

Fig. 7.1 c *Anthracoporus cystidiatus* Yan C. Li & Zhu L. Yang (MHHNU7312, type).
a. Basidiospores; b. Basidia and pleurocystidium; c. Cheilocystidia; d. Pleurocystidia; e. Pileipellis. Scale bars = 10 μm.

National Nature Reserve, alt. 900 m, 14 September 2010, Z.H. Chen 30418 (MHHNU 30418). YUNNAN PROVINCE: Jinghong County, Dadugang Town, alt. 1400 m, 31 July 2008, B. Feng 264 (KUN-HKAS 55375).

Commentary: *Anthracoporus cystidiatus* is characterized by the grayish red to brownish red or ruby red pileus, the black to grayish pink hymenophore, the white to pallid context, the fine hymenophore pores, the initially red and then black discoloration when injured. This species shares the same size of basidiospores, discoloration when injured and absence of the reticulate stipe with *Tylopilus alboater* (Schwein.) Murrill. However, *T. alboater* differs from *An. cystidiatus* in its black to dark grayish brown pileus, and subcutis to trichoderm pileipellis composed of 4–7 μm wide filamentous hyphae (Singer 1947; Smith and Thiers 1971).

7.2 *Anthracoporus holophaeus* (Corner) Yan C. Li & Zhu L. Yang, The Boletes of China: *Tylopilus* s.l. 53 (2021)

MycoBank: MB 834696

Basionym: *Boletus holophaeus* Corner, Trans Br mycol Soc 59: 180 (1972).

Synonyms: *Porphyrellus holophaeus* (Corner) Yan C. Li & Zhu L. Yang, J Fungal Res 9(4): 207 (2011); *Tylopilus holophaeus* (Corner) E. Horak, Malay Fores Rec 51: 235 (2011).

Basidioma small to medium-sized. Pileus 4–10 cm, subhemispherical to applanate, dry, with tomentose squamules, umbrous purple to umbrous red when young, then blackish brown to blackish red with age; context white to grayish white, becoming pallid brownish red to red then black when bruised. Hymenophore adnate to depressed around apex of stipe; surface black to grayish black when young, dingy pink or grayish pink when mature, becoming at first red and then black red to black when injured; pores angular to roundish, 0.3–2 mm wide; tubes 6–15 mm long, concolorous or a little paler than hemenophoral surface, staining at first red and then black red to black when injured. Spore print grayish red. Stipe 4–6 × 1–2.5 cm, subcylindrical to clavate, sometimes attenuate downwards; surface furfuraceous to pruinose, concolorous with pileal surface, whole stipe with white to grayish reticulum but only the upper half is distinct; mycelium on the base of stipe white, becoming at first red then black when bruised; context white to grayish white, becoming at first brownish red to red then black red to black when injured. Taste and odor mild.

Basidia 25–30 × 7–9.5 μm, clavate, hyaline in KOH and yellowish in Melzer's reagent, 4-spored. Basidiospores [60/3/2] (9) 10.5–13.5 × (3.5) 4–5 μm (Q = 2.57–3.29, Q_m = 2.87 ± 0.3), subfusiform

Fig. 7.2 a *Anthracoporus holophaeus* (Corner) Yan C. Li & Zhu L. Yang. Photo by Y.C. Li (**KUN-HKAS 59407**).

Chapter 7 *Anthracoporus* Yan C. Li & Zhu L. Yang   49

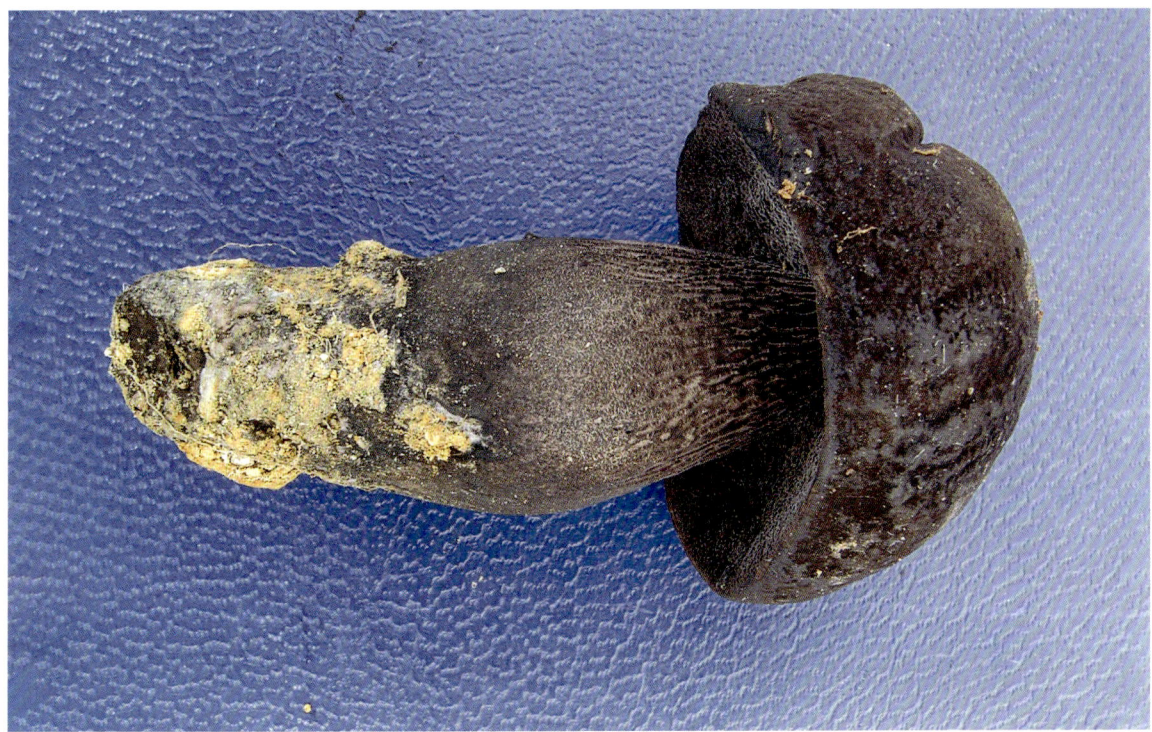

Fig. 7.2 b *Anthracoporus holophaeus* (Corner) Yan C. Li & Zhu L. Yang. Photo by Z.L. Yang (KUN-HKAS 50508).

to subcylindrical in side view with slight suprahilar depression, fusiform to cylindrical in ventral view, smooth, somewhat thick-walled (0.5–1 μm thick), nearly hyaline to light pinkish in KOH and yellowish brown in Melzer's reagent. Pleuro- and cheilocystidia 35–65 × 8–11 μm, subfusiform to ventricose, hyaline in KOH and yellowish to yellow in Melzer's reagent. Pileipellis a palisadoderm consisting of 4–9.5 μm wide vertically arranged hyphae, purplish umber to purplish brown in KOH and brown to dark brown in Melzer's reagent; terminal cells 16–42 × 5.5–9.5 μm, clavate to cystidioid. Pileal trama composed of 4–6 μm wide filamentous hyphae, yellowish brown to brownish in KOH and yellow-brown to dark brown in Melzer's reagent. Clamp connections absent in all tissues.

Habitat: Solitary to scattered, in tropical and subtropical forests dominated by plants of the family Fagaceae.

Known distribution: Currently known from Brunei and China.

Specimens examined: CHINA, YUNNAN PROVINCE: Yingjiang County, on the border between China and Myanmar, alt. 1500 m, 17 July 2009, Y.C. Li 1660 (KUN-HKAS 59407); Jingdong County, Ailao Mountain, alt. 2450 m, 21 July 2006, Z.L. Yang 4711 (KUN-HKAS 50508).

Commentary: *Anthracoporus holophaeus* was originally described as *Boletus holophaeus* by Corner (1972). Horak (2011) transferred it to *Tylopilus* based on its grayish pink hymenophore, while Li and Yang (2011) treated it as *Porphyrellus* based on its umber basidiomata and palisadoderm pileipellis. However, this species stains at first red and then black when bruised or cut, a feature that could be used to distinguish it from *Tylopilus* and *Porphyrellus*.

Phylogenetically, *An. holophaeus* nests within the genus *Anthracoporus* and is closely related to *An. cystidiatus*. However, *An. cystidiatus* differs from *An. holophaeus* in its nearly glabrous stipe, epithelium pileipellis composed of 8–21 μm wide inflated concatenated cells, and relatively small basidiospores measuring (8) 9–10.5 (12) × 4–5 μm.

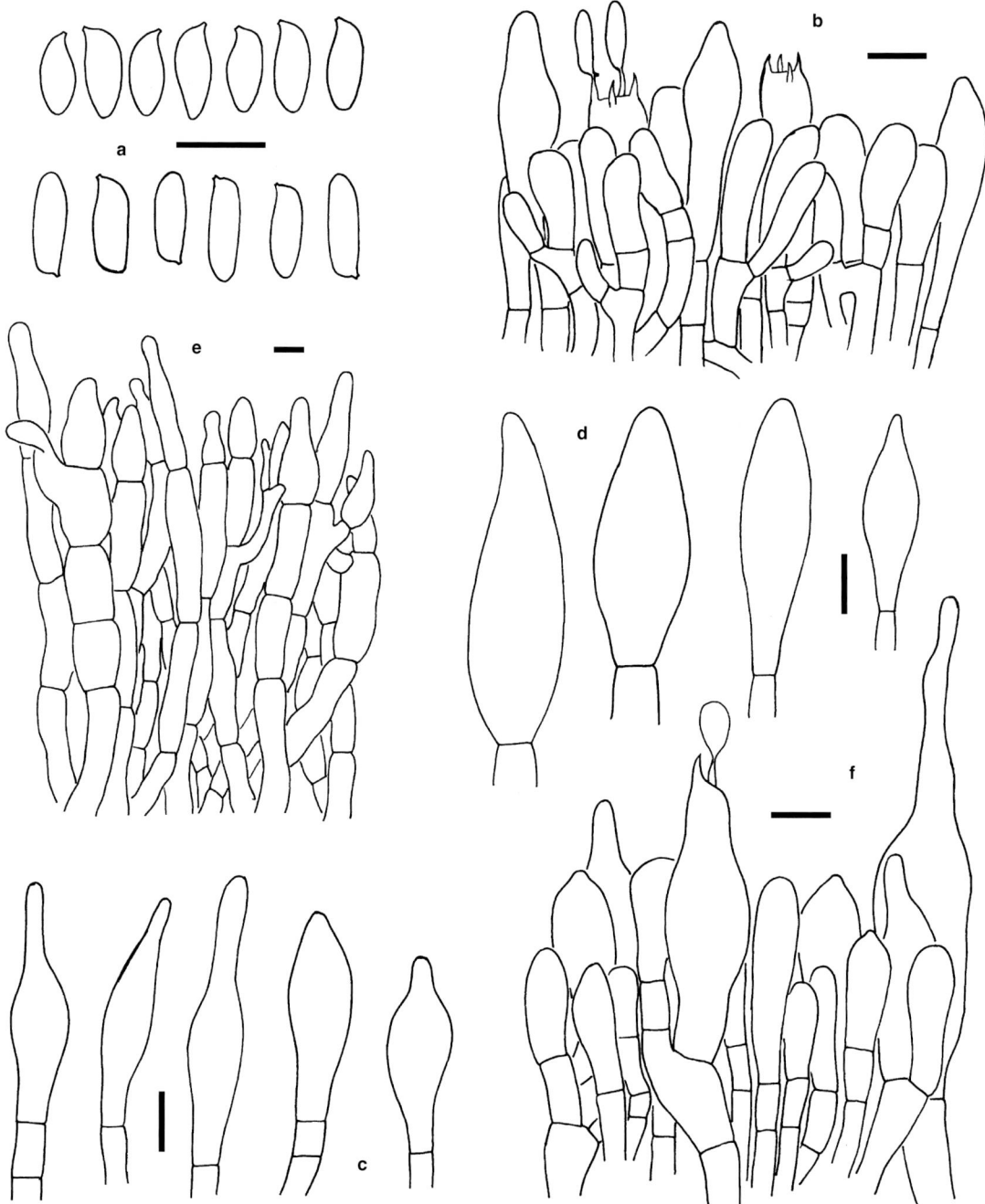

Fig. 7.2 c *Anthracoporus holophaeus* **(Corner) Yan C. Li & Zhu L. Yang (KUN-HKAS 59407).**
a. Basidiospores; b. Basidia and cheilocystidia; c. Cheilocystidia; d. Pleurocystidia; e. Pileipellis; f. Stiptipellis. Scale bars = 10 μm.

7.3 *Anthracoporus nigropurpureus* (Hongo) Yan C. Li & Zhu L. Yang, The Boletes of China: *Tylopilus* s.l. 57 (2021)

MycoBank: MB 834697

Type: SINGAPORE, August 1940, Corner s.n. 27 (CGE).

Basionym: *Tylopilus nigropurpureus* Hongo, Memoirs of Shiga University 23: 40 (1973).

Synonyms: *Boletus nigropurpureus* Corner, *Boletus* in Malaysia (Singapore) 178 (1972) (non Schwein. 1822); *Tylopilus purpureoniger* E. Horak, Malay Fores Rec 51: 61, Fig. 22 (2011); *Porphyrellus nigropurpureus* (Hongo) Yan C. Li & Zhu L. Yang, J Fungal Res 9 (4): 208 (2011).

Basidioma medium-sized to large. Pileus 5–15 cm, subhemispherical to applanate; surface dry, fibrillose to tomentose, dark purplish to blackish purple; context white to grayish, becoming at first red to incarnadine and then black when hurt. Hymenophore depressed around apex of stipe; surface black to grayish black when young, grayish pink to dingy pinkish when mature, becoming at first reddish to incarnadine and then black when injured, pores round to roundish or angular, 0.5–1.5 mm wide; tubes 8–20 mm long, concolorous or a little paler than hymenophoral surface. Spore print grayish red (Fig. 3.4 b). Stipe 4–9 × 2–3 cm, cylindrical to subclavate, concolorous with pileal surface, entirely reticulate; context white to grayish, becoming at first red to incarnadine, then black when injured; mycelium on the base of stipe white, becoming at first red to incarnadine and then black when bruised. Taste bitter and odor mild.

Basidia 25–30 × 8–11 μm, clavate, hyaline in KOH and yellowish in Melzer's reagent, 4-spored. Basidiospores (8) 8.5–9.5 (10) × (4) 4.5–5.5 μm [Q = (1.78) 1.89–2.25 (2.4), Q_m = 2.02 ± 0.14], subcylindrical in side view with slight suprahilar depression, elongated to cylindrical in ventral view (Fig. 3.3 c), smooth, somewhat thick-walled (up to 0.5 μm thick), almost hyaline to light pinkish in KOH and yellowish brown in Melzer's reagent. Pleurocystidia 45–68 × 10–15 μm and cheilocystidia 39–47 × 11–17 μm, subfusiform to ventricose, hyaline in KOH and yellowish to yellow in Melzer's reagent. Pileipellis a palisadoderm composed of 8–10 μm wide filamentous hyphae, dark brown to blackish brown in KOH and yellow-brown to dark yellow-brown in Melzer's reagent; surface always covered with brown to dark brown encrustations; terminal cells 22–58 × 8–11 μm, clavate to subcylindrical. Pileal trama composed of 7–11 μm wide interwoven hyphae, brownish yellow to pinkish brown in KOH and brown-yellow to dark brown in Melzer's reagent. Clamp connections absent in all tissues.

Fig. 7.3 a *Anthracoporus nigropurpureus* (Hongo) Yan C. Li & Zhu L. Yang. Photo by K. Zhao (KUN-HKAS 80653).

Fig. 7.3 b *Anthracoporus nigropurpureus* (Hongo) Yan C. Li & Zhu L. Yang. Photo by Y.C. Li (KUN-HKAS 53358).

Habitat: Solitary to scattered, in tropical and subtropical forests dominated by plants of the family Fagaceae.

Known distribution: Currently known from Singapore, China, and Japan.

Specimens examined: CHINA, FUJIAN PROVINCE: Sanming, Sanyuan National Forest Park, alt. 260 m, 24 August 2007, Y.C. Li 998 (KUN-HKAS 52685), the same location, 25 August 2007, Y.C. Li 1013 (KUN-HKAS 53358), the same location, 26 August 2007, Y.C. Li 1025 (KUN-HKAS 53370). HAINAN PROVINCE: Ledong County, Jianfengling, alt. 850 m, 6 August 2009, N.K. Zeng 486 (KUN-HKAS 59827). GUANGDONG PROVINCE: Guangzhou, Baiyun Mountain, alt. 350 m, 28 May 2013, K. Zhao 228 (KUN-HKAS 80653), the same location and date, G. Wu 1105 (KUN-HKAS 80479). YUNNAN PROVINCE: Ruili, Ruili Botanic Garden, alt. 900 m, 7 October 2010, Y.C. Li 2164 (KUN-HKAS 63663); Nanhua County, Wild Mushroom Market, altitude unknown, 2 August 2009, Y.C. Li 1954 (KUN-HKAS 59702). SICHUAN PROVINCE: Pujiang County, Datang Town, alt. 650 m, 4 September 1986, M.S. Yuan 1294 A (KUN-HKAS 18426).

Commentary: *Anthracoporus nigropurpureus* was originally described as *Boletus nigropurpureus* from Singapore by Corner in 1972. Nomenclaturally, the name of this species was illegitimate, because "*Boletus nigropurpureus* Schwein." (1822) is available. Hongo (1973) and Li and Yang (2011) were unaware of its illegitimacy in their studies related to this species. Horak (2011) recognized the illegitimacy, and proposed the epithet "*purpureoniger*" for the Asian species. However, *Tylopilus nigropurpureus* was published by Hongo (1973) as a new combination ("comb. nov.") citing *Boletus nigropurpureus* Corner as basionym. Since *B. nigropurpureus* Corner is a later homonym of *B. nigropurpureus* Schwein., *T. nigropurpureus* Hongo is therefore validly published as a replacement name under Art. 6.14 leaving *Porphyrellus nigropurpureus* (Corner) Yan C. Li & Zhu L. Yang (Li and Yang 2011) and *T. purpureoniger* E. Horak (Horak 2011) its synonyms. This species is characterized by the dark purplish to blackish purple basidiomata, the white to pinkish gray or dingy pinkish hymenophore, the change from a reddish to a blackish color when injured, the distinct reticulum around stipe, and the short basidiospores (up to 10 μm long). *Anthracoporus nigropurpureus* is morphologically similar to *An. holophaeus*, however, *An. holophaeus* has a umbrous purple to umbrous red or blackish brown to blackish red basidioma, indistinct reticulum on the upper half of the stipe, and relatively long basidiospores (up to 13.5 μm long).

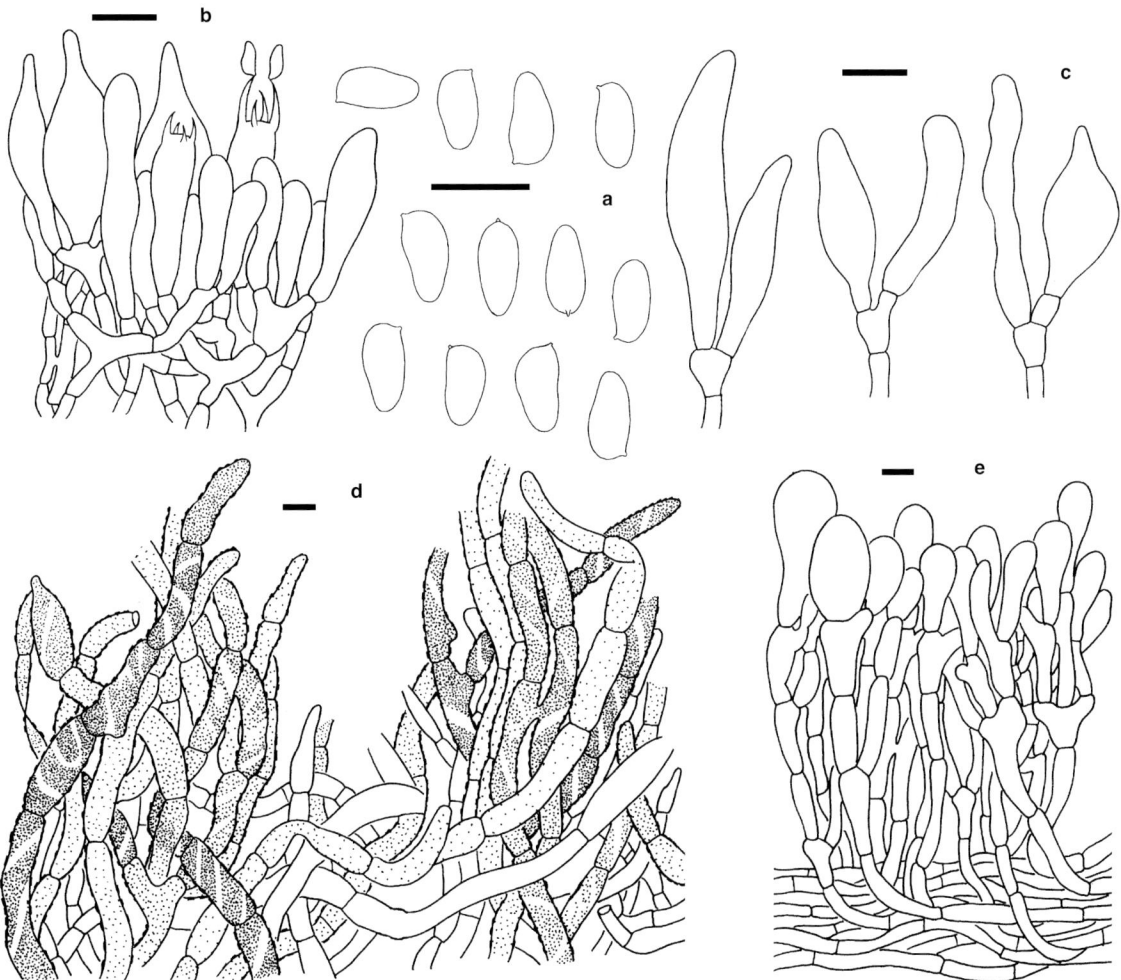

Fig. 7.3 c *Anthracoporus nigropurpureus* **(Hongo) Yan C. Li & Zhu L. Yang (KUN-HKAS 80479).**
a. Basidiospores; b. Basidia and pleurocystidia; c. Cheilocystidia; d. Pileipellis. e. Stiptipellis. Scale bars = 10 μm.

Chapter 8

Austroboletus (Corner) Wolfe

Austroboletus (Corner) Wolfe, Biblthca Mycol 69: 64 (1979a)

Type species: *Porphyrellus dictyotus* Boedijn, Persoonia 1(3): 316 (1960).

Diagnosis: This genus is different from other genera in the Boletaceae in its distinctly extended pileal margin which embracing the stipe in younger basidioma and then breaking into pieces and hanging on the pileal margin in mature basidioma, white to pallid context without color change when injured, pink to purplish pink or purple hymenophore without color change when injured, distinctly reticulate stipe, and conspicuously ornamented basidiospores.

Pileus subhemispherical or convex when young, subhemispherical to applanate or convex to plano-convex when mature; surface viscid, strongly gelatinous or dry, nearly glabrous or with fibrillose to tomentose or scaly squamules; margin extended, forming membranous veil and embracing the stipe when young and then breaking into pieces and hanging on the pileal margin; context soft, white to pallid, without discoloration when bruised. Hymenophore adnate when young, adnate to depressed around apex of stipe when mature; surface pink to purplish when young, purplish pink to purple when mature; pores angular to roundish, without discoloration when injured; tubes concolorous with hymenophoral surface, without discoloration when injured. Stipe central, clavate to subcylindrical; surface distinctly reticulate; basal mycelium white to pallid. Spore print pinkish to purplish or purple-white. Basidiospores ellipsoid, elongated, subfusiform or subcylindrical, conspicuously ornamented with warts, striate reticulum, reticulate ridges or shallow and irregularly furrowed pits (Fig. 8). Hymenial cystidia abundant, subfusiform without any septum, or subcylindrical with 1–2 secondary septa, or subclavate composed of 2 cells with upper cells cylindrical and lower cells broadly clavate; cheilo- and pleurocystidia similar to or greatly different from each other. Pileipellis a cutis to ixocutis, or a trichoderm. Clamp connections absent in all tissues.

Commentary: *Austroboletus* is characterized by the subhemispherical to applanate or convex pileus, the extended pileal margin, the white to pallid context without color change when bruised, the pink to purplish pink or purple-pink hymenophore without color change when bruised, the distinctly reticulate stipe, the conspicuously ornamented basidiospores, and the subfusiform or subcylindrical or subclavate cheilo- and pleurocystidia which are morphologically similar to or greatly different from each other (Wolfe 1979a; Singer 1986; Fulgenzi *et al.* 2010). These traits are significantly different from those of the other genera in the Boletaceae. Currently, seven species including three new species are known from China.

Chapter 8 *Austroboletus* (Corner) Wolfe 55

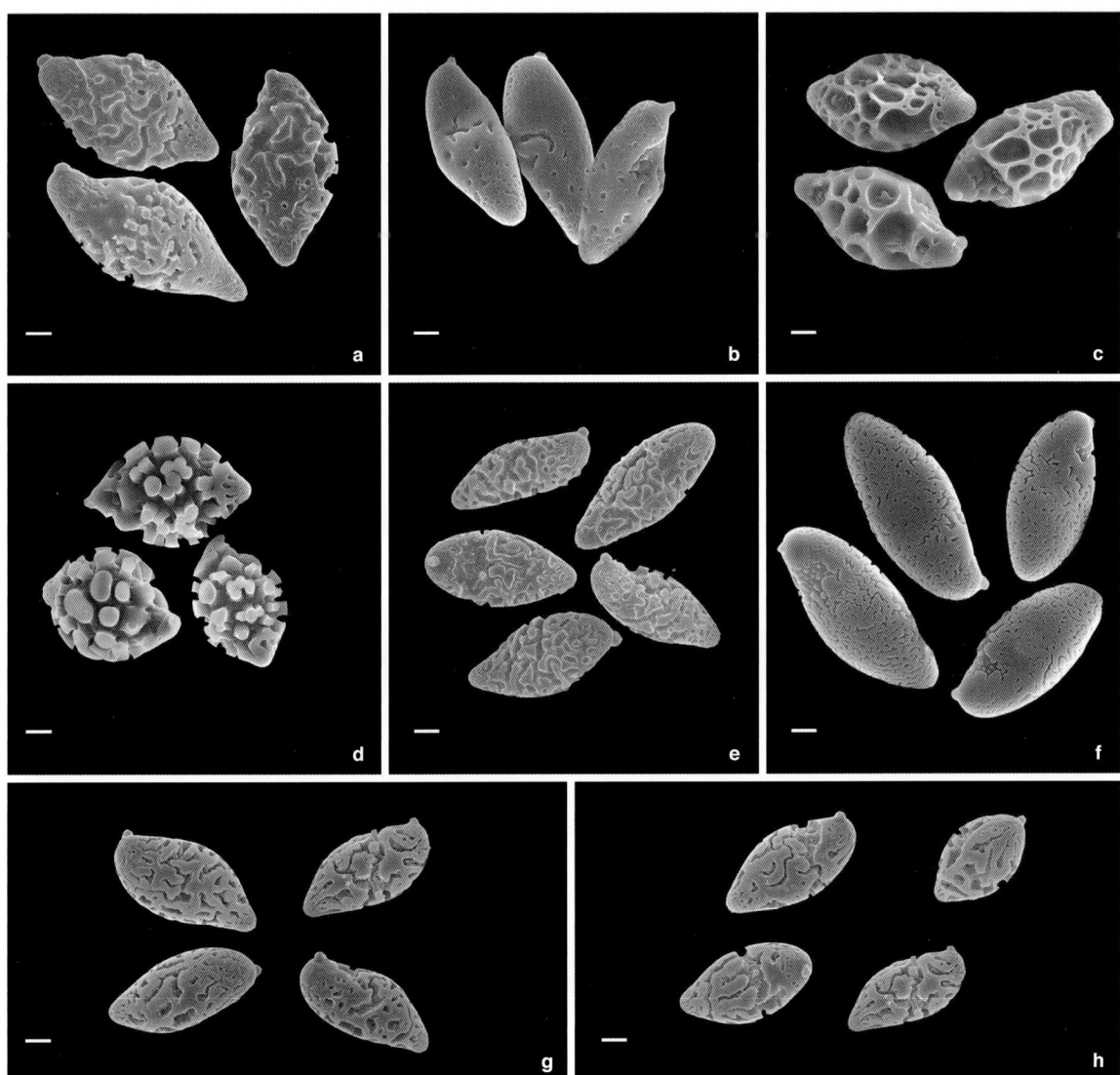

Fig. 8 Characteristics of basidiospore ornamentations in *Austroboletus*.
a. *A. albidus* (KUN-HKAS 107148, type); b. *A. albovirescens* (KUN-HKAS 85661); c. *A. dictyotus* (KUN-HKAS 53450); d. *A. fusisporus* (KUN-HKAS 53461); e. *A. olivaceobrunneus* (KUN-HKAS 92428, type); f. *A. olivaceoglutinosus* (KUN-HKAS 57756); g, h. *A. subvirens* (g. KUN-HKAS 107142, h. KUN-HKAS 107149). Scale bars = 2 μm.

Key to the species of *Austroboletus* in China

1. Pileus with green, dark green or greenish tinge ·· 2
1. Pileus without green, dark green or greenish tinge ·· 5
2. Basidiomata strongly gelatinous; pileipellis an ixocutis composed of gelatinizing filamentous hyphae ·· 3
2. Basidiomata non-gelatinous; pileipellis a trichoderm or an intricate trichoderm composed of filamentous hyphae ·· 4
3. Basidiomata distributed in subtropical forests; pileus matte green to grayish green; surface of basidiospores with unequal pits; pileipellis an ixocutis composed of 3–6 μm wide filamentous hyphae ·· *A. albovirescens*
3. Basidiomata distributed in subalpine forests; pileus dark green to olive and then pale olive to greenish yellow; surface of basidiospores with irregular shallow reticulum; pileipellis an ixocutis composed of 1.5–3 μm wide filamentous hyphae ··· *A. olivaceoglutinosus*
4. Pileus yellowish green to brownish green when young, becoming brownish to grayish brown when mature; basidiospores 12–15 × 5.5–6.5 μm ·· *A. olivaceobrunneus*
4. Pileus olivaceous green to grayish green when young, and then dull green to yellowish green in the center and yellowish brown to yellow-brown elsewhere; basidiospores relatively broad 11–16 (17) × (6) 7–8 (9) μm ·· *A. subvirens*
5. Basidioma very small to small; pileus viscid, without yellowish brown to chestnut-brown tinge; basidiospores ornamented with regular to irregular cylindrical protuberances, or with large tubercles ·· 6
5. Basidioma medium-sized to large; pileus dry, yellowish brown to chestnut-brown; basidiospores ornamented with high ridges forming relatively large and deep pits on the surface ··· *A. dictyotus*
6. Pileus dark apricot to apricot; basidiospores 9.5–12.5 × 8–9 μm, surface with regular to irregular cylindrical protuberances ·· *A. fusisporus*
6. Pileus white to cream; basidiospores relatively long 13–18 × 7.5–9 μm, surface with large tubercles ·· *A. albidus*

8.1 *Austroboletus albidus* Yan C. Li & Zhu L. Yang, The Boletes of China: *Tylopilus* s.l. 63 (2021)

MycoBank: MB 834698

Etymology: The epithet "*albidus*" refers to the color of the basidiomata.

Type: CHINA, YUNNAN PROVINCE: Maguan County, Dalishu Town, Adushangba Village, alt. 1780 m, 7 August 2017, S.F. Shi 238 (KUN-HKAS 107148, GenBank Acc. No.: MT154756 for nrLSU).

Basidioma small. Pileus 3–5 cm in diam., conical to convex; surface white (1A1) to cream (2B2–3) when young, cream to grayish yellow (1B2–3) when mature, covered with light orange (4C2–3) to brownish (6D4–5) nubby squamules; margin extended; context solid when young then spongy, white, without color change when bruised. Hymenophore depressed around apex of stipe; surface grayish white (13B1) to purplish gray (13D2) when young, grayish purple (13E2–3) to gray-purple (14E2–3) or dark purple (14F2–3) when mature, without discoloration when injured; pores roundish to angular, 0.5–1 mm wide; tubes 8–15 mm long, concolorous or a little paler than hymenophoral surface, without color change when bruised. Spore print pinkish (13B2) to purplish white (14B2). Stipe 4–7 × 0.4–0.6 cm, subcylindrical to clavate, pallid (1C2) to white (1A1), staining grayish yellow (1B3–4) to reddish brown (8E6–7) when bruised or aged; surface entirely reticulate with elongated meshes up to 5 mm high; context soft, whitish to pallid, without color change when injured; basal mycelium whitish to pallid. Taste and odor mild.

Basidia 28–39 × 14–20 μm, clavate to broadly clavate, 4-spored, hyaline in KOH. Basidiospores [60/3/3] 13–18 × 7.5–9 μm (Q = 1.63–2.13, Q_m = 1.89 ± 0.18) including ornamentation, elongated to subfusiform with a pronounced suprahilar depression and acute base; surface with large tubercles (0.2–1.8 μm high, 0.4–1.4 μm wide) (Fig. 8 a), forming intricate reticulum and shallow pits at apex and base, yellowish to

Fig. 8.1 a *Austroboletus albidus* Yan C. Li & Zhu L. Yang. Photo by S.F. Shi (KUN-HKAS 107148, type).

pale olivaceous in KOH, yellow to brown in Melzer's reagent. Hymenophoral trama boletoid composed of 3–12 μm wide filamentous hyphae, hyaline to yellowish in KOH, yellowish to yellow in Melzer's reagent. Cheilocystidia 27–48 × 11–22 μm, broadly clavate to subfusiform. Pleurocystidia composed of 2 cells with upper cells cylindrical to finger-like 27–44 × 8–11 μm, lower cells broadly clavate 46–56 × 17–30 μm, colorless to yellowish in KOH, thin-walled. Caulocystidia forming the reticulum over the surface of stipe, morphologically similar to hymenial cystidia. Pileipellis an ixocutis, composed of 3–6.5 μm wide gelatinizing filamentous hyphae, hyaline to yellowish in KOH and yellowish to brownish in Melzer's reagent; terminal cells 28–56 × 3.5–6.5 μm, subcylindrical. Pileal trama composed of 5–15 μm wide interwoven hyphae, colorless to yellowish in KOH and yellow to yellowish brown in Melzer's reagent. Clamp connections absent in all tissues.

Habitat: Scattered on soil in tropical and southern subtropical forests dominated by plants of the family Fagaceae.

Known distribution: Currently known from southeastern and southwestern China.

Additional specimens examined: CHINA, TAIWAN PROVINCE: Taizhong, Daxueshan Forest Park, alt. 1900 m, 19 September 2012, B. Feng 1321 (KUN-HKAS 82463). YUNNAN PROVINCE: Hekou County, alt. 150 m, 14 September 2019, X. Xu 532532MF0489 (KUN-HKAS 107158).

Commentary: *Austroboletus albidus* is characterized by the viscid and white to cream pileus which is covered with light orange to brownish nubby squamules, the grayish white to purplish gray or grayish purple to dark purple hymenophore, the pallid to white reticulate stipe staining grayish yellow to reddish brown at base when bruised, the elongated to subfusiform basidiospores ornamented with large tubercles, and the ixocutis pileipellis.

Phylogenetically, *A. albidus* formed a distinct lineage within *Austroboletus*. Its phylogenetic relationships with other species in this genus remain unresolved (Fig. 4.1). Morphologically, the white to cream pileus of *A. albidus* is similar to that of *A. longipes* var. *albus* (Corner) E. Horak. However, *A. longipes* var. *albus* differs from *A. albidus* in its small and slender basidiospores measuring 12.5–16.5 × 4–5 μm, fusiform to sublanceolate hymenial cystidia, and palisadoderm pileipellis composed of 6–12 μm wide vertically arranged hyphae (Corner 1974; Horak 2011).

Fig. 8.1 b *Austroboletus albidus* Yan C. Li & Zhu L. Yang. Photo by B. Feng (KUN-HKAS 82463).

Chapter 8 *Austroboletus* (Corner) Wolfe

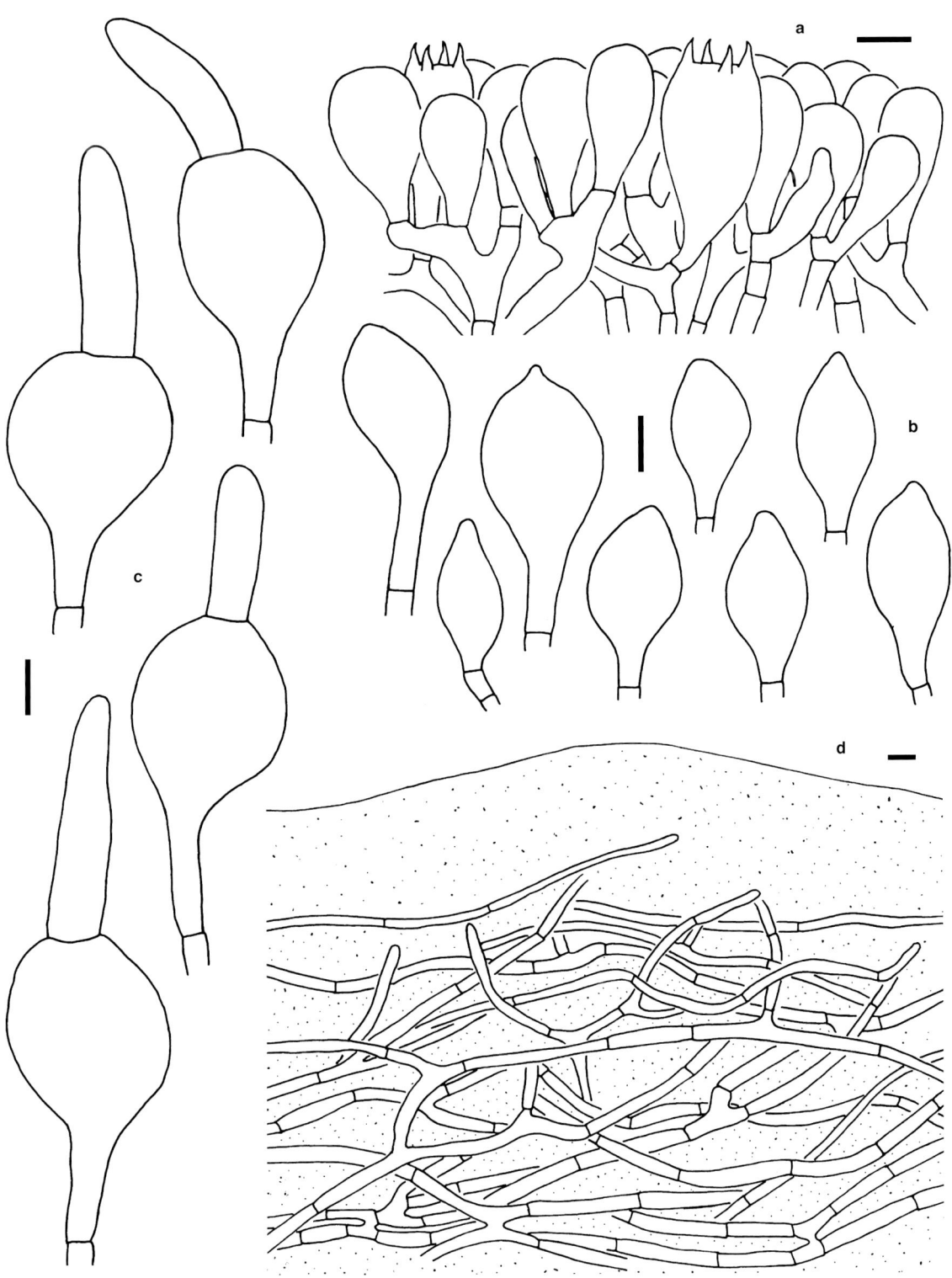

Fig. 8.1 c *Austroboletus albidus* Yan C. Li & Zhu L. Yang (KUN-HKAS 107148, type).
a. Basidia; b. Cheilocystidia; c. Pleurocystidia; d. Pileipellis. Scale bars = 10 μm.

8.2 *Austroboletus albovirescens* Yan C. Li & Zhu L. Yang, The Boletes of China: *Tylopilus* s.l. 66 (2021)

MycoBank: MB 834699

Etymology: The epithet "*albovirescens*" refers to the color of the pileus.

Type: CHINA, YUNNAN PROVINCE: Lvchun County, alt. 1650 m, 23 June 2019, Q. Cai 1508 (KUN-HKAS 107171, GenBank Acc. No.: MW114850 for nrLSU).

Basidioma small to medium-sized. Pileus 3–6 cm in diam., subhemispherical to convex or plano-convex; surface dry, matte green (30E3–4) to gray-green (30E5), slightly darker in the center, covered with concolorous floccose squamules, usually cracked with white background exposed when mature, without discoloration when touched; margin extended and forming membranous veil and embracing the stipe when young and then breaking into pieces and hanging on the pileal margin; context solid when young and then spongy when mature, white to pallid, without color change when bruised. Hymenophore adnate when young, and then adnate to depressed around apex of stipe; surface pinkish (4A2) to grayish pink (14C2) when young and pink (8A3) to purplish pink (12A3) when mature, without discoloration when bruised; pores subangular to roundish, 0.5–1 mm wide; tubes 6–10 mm long, concolorous or a little paler than hymenophoral surface, without discoloration when bruised. Spore print pinkish (13B2) to dark purplish (14B2). Stipe 4–9 × 0.5–1.2 cm, clavate; surface pallid to white, but becoming cream to yellowish brown when bruised, entirely reticulate with elongated meshes up to 3 mm high; context solid when young, spongy when old without discoloration when bruised; basal mycelium whitish to pallid. Taste and odor mild.

Basidia 26–35 × 13–15 µm, clavate to narrowly clavate, 4-spored, rarely 2-spored, hyaline in KOH. Basidiospores [100/5/5] 14.5–17.5 × 6–7.5 µm [Q = (2) 2.07–2.54 (2.62), Q_m = 2.3 ± 0.13], elongated to subfusiform, with a indistinct suprahilar depression; surface with unequal pits (Fig. 3.3 j, Fig. 8 b), apex and base forming intricate reticulum or with relatively dense and shallow pits, yellowish to brownish yellow in KOH, yellow to yellow-brown in Melzer's reagent. Hymenophoral trama boletoid composed of 3.5–8 µm wide filamentous hyphae, hyaline to yellowish in KOH, yellowish to yellow in Melzer's reagent. Cheilocystidia subcylindrical, composed of 2–3 cylindrical cells with terminal cells 33–48 × 4.5–6 µm, narrowly clavate to cylindrical, brownish to brownish yellow in KOH, somewhat thick-walled (up to 1.5 µm thick). Pleurocystidia 34–56 × 5–10 µm, subfusiform to subfusoid-mucronate or subfusoid-ventricose, sometimes narrowly mucronate, rostrate, thin-walled. Caulocystidia forming the reticulum over the surface of stipe, morphologically similar to hymenial cystidia. Pileipellis an ixocutis composed of 3–6 µm wide gelatinizing filamentous hyphae, hyaline to yellowish in KOH and yellowish to brownish in Melzer's reagent; terminal cells 14–65 × 4–6 µm, cylindrical. Pileal trama composed of 4–8 µm wide interwoven hyphae, hyaline to yellowish in KOH and yellowish to brownish in Melzer's reagent. Clamp connections absent in all tissues.

Habitat: Scattered to solitary on soil in subtropical forests dominated by plants of the family Fagaceae.

Known distribution: Currently known from southwestern China.

Chapter 8 *Austroboletus* (Corner) Wolfe 61

Fig. 8.2 a *Austroboletus albovirescens* Yan C. Li & Zhu L. Yang. Photos by Q. Cai (KUN-HKAS 107171, type).

Fig. 8.2 b *Austroboletus albovirescens* Yan C. Li & Zhu L. Yang. Photo by X.X. Ding (KUN-HKAS 107172).

Additional specimens examined: YUNNAN PROVINCE: Yongping County, near Milestone 3260 KM of the National Road 320#, alt. 2200 m, 30 July 2009, Y.C. Li 1876 (KUN-HKAS 59624); Gongshan County, Bingzhongluo Town, alt. 2000 m, 29 July 2011, G. Wu 432 (KUN-HKAS 74743); Nanjian County, Lingbaoshan National Forest Park, alt. 2100 m, 4 August 2014, C. Yan 63 (KUN-HKAS 85661); Lvchun County, alt. 1650 m, 19 June 2019, X.X. Ding 433 (KUN-HKAS 107172).

Commentary: *Austroboletus albovirescens* is characterized by the dry matte green to gray-green pileus, the white context without discoloration when injured, the pinkish to grayish pink when young and then pink to purplish pink hymenophore, the distinctly reticulate stipe, the elongated to subfusiform basidiospores which are ornamented with unequal pits, and the ixocutis pileipellis.

Morphologically, *A. albovirescens* shares the color of the pileus and hymenophore with *A. olivaceoglutinosus* K. Das & Dentinger. However, *A. olivaceoglutinosus* differs from *A. albovirescens* in its glutinous pileus, intricately reticulate basidiospores, ixocutis pileipellis composed of 1.5–3 μm wide filamentous hyphae, and distribution in subalpine forests.

Chapter 8 *Austroboletus* (Corner) Wolfe

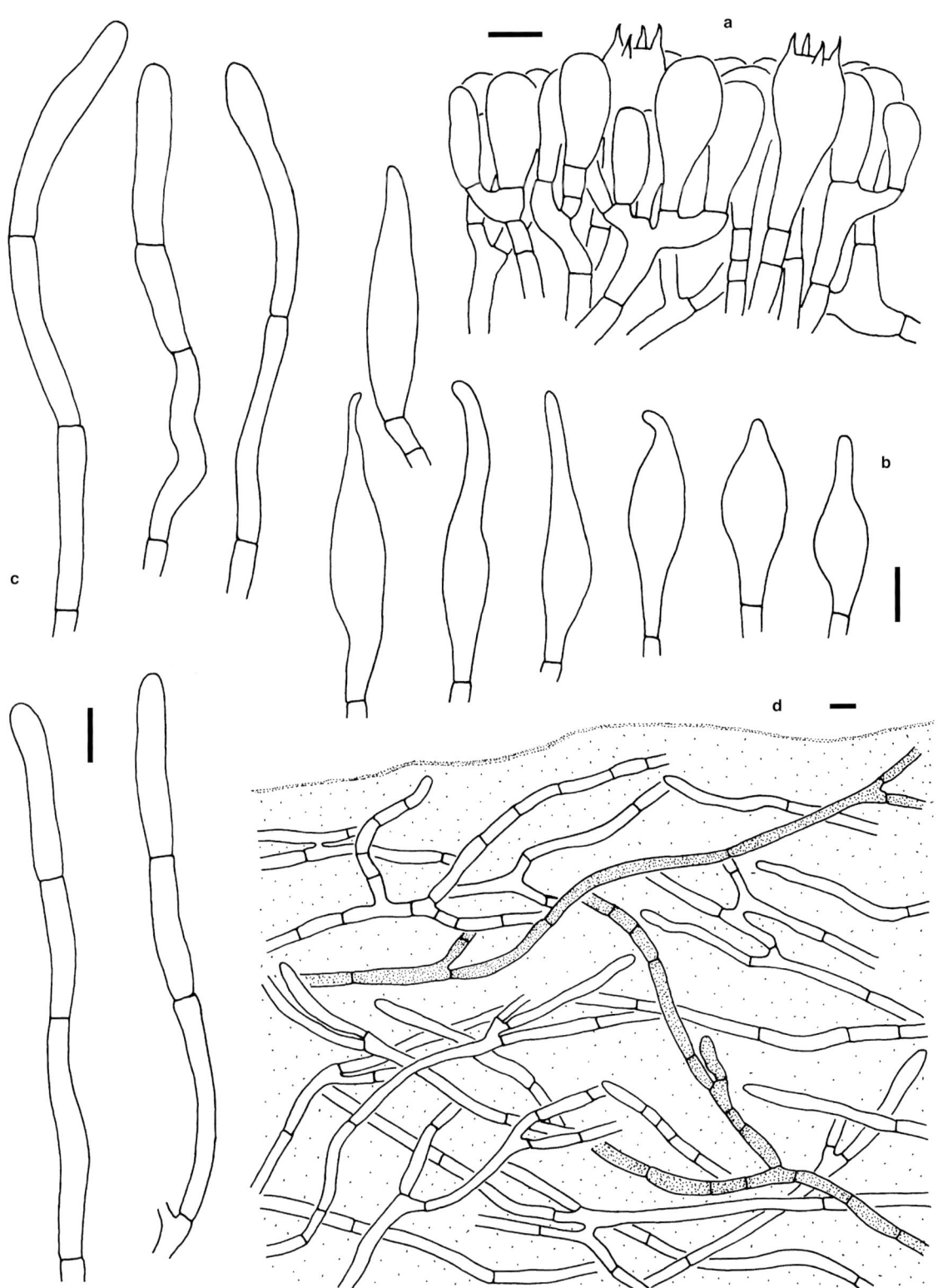

Fig. 8.2 c *Austroboletus albovirescens* Yan C. Li & Zhu L. Yang (KUN-HKAS 107171, type).
a. Basidia; b. Pleurocystidia; c. Cheilocystidia; d. Pileipellis. Scale bars = 10 μm.

8.3 *Austroboletus dictyotus* (Boedijn) Wolfe, Biblthca Mycol 69: 92 (1979a)

Basionym: *Porphyrellus dictyotus* Boedijn, Persoonia 1(3): 316 (1960).
Synonym: *Boletus dictyotus* (Boedijn) Corner, *Boletus* in Malaysia 80 (1972).

Basidioma medium-sized to large. Pileus 5–15 cm in diam., subhemispherical to convex when young, plano-convex to applanate when mature; surface tomentose to floccose, reddish brown to chestnut-brown when young, yellowish brown to yellow-brown when mature, without discoloration when touched; margin extended, forming membranous veil and embracing the stipe when young and then breaking into pieces and hanging on the pileal margin when mature; context soft, white to pallid, without discoloration when bruised. Hymenophore adnate when young, adnate to depressed around apex of stipe when mature; surface pinkish to grayish pink and then pink to purplish pink, without discoloration when bruised; pores roundish or angular, 0.3–1 mm wide; tubes 10–20 mm long, concolorous or a little paler than hymenophoral surface, without discoloration when touched. Spore print pink to purplish. Stipe 3–8 × 1.2–2.5 cm, clavate to subcylindrical, pallid to white, staining orange-yellow to brownish when aged or bruised; surface entirely covered with elongated meshes up to 8 mm high; context solid, white, without discoloration when hurt; basal mycelium whitish to pallid. Taste and odor mild.

Basidia 28–48 × 9–16 μm, clavate, 4-spored, hyaline in KOH. Basidiospores 11.5–15.5 × 6.5–8.5 μm ($Q = 1.53–2$, $Q_m = 1.78 \pm 0.15$) including ornamentation, subfusiform with a shallow suprahilar depression; surface with high ridges forming reticulum, apex and base forming intricate reticulum and shallow pits (Fig. 8 c), hyaline to yellowish in KOH, yellow to brownish yellow in Melzer's reagent. Hymenophoral trama boletoid composed of 3–8 μm wide filamentous hyphae, hyaline to yellowish in KOH, yellowish to yellow in Melzer's reagent. Cheilocystidia 50–88 × 5–9 μm, subcylindrical, composed of 1–3 cylindrical cells, brown to yellowish brown in KOH, thin-walled; terminal cells 15–25 × 4.5–8 μm. Pleurocystidia morphologically different from cheilocystidia, composed of 2 cells; upper cells cylindrical, 16–36 × 6–9

Fig. 8.3 a *Austroboletus dictyotus* (Boedijn) Wolfe. Photos by Y.C. Li (KUN-HKAS 53399).

Fig. 8.3 b *Austroboletus dictyotus* (Boedijn) Wolfe. Photo by N.K. Zeng (KUN-HKAS 59804).

μm; lower cells broadly clavate, 39–60 × 14–22 μm, yellowish to brownish yellow in KOH, thin-walled. Caulocystidia forming the reticulum over the surface of stipe, morphologically similar to hymenial cystidia. Pileipellis a trichoderm to palisadoderm composed of 4–10 μm wide filamentous hyphae, yellowish to pale brownish in KOH and yellow to brownish in Melzer's reagent; terminal cells 20–48 × 3.5–6 μm, clavate to subcylindrical. Pileal trama composed of 4–8 μm wide interwoven hyphae, hyaline to yellowish in KOH and yellowish to brownish in Melzer's reagent. Clamp connections absent in all tissues.

Habitat: Scattered on soil in tropical and subtropical forests dominated by plants of the family Fagaceae.

Known distribution: Currently known from Indonesia, Malaysia, and southern and central China.

Specimens examined: CHINA, HAINAN PROVINCE: Qiongzhong County, Limu Mountain, alt. 870 m, 4 August 2010, N.K. Zeng 830 (KUN-HKAS 59804). HUNAN PROVINCE: Yizhang County, Mangshan National Forest Park, alt. 880 m, 2 September 2007, Y.C. Li 1054 (KUN-HKAS 53399), the same location, 4 September 2007, Y.C. Li 1088 (KUN-HKAS 53450).

Commentary: *Austroboletus dictyotus* was originally described from Indonesia and then reported from Singapore and China (Boedijn 1960; Corner 1972; Wu *et al.* 2016a). Four additional varieties under *A. dictyotus* were described by Corner (1972), based on basidiospore ornamentation and the height of the ridges on the stipe. The taxonomy of these taxa need to be clarified based on properly documented materials (including SEM pictures of basidiospores and DNA sequences), preferably from type localities.

8.4 *Austroboletus fusisporus* (Imazeki & Hongo) Wolfe, Biblthca Mycol 69: 96 (1979a)

Basionym: *Porphyrellus fusisporus* Kawam. ex Imazeki & Hongo, Acta Phytotax Geobot, Kyoto 18(4): 110 (1960).

Basidioma very small to small. Pileus 1.5–4 cm in diam., conical to subconical or convex, reddish brown to cinnamon brown, paler towards margin; surface glutinous and very sticky when wet, shiny when dry, without discoloration when touched; margin extended, forming membranous veil and embracing the stipe in younger basidioma, and then breaking into pieces and hanging on the pileal margin; context soft, white to pallid, without discoloration when injured. Hymenophore adnate when young, adnate to depressed around apex of stipe when mature; surface initially pinkish to pink and then pink to purplish pink when mature, without discoloration when bruised; pores roundish

Fig. 8.4 *Austroboletus fusisporus* (Imazeki & Hongo) Wolfe. Photos by G. Wu (KUN-HKAS 99912).

or angular, 0.3–1 mm wide; tubes short, 4–7 mm long, concolorous or a little paler than hymenophoral surface, without discoloration when bruised. Spore print pinkish to purplish (Fig. 3.4 c). Stipe 2–5 × 0.3–0.8 cm, clavate to subcylindrical, always enlarged downwards, whitish to pallid; surface entirely reticulate with elongated meshes up to 4 mm high, staining reddish brown to cinnamon brown when injured; basal mycelium whitish to pallid. Taste and odor mild.

Basidia 24–35 × 10–16 μm, clavate, 4-spored, hyaline in KOH. Basidiospores 9.5–12.5 × 8–9 μm (Q = 1.17–1.5, Q_m = 1.33 ± 0.13) including ornamentation, broadly ellipsoid to ellipsoid with a distinct suprahilar depression; surface ornamented with large regular to irregular subcylindrical tubercles which are 0.5–1.5 μm high and 1–2.5 μm wide, apex and base remaining smooth or with shallow pits (Figs. 3.3 l and 8 d), yellowish to brownish yellow in KOH, yellow to yellow-brown in Melzer's reagent. Hymenophoral trama boletoid composed of 3–8 μm wide filamentous hyphae, hyaline to yellowish in KOH, yellowish to yellow in Melzer's reagent. Cheilo- and pleurocystidia composed of 2 cells with upper cells cylindrical to finger-like 19–44 × 3.5–6.5 μm, lower cells clavate to broadly clavate 31–46 × 13–18 μm, pale brownish to yellowish brown in KOH, thin-walled. Caulocystidia forming the reticulum over the surface of stipe, morphologically similar to hymenial cystidia. Pileipellis an ixotrichoderm composed of strongly gelatinous 2.5–5 μm wide interwoven hyphae, yellowish to pale brownish in KOH and yellow to brownish in Melzer's reagent; terminal cells 28–95 × 3–5 μm, subcylindrical. Pileal trama composed of 4–8 μm wide filamentous hyphae, hyaline to yellowish in KOH and yellowish to brownish in Melzer's reagent. Clamp connections absent in all tissues.

Habitat: Scattered on soil in tropical and subtropical forests dominated by plants of the family Fagaceae.

Known distribution: Currently known from Japan, Korea, and China.

Specimens examined: CHINA, GUANGDONG PROVINCE: Fengkai County, Heishiding Nature Reserve, alt. 250 m, 1 June 2013, Y.J Hao 820 (KUN-HKAS 80100), the same location, 2 June 2013, Y.J Hao 833 (KUN-HKAS 80113), the same location, 3 June 2013, Y.J Hao 842 (KUN-HKAS 80122). HAINAN PROVINCE: Changjiang County, Bawangling National Nature Reserve, alt. 750 m, 23 August 2009, N.K. Zeng 567 (KUN-HKAS 59805); Lingshui County, Diaoluo Mountain, alt. 850 m, 28 July 2010, N.K. Zeng 750 (KUN-HKAS 59806). HUNAN PROVINCE: Yizhang County, Mangshan National Forest Park, alt. 880 m, 5 September 2007, Y.C. Li 1099 (KUN-HKAS 53461), the same location, 12 September 2016, G. Wu 1793 (KUN-HKAS 99912). YUNNAN PROVINCE: Mengla County, Menglun Town, alt. 550 m, 9 July 2006, Y.C. Li 484 (KUN-HKAS 50238); Jinghong County, Dadugang Town, alt. 1000 m, 14 July 2006, Y.C. Li 526 (KUN-HKAS 50280), the same location, 21 July 2007, Y.C. Li 920 (KUN-HKAS 52607) and 23 July 2007, Y.C. Li 944 (KUN-HKAS 52631).

Commentary: *Austroboletus fusisporus* originally described from Japan and then reported from Korea and China (An *et al.* 1998; Wu *et al.* 2016a) is characterized by the very small to small basidiomata, the glutinous and very sticky basidiomata, the reddish brown to cinnamon brown pileus, the broadly ellipsoid to ellipsoid basidiospores with large regular to irregular subcylindrical ornamentations, and the ixotrichoderm pileipellis. *Austroboletus fusisporus* is morphologically similar to *A. mucosus* (Corner) Wolfe in the color of the pileus, but they can be distinguished from each other by the dimensions of the basidiospores, and the ornamentations of the basidiospores (Corner 1972; Wolfe 1979a).

8.5 *Austroboletus olivaceobrunneus* Yan C. Li & Zhu L. Yang, The Boletes of China: *Tylopilus* s.l. 75 (2021)

MycoBank: MB 834701

Etymology: The epithet "*olivaceobrunneus*" refers to the color of the basidiomata.

Type: CHINA, YUNNAN PROVINCE: Nanjian County, Gonglang Town, Yangdianhe Village, alt. 1400 m, 1 July 2015, K. Zhao 798 (KUN-HKAS 92428, GenBank Acc. No.: MT154757 for nrLSU, MT110363 for *tef1-α*, MT110400 for *rpb1*, MT110434 for *rpb2*).

Basidioma small to medium-sized. Pileus 5–8 cm in diam., hemispherical, subhemispherical, or convex; surface dry, yellowish green (29C8) to brownish green (30D7–8) when young, grayish brown (6E3–4) to brownish (6D4–5) with age, without discoloration when bruised; surface dry, covered with concolorous floccose squamules; margin extended, forming membranous veil and embracing the stipe when young and then breaking into pieces and hanging on the pileal margin; context 4 mm thick in the center, soft, white to pallid but with yellowish to orange-yellow tinge beneath pileipellis, without discoloration when bruised. Hymenophore adnate to depressed around apex of stipe; surface whitish (4A1) to pinkish (4A2) when young, purplish (14B2) when mature, without discoloration when bruised; pores roundish or angular, 0.5–1 mm wide; tubes 8 mm long, concolorous or a little paler than hymenophoral surface, without discoloration when bruised. Spore print pinkish (13B2) to purplish (14B2). Stipe 6–10 × 0.4–1.2 cm, clavate to subcylindrical, always enlarged downwards, pallid to white, staining orange-yellow (5B2–5) to brownish (7D5–6) when bruised; surface distinctly reticulate with elongated meshes up to 3 mm high; context soft, white to pallid, without discoloration when hurt; basal mycelium whitish to pallid. Taste and odor mild.

Basidia 18–38 × 8–12 μm, clavate to narrowly clavate, 4-spored, sometimes 2-spored, hyaline to yellowish in KOH. Basidiospores [60/3/2] 12–15 × (5.5) 6–6.5 μm [Q = 2–2.42 (2.5), Q_m = 2.18 ± 0.15] including ornamentation, subcylindrical to subfusiform with a suprahilar depression; surface reticulate with

Fig. 8.5 a *Austroboletus olivaceobrunneus* Yan C. Li & Zhu L. Yang. Photos by K. Zhao (KUN-HKAS 92428, type).

Fig. 8.5 b *Austroboletus olivaceobrunneus* Yan C. Li & Zhu L. Yang. Photo by J.W. Liu (KUN-HKAS 107144).

high ridges, apex and base forming intricate reticulum and shallow pits (Fig. 8 e), yellowish to brownish yellow in KOH, yellow to yellow-brown in Melzer's reagent. Hymenophoral trama boletoid composed of 3–12 μm wide filamentous hyphae, hyaline to yellowish in KOH, yellowish to yellow in Melzer's reagent. Cheilo- and pleurocystidia 26–69 × 7–11 μm, subfusiform or fusiform, thin-walled, yellowish brown to brownish in KOH, yellow-brown to brown in Melzer's reagent. Caulocystidia forming the reticulum over the surface of stipe, morphologically similar to hymenial cystidia. Pileipellis a trichoderm to palisadoderm or ixopalisadoderm composed of 4–8.5 μm wide interwoven hyphae, yellowish to pale brownish in KOH and yellow to brownish in Melzer's reagent; terminal cells 23–60 × 4–8.5 μm, subcylindrical. Pileal trama composed of 5–7 μm wide filamentous hyphae, hyaline to yellowish in KOH and yellowish to brownish in Melzer's reagent. Clamp connections absent in all tissues.

Habitat: Scattered on soil in subtropical mixed forests dominated by plants of the families Fagaceae and Pinaceae.

Known distribution: Currently known from southwestern China.

Additional specimen examined: CHINA, YUNNAN PROVINCE: Lancang County, alt. 1000 m, 28 June 2017, J.W. Liu 530828MF0042 (KUN-HKAS 107144).

Commentary: *Austroboletus olivaceobrunneus* is characterized by the unviscid yellowish green to brownish green and then grayish brown to brownish pileus, the whitish to pinkish hymenophore, the distinctly reticulate stipe, the subfusiform basidiospores with distinct reticulum, the subfusiform or fusiform hymenial cystidia, and the trichoderm to palisadoderm pileipellis composed of 4–8 μm wide interwoven or somewhat vertically arranged hyphae. This species is morphologically similar and phylogenetically related to *A. subvirens* (Hongo) Wolfe, originally described from Japan (Fig. 4.1). However, *A. subvirens* has a green to olivaceous pileus lacking brownish tinge, and relatively big basidiospores 13–19 × 6–7.5 (8) μm (Hongo 1960; Wolfe 1979a).

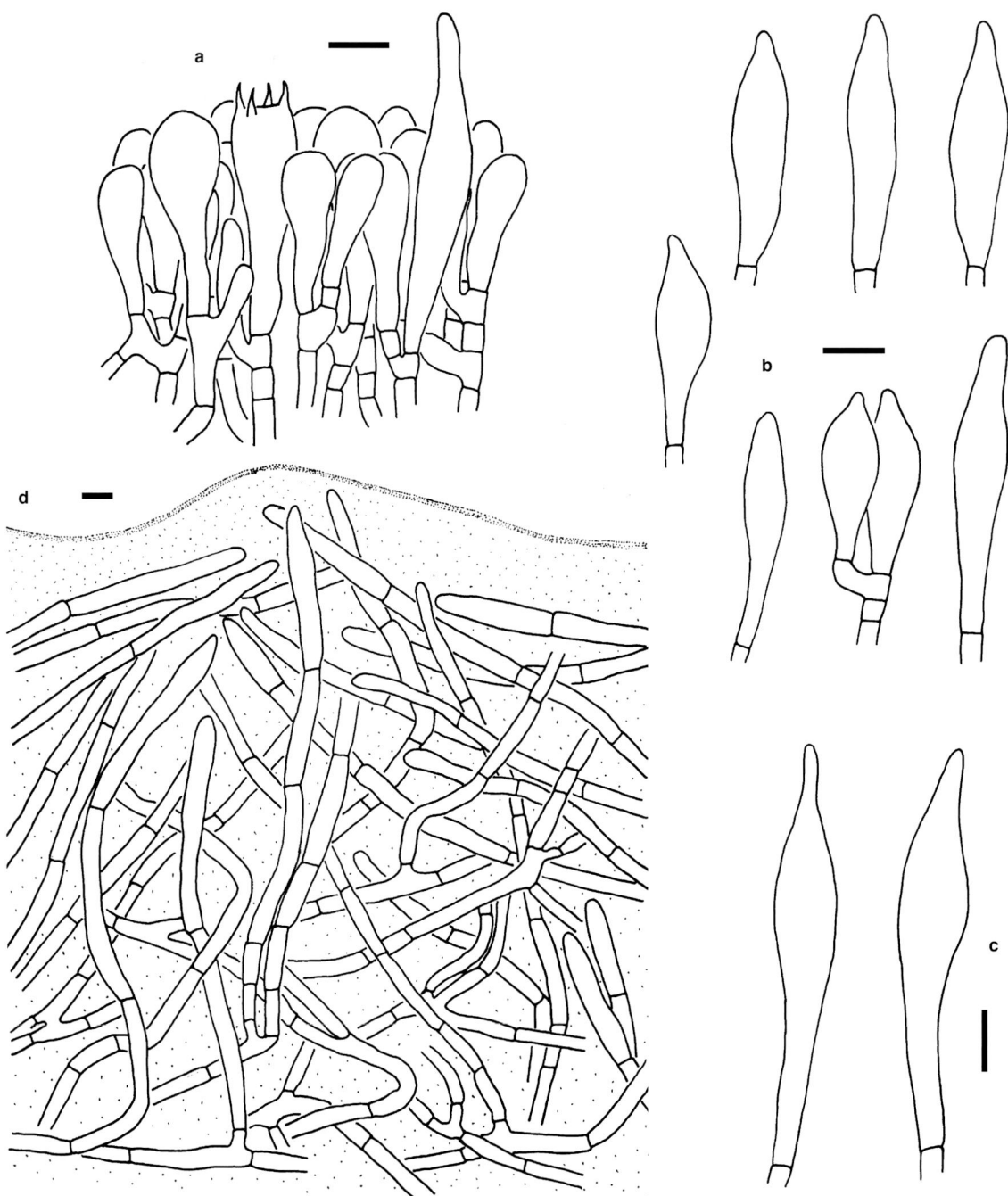

Fig. 8.5 c *Austroboletus olivaceobrunneus* Yan C. Li & Zhu L. Yang (KUN-HKAS 92428, type).
a. Basidia and pleurocystidium; b. Cheilocystidia; c. Pleurocystidia; d. Pileipellis. Scale bars = 10 μm.

8.6 *Austroboletus olivaceoglutinosus* K. Das & Dentinger, Kew Bull 70: 15 (2015)

Basidioma very small to small. Pileus 2.5–5 cm in diam., conical, convex to plano-convex, glutinous to sticky when wet, shiny when dry, dark green to olive when young, and then pale olive to greenish yellow when mature; surface nearly glabrous to finely tomentose when dry, margin extended and embracing the stipe when young and then breaking into pieces and hanging on the pileal margin; context soft, white to pallid, without discoloration when bruised. Hymenophore adnate when young, adnate to slightly depressed around apex of stipe; surface pinkish to grayish pink when young and then pink to purplish pink when mature, without color change when injured; pores roundish to angular, 0.5–1 mm wide; tubes 4–7 mm long, concolorous or a little paler than hymenophoral surface, without discoloration when bruised. Spore print pinkish to purplish. Stipe 6–12 × 1–1.2 cm, clavate to subcylindrical, always enlarged downwards, white to cream, staining brownish yellow to yellowish when injured; surface viscid when wet, distinctly reticulate; context soft, white to pallid, without discoloration when injured; basal mycelium whitish to pallid. Taste and odor mild.

Fig. 8.6 a *Austroboletus olivaceoglutinosus* K. Das & Dentinger. Photo by X. Xu (KUN-HKAS 106465).

Fig. 8.6 b *Austroboletus olivaceoglutinosus* K. Das & Dentinger. Photos by J.W. Liu (KUN-HKAS 91167).

Basidia 35–50 × 11–17 μm, clavate, 4-spored, hyaline in KOH. Basidiospores 12–17 × 6–7.5 μm (Q = 2.1–2.33, Q_m = 2.15 ± 0.11), subcylindrical to subfusiform, inequilateral in side view; surface with intricate reticulum or irregular pits (Fig. 3.3 k, Fig. 8 f), yellowish to pale olivaceous in KOH, yellow to olivaceous brown in Melzer's reagent. Hymenophoral trama boletoid composed of 2.5–11 μm wide filamentous hyphae, hyaline to yellowish in KOH, yellowish to yellow in Melzer's reagent. Cheilocystidia cylindrical or finger-like composed of 2–3 cells, with terminal cells 22.5–47 × 5–7.5 μm, yellowish to brownish yellow in KOH, somewhat thick-walled (up to 1 μm thick). Pleurocystidia not observed. Caulocystidia forming the reticulum over the surface of stipe, morphologically similar to cheilocystidia. Pileipellis an ixocutis, composed of 1.5–3 μm wide filamentous hyphae, embedded in thick gelatinous matrix, yellowish to pale brownish in KOH and yellow to brownish in Melzer's reagent; terminal cells 6.5–40.5 × 1.5–3 μm, cylindrical. Clamp connections absent in all tissues.

Habitat: Scattered on soil in subalpine forests dominated by plants of the family Pinaceae.

Known distribution: Currently known from India and southwestern China.

Specimens examined: CHINA, XIZANG AUTONOMOUS REGION: Milin County, alt. 3350 m, 8 August 2015, J.W. Liu 359 (KUN-HKAS 91167); Bomi County, alt. 3300 m, 8 August 2018, X. Xu 132 (KUN-HKAS 106456). YUNNAN PROVINCE: Yulong County, Laojun Mountain, alt. 3600 m, 2 September 2009, G. Wu 224 (KUN-HKAS 57756).

Commentary: *Austroboletus olivaceoglutinosus* was originally described from subalpine coniferous forests in India, and then reported from China (Wu *et al.* 2016a). This species is characterized by the sticky basidioma, the dark green to olive pileus, the initially pinkish to grayish pink and then pink to pinkish purple hymenophore, the reticulate stipe, the subcylindrical to subfusiform basidiospores with intricate reticulum or irregular pits, and the ixocutis pileipellis. *Austroboletus olivaceoglutinosus* can be confused with *A. albovirescens* because of the similar colors of their basidiomata. However, our multi-locus phylogenetic analysis (Fig. 4.1) reveals that they cluster in two independent lineages and represent two distinct species. Moreover, *A. albovirescens* differs from *A. olivaceoglutinosus* in its dry pileus, pitted basidiospores, ixocutis pileipellis composed of 3–6 μm wide gelatinizing filamentous hyphae, and distribution in subtropical forests dominated by plants of the family Fagaceae.

8.7 *Austroboletus subvirens* (Hongo) Wolfe, Biblthca Mycol 69: 125 (1979a)

Basionym: *Porphyrellus subvirens* Hongo, Acta Phytotax Geobot, Kyoto 18(4): 110 (1960).

Basidioma small to medium-sized. Pileus 3–8 cm in diam., hemispherical to subhemispherical or convex to plano-convex, olivaceous green to grayish green when young, and then dull green to yellowish green in the center and yellowish brown to yellow-brown elsewhere; pileal margin extended and embracing the stipe when young and then breaking into pieces and hanging on the pileal margin; surface dry, densely covered with concolorous appressed scaly squamules; context spongy, white, without discoloration when bruised. Hymenophore adnate when young and then depressed around apex of stipe when mature; surface grayish to pinkish or pinkish gray when young and pink to purple-gray when mature, without color change when injured; pores angular to roundish, 0.5–1 mm wide; tubes 3–12 mm long, concolorous or a little paler than hymenophoral surface, without discoloration when bruised. Spore print pinkish to purplish. Stipe 9–12 × 1.2–2 cm, clavate to subcylindrical, always enlarged downwards, white to cream, staining yellowish brown to brownish when touched or aged; surface dry, deeply reticulate, sometimes with olivaceous tinge, becoming yellowish or pale yellow to grayish yellow when bruised or mature; context white, without

Fig. 8.7 a *Austroboletus subvirens* (Hongo) Wolfe. Photos by J.W. Liu (KUN-HKAS 107142).

Fig. 8.7 b *Austroboletus subvirens* (Hongo) Wolfe. Photos by G. Wu (KUN-HKAS 107149).

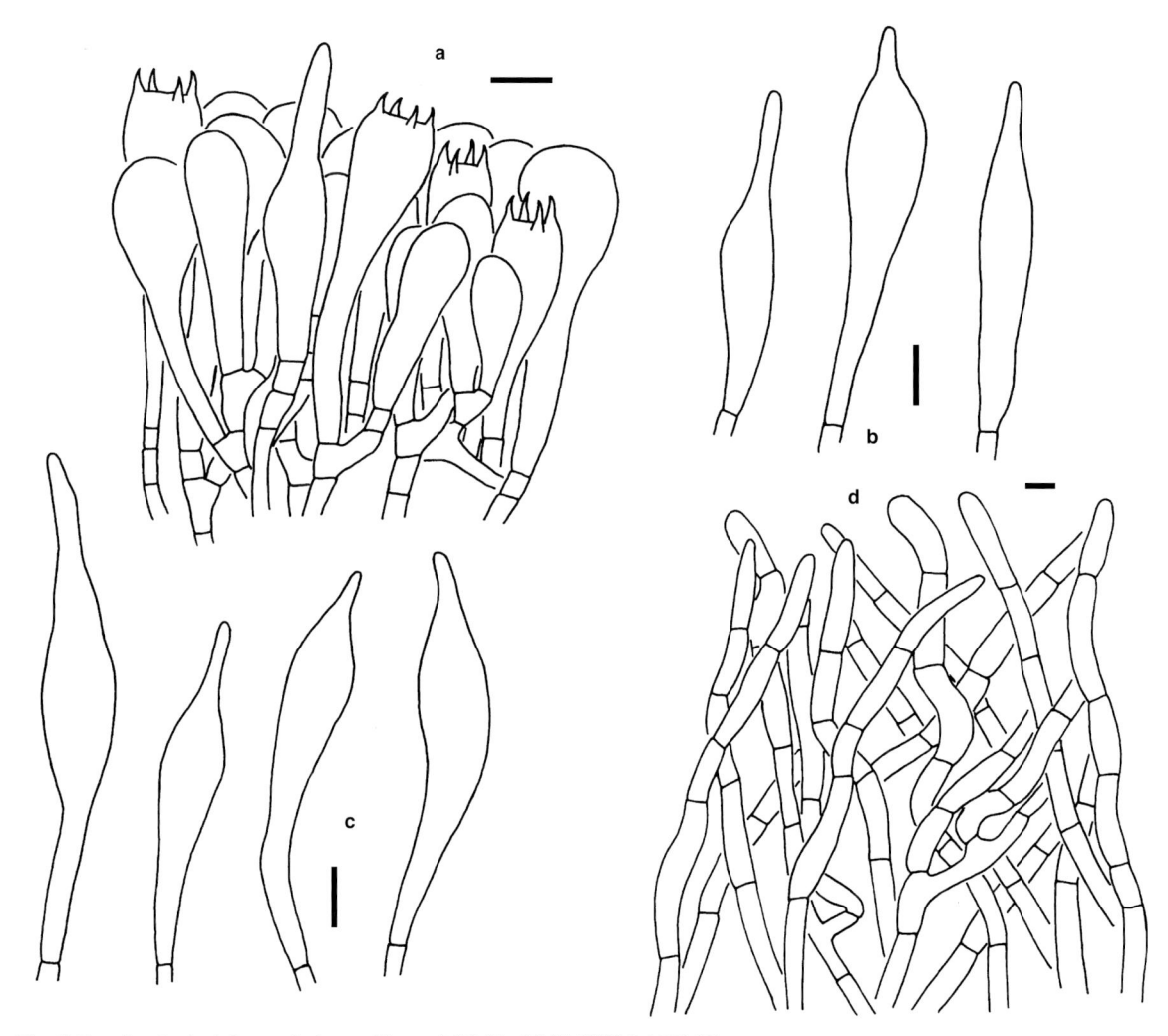

Fig. 8.7 c *Austroboletus subvirens* (Hongo) Wolfe (KUN-HKAS 107142).
a. Basidia and cheilocystidium; b. Cheilocystidia; c. Pleurocystidia; d. Pileipellis. Scale bars = 10 μm.

discoloration when injured; basal mycelium whitish to pallid. Taste and odor mild.

Basidia 25–55 × 11–14 μm, clavate, 4-spored, hyaline in KOH. Basidiospores 11–16 (17) × (6) 7–8 (9) μm [Q = (1.57) 1.63–2.2 (2.42), Q_m = 1.86 ± 0.15], ellipsoid to elongated or subfusiform, inequilateral in side view; surface with intricate reticulum and irregular pits (Fig. 8 g, h), yellowish to pale olivaceous in KOH, yellow to brownish or olivaceous in Melzer's reagent. Hymenophoral trama boletoid composed of 2–16 μm wide filamentous hyphae, hyaline to yellowish in KOH, yellowish to yellow in Melzer's reagent. Cheilo- and pleurocystidia rare, 31–85 × 8–13 μm, subfusiform to fusoid-ventricose, brownish to yellow-brown in KOH, somewhat thick-walled (up to 1 μm thick). Caulocystidia forming the reticulum over the surface of stipe, morphologically similar to hymenial cystidia. Pileipellis a trichoderm to palisadoderm composed of 4–10 μm wide filamentous hyphae, pale yellow to brownish green in KOH and yellow-brown to greenish brown in Melzer's reagent; terminal cells 10–72 × 4–10 μm, subcylindrical. Pileal trama compose of 2–20 μm wide filamentous hyphae, hyaline to yellowish in KOH, yellowish to yellow in Melzer's reagent. Clamp connections absent in all tissues.

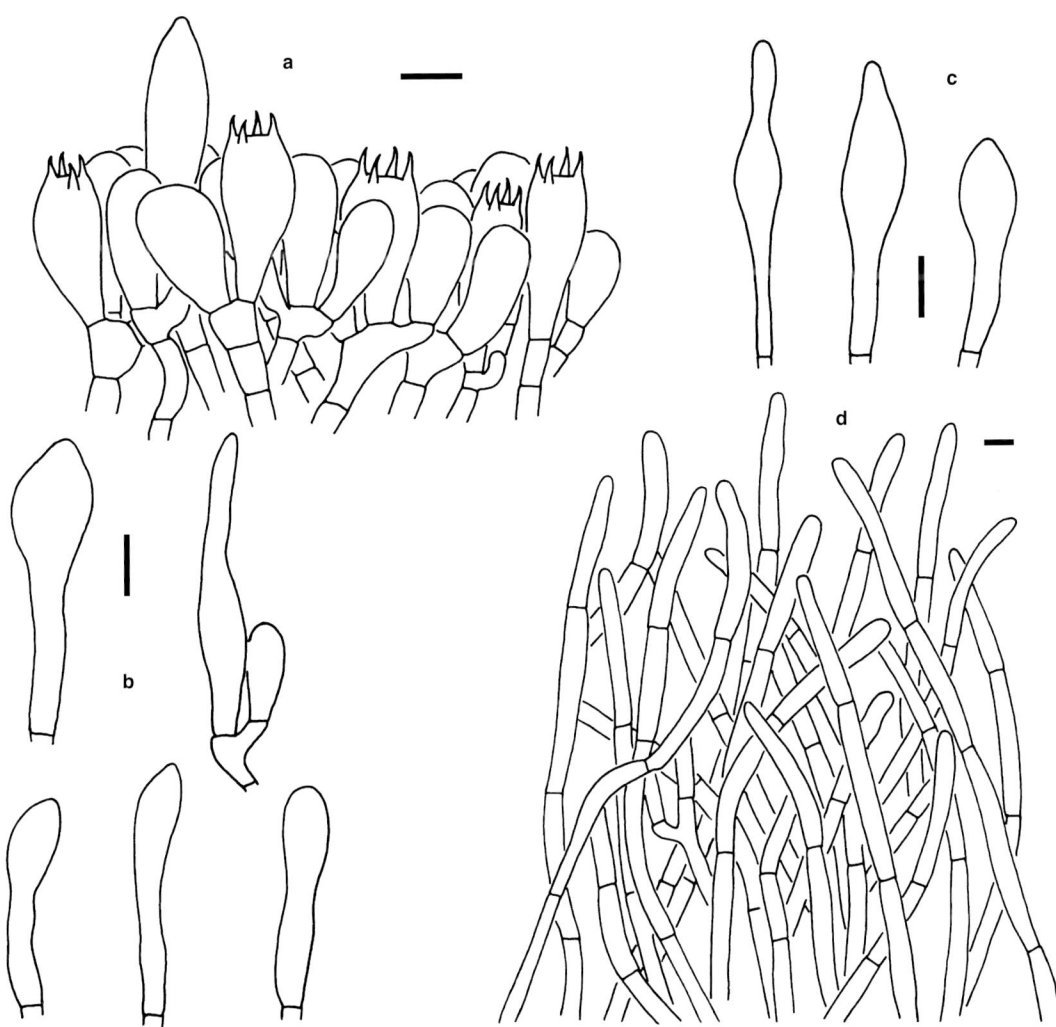

Fig. 8.7 d *Austroboletus subvirens* (Hongo) Wolfe (KUN-HKAS 107149).
a. Basidia and cheilocystidium; b. Cheilocystidia; c. Pleurocystidia; d. Pileipellis. Scale bars = 10 μm.

Habitat: Scattered on soil in subtropical forests dominated by plants of the family Fagaceae.

Known distribution: Currently known from Japan southwestern China.

Specimens examined: CHINA, YUNNAN PROVINCE: Malipo County, Shuitangzi Village, alt. 1200 m, 24 June 2017, G. Wu 2027 (KUN-HKAS 107149); Lancang County, alt. 1000 m, 26 June 2017, J.W. Liu 530828MF0013 (KUN-HKAS 107142).

Commentary: *Austroboletus subvirens* was originally reported from Japan, and then reported from China (Hongo 1960; Li and Song 2003). This species resembles the species with olivaceous or greenish tinge and non-gelatinous pileus, namely *A. olivaceobrunneus*. However, *A. olivaceobrunneus* has a yellow-green to brownish green pileus, and marrow basidiospores 12–15 × 5.5–6.5 μm. *Austroboletus subvirens* is also comparable to *A. dictyotus* in that they share a yellowish brown to yellow-brown pileus in aged basidiomata. However, *A. dictyotus*, originally described from Indonesia, differs from *A. subvirens* in its big basidiomata, reddish brown tinged pileus when young, and irregular reticulate basidiospores (Boedijn 1960; Corner 1972; Horak 2011).

Chapter 9
Chiua Yan C. Li & Zhu L. Yang

Chiua Yan C. Li & Zhu L. Yang, Fungal Divers 81: 76 (2016)

Type species: *Boletus virens* W.F. Chiu, Mycologia 40(2): 206 (1948).

Diagnosis: The genus differs from the other genera in the Boletaceae in its dark green to green or yellow-green pileus, yellow to bright yellow stipe covered with yellow or pink to red scabrous squamules, yellow to chrome-yellow context in pileus and stipe without discoloration when injured, cutis to trichoderm, epithelium or hyphoepithelium pileipellis.

Pileus subhemispherical to convex when young, subhemispherical to applanate or plano-convex when mature; surface dark green to green or yellow-green, dry, fibrillose or tomentose; context yellow to bright yellow, without discoloration when injured. Hymenophore adnate when young, adnate to depressed around apex of stipe when mature; surface white when young and becoming pinkish or pink when mature, without discoloration when injured; pores angular to roundish, 0.3–1 mm wide; tubes concolorous or much paler than hymenophoral surface, without discoloration when injured. Spore print brownish red to red. Stipe central, yellow to bright yellow, but golden yellow to chrome-yellow at base; basal mycelium chrome-yellow to golden yellow. Basidiospores subfusiform to elongated or cylindrical; surface smooth. Hymenial cystidia subfusiform to ventricose or clavate. Pileipellis an epithelium with terminal cells subfusoid-mucronate to mucronate or flagelliform, or a cutis composed of filamentous hyphae or a hyphoepithelium with an upper layer composed of filamentous hyphae and an inner layer composed of erect moniliform hyphae. Clamp connections absent in all tissues.

Commentary: *Chiua* generally shares the color of the hymenophore, stipe base and basidiospore deposit with *Harrya*, *Hymenoboletus*, *Royoungia* Castellano, Trappe & Malajczuk, and *Zangia*. However, *Harrya* differs from *Chiua* in its dry and subtomentose pileus usually with wine-red tinge when bruised, white to cream pileal context which is somewhat with pinkish purple tinge beneath pileipellis, white to cream or yellowish surface of stipe which is covered with pink to reddish scabrous squamules (Halling *et al.* 2012b; Wu *et al.* 2016a). *Hymenoboletus* differs from *Chiua* in its purplish red stipe which is yellowish to yellow at apex and bright yellow to chrome-yellow at base, and white to pallid context in pileus (Wu *et al.* 2016a). *Royoungia* differs from *Chiua* in its whitish to white context in pileus which is without discoloration when injured, yellowish to yellow surface of stipe covered with pink to reddish scabrous squamules (Wu *et al.* 2016a). *Zangia* differs from *Chiua* in its whitish to pallid pileal context, ixohyphoepithelium pileipellis, and bluish discoloration when injured (Li *et al.* 2011; Wu *et al.* 2016a). Currently four species are known from China.

Chapter 9 *Chiua* Yan C. Li & Zhu L. Yang

Key to the species of the genus *Chiua* in China

1. Surface of stipe always covered with pink or red scabrous squamules; pileipellis an epithelium composed of vertically arranged moniliform hyphae or a hyphoepithelium with an outer layer composed of filamentous hyphae and an inner layer composed of inflated concatenated cells ····· ·· 2
1. Surface of stipe nearly glabrous or with yellow scabrous squamules; pileipellis a cutis to intricate trichoderm composed of 3.5–7 μm wide interwoven filamentous hyphae ········ *C. virens*
2. Pileipellis a hyphoepithelium, with outer layer composed of filamentous hyphae and inner layer composed of concatenated globose to subglobose cells ·· 3
2. Pileipellis an epithelium composed of concatenated globose to subglobose cells, with terminal cells subfusoid-mucronate to mucronate or flagelliform ······························· *C. angusticystidiata*
3. Basidiomata distributed in subtropical to temperate forests and associated with plants of the family Pinaceae; basidiospores 11–14 × 4–5 μm ·· *C. olivaceoreticulata*
3. Basidiomata distributed in subtropical forests and associated with plants of the family Fagaceae; basidiospores 10–12 × 4–5 μm ··· *C. viridula*

9.1 *Chiua angusticystidiata* Yan C. Li & Zhu L. Yang, Fungal Divers 81: 77 (2016)

Basidioma small to medium-sized. Pileus 3–6 cm, subhemispherical, convex, applanate or plano-convex, deep olivaceous or olivaceous green and then olive-yellow to yellowish green; surface dry, covered with subtomentose to fibrillose squamules; context yellow to bright yellow, without discoloration when cut. Hymenophore adnate when young, depressed around apex of stipe; surface initially white to pallid and then pinkish or dingy pink, without discoloration when injured; pores angular to roundish, 0.3–1 mm wide; tubes 5–15 mm long, concolorous or a little paler than hymenophoral surface, without discoloration when injured. Spore print pinkish to pale brownish red. Stipe 3–6 × 0.6–1.3 cm, clavate, always enlarged downwards, yellowish to yellow at upper part and bright yellow to chrome-yellow at base; surface covered with pink to pinkish red scabrous squamules; context yellow; but bright yellow to chrome-yellow at base, without discoloration when injured; basal mycelium bright yellow to chrome-yellow, without discoloration when bruised. Taste and odor mild.

Basidia 20–36 × 7–10 μm, clavate, 4-spored, hyaline in KOH and yellowish in Melzer's reagent. Basidiospores (9) 10.5–12.5 (14) × 4.5–5.5 (6) μm (Q = 1.09–2.8, Q_m = 2.4 ± 0.2), elongated to subcylindrical with slight suprahilar depression, smooth, somewhat thick-walled (0.5–1 μm thick), yellowish to light pinkish in KOH and brownish to yellowish brown in Melzer's reagent. Pleuro- and cheilocystidia 30–51 × 5–7 μm, clavate, lanceolate, ventricose or mucronate, thin-walled, hyaline in KOH and yellowish to yellow in Melzer's reagent. Pileipellis an epithelium composed of broad (up to 20 μm wide) vertically arranged moniliform hyphae; terminal cells 27–35 × 3–11 μm, subfusoid-mucronate to mucronate or flagelliform; pileal trama composed of 6–8 μm wide filamentous hyphae, colorless or hyaline in KOH, yellowish to yellowish brown in Melzer's reagent. Clamp connections absent in all tissues.

Habitat: Gregarious to solitary to scattered on soil in subtropical forests dominated by plants of the family Fagaceae.

Known distribution: Currently known from central, southeastern, southwestern, and southern China.

Specimens examined: CHINA, FUJIAN PROVINCE: Anxi County, Changqin Town, Nanyang Village, alt. 240 m, 28 August 2009, N.K. Zeng 592 (KUN-HKAS 63664). HAINAN PROVINCE: Qiongzhong County, Limu Mountain, alt. 820 m, 3 August 2010, N.K. Zeng 812 (KUN-HKAS 59840). HUNAN PROVINCE: Chengbu County, Erbaoding, alt. 1300 m, 9 August 2009, Z.H. Chen 30372 (MHHNU 30372). GUANGDONG PROVINCE: Fengkai County, Heishiding Nature Reserve, alt. 800 m, 3 June 2013, Y.J. Hao 840 (KUN-HKAS 80120). YUNNAN PROVINCE: Jinghong County, Dadugang Town, alt. 1000 m, 14 July 2006, Y.C. Li 528 (KUN-HKAS 50282), the same location, 10 August 2006, Y.C. Li 706 (KUN-HKAS 50460, type); Nanhua County, Wild Mushroom Market, altitude unknown, 25 August 2007, Z.L. Yang 4922 (KUN-HKAS 52239).

Commentary: *Chiua angusticystidiata* was originally described by Wu *et al.* (2016a) from China, and is characterized by the deep olivaceous or olivaceous green and then olive-yellow to yellowish green pileus, the pinkish to pink or dingy pink hymenophore, and the yellow to chrome-yellow context of pileus and stipe. These traits are very similar to those of *C. olivaceoreticulata* Yan C. Li & Zhu L. Yang,

C. virens (W.F. Chiu) Yan C. Li & Zhu L. Yang and *C. viridula* Yan C. Li & Zhu L. Yang. However, *C. olivaceoreticulata* differs in its relatively slender basidiospores measuring 10–14 × 4–5.5 μm, hyphoepithelium pileipellis with an outer layer composed of filamentous hyphae and an inner layer composed of subglobose concatenated cells, distribution in subtropical to temperate forests, and association with plants of the family Pinaceae (Wu *et al.* 2016a). *Chiua virens* differs in its initially dark green to green or olive-green and then becoming mustard-yellow to greenish yellow pileus, yellow to chrome-yellow scabrous stipe but without any pink or reddish tinge, and cutis to intricate trichoderm pileipellis. *Chiua viridula* differs in its relatively slender basidiospores measuring 9.5–12 × 4–5 μm, and hyphoepithelium pileipellis (Wu *et al.* 2016a).

Fig. 9.1 *Chiua angusticystidiata* Yan C. Li & Zhu L. Yang. Photos by Y.J. Hao (KUN-HKAS 80120).

9.2 *Chiua olivaceoreticulata* Yan C. Li & Zhu L. Yang, Fungal Divers 81: 78 (2016)

Basidioma small to medium-sized. Pileus 4–8 cm in diam., subhemispherical to convex and then plano-convex to applanate when mature, dark green to grayish green and then green to yellowish green, much paler towards margin; surface dry, fibrillose to tomentose; context yellow to chrome-yellow, color unchanged when injured. Hymenophore adnate when young, adnate to depressed around apex of stipe when mature; surface initially white to pallid and then pinkish to pink, without discoloration when injured; pores angular to roundish, 0.3–1 mm wide; tubes up to 10 mm long, concolorous or much paler than hymenophoral surface, without discoloration when bruised. Spore print pinkish to pale brownish red. Stipe 3–7 × 1.2–2.5 cm, clavate, always attenuate downwards, pink to reddish on the upper part, yellow downwards, but chrome-yellow to golden yellow at base; surface covered with pink to red scabrous squamules, sometimes with distinct reticulum, without color change when injured; context yellow in the upper part, chrome-yellow or golden yellow downwards, without color change when injured; basal mycelium chrome-yellow or golden yellow. Taste and odor mild.

Basidia 25–43 × 8–10.5 μm, clavate, 4-spored, hyaline or colorless in KOH. Basidiospores 10–14 × 4–5.5 μm (Q = 2.18–2.88, Q_m = 2.58 ± 0.15), subfusiform to subcylindrical, smooth, hyaline to yellowish in KOH, yellow to brownish yellow in Melzer's reagent. Hymenophoral trama boletoid composed of 4–8 μm wide filamentous hyphae. Cheilo- and pleurocystidia 35–60 × 5–7.5 μm, lanceolate to subfusiform, with 1–2 constrictions in the middle parts, hyaline to yellowish in KOH, thin-walled. Caulocystidia forming the scabrous squamules over the surface of stipe, clavate to lanceolate or subfusiform. Pileipellis a hyphoepithelium: outer layer composed of 4–7 μm wide interwoven filamentous hyphae with cylindrical terminal cells 12–40 × 3.5–6 μm, yellowish to pale brownish in KOH and yellow to brownish in Melzer's reagent; inner layer composed of globose to subglobose (up to 19 μm wide) concatenated cells arising from 4–9 μm wide radially arranged filamentous hyphae, colorless to yellowish in KOH and yellow to brownish yellow in Melzer's reagent. Pileal trama composed of broad (up to 9 μm wide) interwoven hyphae, hyaline

Fig. 9.2 a *Chiua olivaceoreticulata* Yan C. Li & Zhu L. Yang. Photos by Y.C. Li (KUN-HKAS 59675, type).

Fig. 9.2 b *Chiua olivaceoreticulata* Yan C. Li & Zhu L. Yang. Photos by K. Zhao (KUN-HKAS 92480).

to yellowish in KOH and yellowish to brownish in Melzer's reagent. Clamp connections absent in all tissues.

Habitat: Scattered on soil in subtropical to temperate forests dominated by plants of the family Pinaceae or in the mixed forests dominated by plants of the families Fagaceae and Pinaceae.

Known distribution: Currently known from northeastern, central, and southwestern China.

Specimens examined: CHINA, HENAN PROVINCE: Luanchuan County, Laojun Mountain, alt. 2000 m, 13 August 2015, B. Li 48 (KUN-HKAS 89814). HUBEI PROVINCE: Shennongjia Forestry District, Muyu Town, alt. 1800 m, 8 August 2015, X.B. Liu 749 (KUN-HKAS 92294). LIAONING PROVINCE: Dandong County, Jinshan Town, Baihuagu, alt. 110 m, 30 August 2018, H.Y. Liu 3 (HMJAU 55233) and H.Y. Liu 93 (HMJAU 55323), the same town, Wulong Mountain, alt. 110 m, 31 August 2018, H.Y. Liu 97 (HMJAU 55327). SICHUAN PROVINCE: Xichang, Puge County, Luoji Mountain, alt. 2000 m, 12 August 2010, X.F. Shi 677 (KUN-HKAS 76678). YUNNAN PROVINCE: Yongping County, near Milestone 3295 KM of the National Road 320#, alt. 2100 m, 31 July 2009, Y.C. Li 1927 (KUN-HKAS 59675, type); Nanhua County, Wild Mushroom Market, altitude unknown, 3 August 2009, Y.C. Li 1960 (KUN-HKAS 59706); Gongshan County, Bingzhongluo Town, alt. 1800 m, 24 August 2015, K. Zhao 848 (KUN-HKAS 90195), and K. Zhao 850 (KUN-HKAS 92480).

Commentary: *Chiua olivaceoreticulata* was described from southwestern China by Wu *et al.* (2016a). This species is characterized by the dark green to grayish green and then green to yellowish green pileus, the initially white and then pinkish hymenophore, the yellow to bright yellow context in pileus and stipe, and the hyphoepithelium pileipellis. *Chiua olivaceoreticulata* is phylogenetically related and morphologically similar to *C. angusticystidiata* (Fig. 4.1). However, *C. angusticystidiata* differs in its epithelium pileipellis with terminal cells subfusoid-mucronate to mucronate or flagelliform, relatively broad basidiospores measuring 9–14 × 4.5–6 μm, and distribution in tropical or subtropical forests (Wu *et al.* 2016a).

9.3 *Chiua virens* (W.F. Chiu) Yan C. Li & Zhu L. Yang, Fungal Divers 81: 79 (2016)

Basionym: *Boletus virens* W.F. Chiu, Mycologia 40: 206 (1948).

Synonyms: *Tylopilus virens* (W.F. Chiu) Hongo, Mem Shiga Univ 14: 46 (1964); *Tylopilus chromoreticulatus* Wolfe & Bougher, Aust Syst Bot 6(3): 205 (1993); *Tylopilus pinophilus* Wolfe & Bougher, Aust Syst Bot 6 (3): 203 (1993).

Basidioma small to medium-sized. Pileus 3–8 cm in diam., subhemispherical to plano-convex or applanate, dark green, green or olive-green and then mustard-yellow to greenish yellow when mature; surface dry, covered with concolorous subtomentose to fibrillose squamules; context yellow to chrome-yellow, without color change when bruised. Hymenophore adnate when young, adnate to depressed around apex of stipe when mature; surface white when young and then pinkish to pink when mature, without discoloration when injured; pores angular to

Fig. 9.3 *Chiua virens* (W.F. Chiu) Yan C. Li & Zhu L. Yang. Photos by Z.L. Yang (KUN-HKAS 107678).

roundish, 0.5–1 mm wide; tubes 5–20 mm long, concolorous or a little paler than hymenophoral surface, without discoloration when injured. Spore print pinkish red to pale brownish red (Fig. 3.4 d). Stipe 3–8 × 1.2–1.8 cm, clavate to subcylindrical, always enlarged downwards, yellow on the upper part, chrome-yellow to bright yellow downwards, without discoloration when touched; surface nearly glabrous or with indistinct nets, occasionally covered with yellow squamules; context bright yellow to yellow, but chrome-yellow to golden yellow at base, without color change when injured; basal mycelium chrome-yellow or golden yellow. Taste and odor mild.

Basidia 20–30 × 9–14 μm, clavate, 4-spored, hyaline in KOH. Basidiospores 11–13.5 × 5–5.5 μm (Q = 2.3–2.7, Q_m = 2.47 ± 0.12), subfusiform to subcylindrical and inequilateral in side view with slight suprahilar depression, cylindrical to fusiform in ventral view, smooth, somewhat thick-walled (up to 1 μm thick), hyaline to yellowish in KOH, yellow to brownish yellow in Melzer's reagent. Hymenophoral trama boletoid composed of 3–11 μm wide filamentous hyphae, hyaline to yellowish in KOH, yellowish to yellow in Melzer's reagent. Cheilocystidia 20–37 × 7–9 μm, clavate, yellowish to pale yellow in KOH, thin-walled. Pleurocystidia 40–56 × 6–10 μm, fusoid-ventricose to subfusiform, thin-walled. Caulocystidia forming the scabrous squamules over the surface of stipe, ventricose or fusoid-ventricose to clavate or lanceolate. Pileipellis a subcutis to trichoderm composed of 3.5–7 μm wide filamentous hyphae, pale yellowish to yellow in KOH and yellow to brownish yellow in Melzer's reagent; terminal cells 8–31×4–7 μm, clavate to subcylindrical. Pileal trama composed of 4–15 μm wide interwoven hyphae, hyaline to yellowish in KOH and yellowish to yellow in Melzer's reagent. Clamp connections absent in all tissues.

Habitat: Scattered on soil in subtropical to cold-temperate mixed forests dominated by plants of the families Fagaceae and Pinaceae.

Known distribution: Currently known from southwestern China.

Specimens examined: CHINA, GUIZHOU PROVINCE: Weining County, alt. 2200 m, 20 August 2008, X.F. Shi 189 (KUN-HKAS 62606). SICHUAN PROVINCE: Xichang, Puge County, Luoji Mountain, alt. 1800 m, 12 September 2010, X.F. Shi 677 (KUN-HKAS 76678). YUNNAN PROVINCE: Kunming, Kunming Botanic Garden, alt. 1980 m, 14 August 2020, Z.L. Yang 6371 (KUN-HKAS 107678); Shangri-La County, Bukou Mountain, alt. 3000 m, 24 July 2006, Y.C. Li 610 (KUN-HKAS 50364); Ninglang County, Dayan Town, alt. 2300 m, 6 August 2011, Y.C. Li 2606 (KUN-HKAS 107177).

Commentary: *Chiua virens*, originally described as *Boletus virens* by Chiu (1948) from China, is characterized by the initially dark green or olive-green and then mustard-yellow to greenish yellow pileus, the white to pinkish or pink hymenophore, the bright yellow to yellow context in pileus without color change when cut, the yellow to mustard-yellow stipe covered with concolorous scabrous squamules, the subcutis to trichoderm pileipellis, and the distribution in subtropical to cold-temperate forests. This species was then transferred to *Tylopilus* due to the color of the hymenophore and basidiospores (Hongo 1964). However, the bright yellow to yellow context, the chrome-yellow to golden yellow base of stipe, the coarsely verrucose surface of stipe are different from those in the genus *Tylopilus* typified by *T. felleus*. Wu *et al.* (2016a) established the genus *Chiua* to accommodate *B. virens* and its allies. *Chiua virens* is phylogenetically related and macroscopically similar to *C. angusticystidiata*, *C. olivaceoreticulata*, and *C. viridula*. However, they can be separated from *C. virens* in the color of the stipe, and the structure of the pileipellis.

9.4 *Chiua viridula* Yan C. Li & Zhu L. Yang, Fungal Divers 81: 80 (2016)

Basidioma small. Pileus 3–4 cm in diam., hemispherical when young, subhemispherical to plano-convex or applanate when mature, dark green when young, green to yellowish green or grayish green when mature, margin much paler in color; surface dry, fibrillose or finely tomentose; context yellow to bright yellow, without color change when injured. Hymenophore adnate when young, adnate to depressed around apex of stipe when mature; surface white when young and then pinkish to pink when mature, without discoloration when bruised; pores subangular to roundish, 0.3–1 mm wide; tubes up to 8 mm long, concolorous or a little paler than hymenophoral surface, without discoloration when injured. Spore print pinkish red to pale brownish red. Stipe 4–7 × 0.6–1.2 cm, subcylindrical to clavate, yellow to bright yellow with pinkish tinge on the upper part, chrome-yellow to golden yellow at base; surface covered with concolorous scabrous squamules, without discoloration when touched; context yellow to bright yellow in the upper part, but bright yellow to chrome-yellow at base, without discoloration when bruised; basal mycelium bright yellow or chrome-yellow. Taste and odor mild.

Basidia 30–38 × 9–12 μm, clavate, 4-spored, hyaline in KOH. Basidiospores 9.5–12 × 4–5 μm (Q = 2.1–2.63, Q_m = 2.35 ± 0.16), subcylindrical in side view, cylindrical to fusiform in ventral view, smooth, hyaline to yellowish in KOH, yellowish brown to brownish in Melzer's reagent. Hymenophoral trama boletoid composed of 3–8 μm wide filamentous hyphae, hyaline to yellowish in KOH, yellowish to yellow in Melzer's reagent. Cheilo- and pleurocystidia 30–58 × 4–6.5 μm, subfusiform to ventricose or narrowly clavate, with 1–2 constrictions in the middle parts, hyaline to yellowish in KOH, thin-walled. Caulocystidia forming the scabrous squamules over the surface of stipe, clavate, lanceolate, ventricose or subfusoid. Pileipellis a hyphoepithelium: outer layer composed of 4–8 μm wide interwoven filamentous hyphae with cylindrical terminal cells 12–41 × 4–6 μm, yellowish to pale brownish in KOH and yellow to pale brownish in Melzer's reagent; inner layer made up of 20–30 μm wide subglobose to globose concatenated cells arising from 4–7 μm wide radially arranged filamentous hyphae, colorless to yellowish in KOH and yellow to brownish yellow in Melzer's reagent. Pileal trama composed of 4–15 μm wide interwoven hyphae, hyaline to yellowish in KOH and yellowish to yellow in Melzer's reagent. Clamp connections absent in all tissues.

Habitat: Gregarious to scattered on soil in subtropical forests dominated by plants of the family Fagaceae.

Known distribution: Currently known from central and southwestern China.

Specimens examined: HUNAN PROVINCE: Dawei Mountain, Chuandiwo, alt. 1300 m, 21 August 2010, P. Zhang 846 (MHHNU 7346). YUNNAN PROVINCE: Tengchong County, Longteng Road (X193), near Milestone 52 KM, alt. 1200 m, 11 August 2011, G. Wu 614 (KUN-HKAS 74928, type); Gejiu County, Manhao Town, Shazhudi Village, alt. 2000 m, 24 September 2011, Y.C. Li 2777 (KUN-HKAS 89420). GUIZHOU PROVINCE: Leishan County, Leigong Mountain, alt. 1800 m, 26 July 2018, X.H. Wang 4917 (KUN-HKAS 104630).

Commentary: *Chiua viridula*, originally described by Wu *et al.* (2016a) from China, is characterized by the dark green to green and then grayish green or yellowish green pileus, the white to pinkish

Fig. 9.4 *Chiua viridula* Yan C. Li & Zhu L. Yang. Photos by Y.C. Li (KUN-HKAS 89420).

hymenophore, the nearly glabrous to yellow scabrous stipe with a chrome-yellow to bright yellow base, the hyphoepithelium pileipellis, and the distribution in subtropical forests. *Chiua viridula* is macroscopically similar to *Zangia olivacea* Yan C. Li & Zhu L. Yang originally described from southwestern China by Li *et al.* (2011). However, *Z. olivacea* has a distribution in temperate to subalpine forests, a bluish discoloration in the stipe when bruised, an ixohyphoepithelium pileipellis, and broad basidiospores measuring 12–17 × 6–7 μm.

In our phylogenetic analysis (Fig. 4.1), *C. viridula* is closely related to *C. angusticystidiata* and *C. olivaceoreticulata*. However, *C. angusticystidiata* has an epithelium pileipellis with terminal cells subfusiform to mucronate or flagelliform, and relatively broad basidiospores measuring 9–14 × 4.5–6 μm (Wu *et al.* 2016a). *Chiua olivaceoreticulata* always has a reticulate stipe, relatively large basidiospores measuring 10–14 × 4–5.5 μm, and a distribution in subtropical to temperate forests.

Chapter 10
Fistulinella Henn.

Fistulinella Henn., Bot Jb 30: 43 (1901)

Type species: *Fistulinella staudtii* Henn., Bot Jb 30: 44 (1901).

Diagnosis: The genus is different from all the other genera in the Boletaceae in its usually gelatinous basidiomata, white to pallid context without color change when hurt, white to brownish pinkish hymenophore, brown to reddish brown spore print, ixocutis or ixotrichoderm pileipellis.

Pileus subhemispherical to convex; surface glabrous, viscid to gelatinous; margin slightly extended; context white to sordid white, without discoloration when injured. Hymenophore adnate to depressed around apex of stipe; surface white to pinkish when young and grayish pink when mature, without discoloration when touched; pores angular to roundish, fine, 0.3–1 mm wide; tubes concolorous with hymenophoral surface, without discoloration when injured. Spore print brown to reddish brown. Stipe central, clavate or tapering upwards, glabrous to subglabrous or finely velutinous to subpruinose; basal mycelium white. Basidiospores smooth, subfusiform to cylindrical. Pleuro- and cheilocystidia fusoid-ventricose to subfusoid. Pileipellis an ixocutis or an ixotrichoderm composed of strongly gelatinous filamentous hyphae. Clamp connections absent in all tissues.

Commentary: *Fistulinella* was originally described from Cameroon based on the species *F. staudtii* Henn. (Hennings 1901). It is characterized by the gelatinous basidiomata, the slight extended pileal margin, the smooth basidiospores, the brown to reddish brown spore print, and the smooth basidiospores (Redeuilh and Soop 2006; Fulgenzi *et al.* 2010; Vasco-Palacios *et al.* 2014; Magnago *et al.* 2017). These traits are somewhat similar to those of *Mucilopilus*. However, *Mucilopilus* differs from *Fistulinella* in its viscid to mucilaginous pileus and non-gelatinizing stipe, white context unchanging in pileus but staining yellow at the base of stipe when injured, pink to grayish pink spore print, and only ixotrichoderm pileipellis (Wolfe 1979b, 1981, 1982). Phylogenetically, this genus is polyphyletic (Vasco-Palacios *et al.* 2014; Magnago *et al.* 2017; this study). Since the type specimen for this genus, is conserved in alcohol, neither phylogenetic nor morphological study is possible. Further studies that include thorough sequences analysis of more species, including type specimens or specimens from the type locality, are required to assess the phylogenetic relationships among these fistulinoid boletes. In this study, two species from China match the morphological features of the genus *Fistulinella*.

10.1 *Fistulinella olivaceoalba* T.H.G. Pham, Yan C. Li & O.V. Morozova, Persoonia 41: 361 (2018)

Basidioma very small to small. Pileus 2.5–5 cm in diam., subhemispherical to convex or applanate, dark green to grayish green or olive-green when young, grass-green to yellowish green when mature, slightly darker in the center; margin sometimes extended, strongly gelatinous when wet; context whitish to grayish white, without discoloration when injured. Hymenophore adnate to slightly depressed around apex of stipe; surface initially white to pallid and then pinkish to grayish pink, without discoloration when injured; pores fine 0.3–1 mm wide; tubes 1–3 mm long, concolorous or a little paler than hymenophoral surface, without discoloration when bruised; spore print brown to reddish brown. Stipe strongly gelatinous, clavate to cylindrical 4–9 × 0.5–0.7 cm; surface white, sparsely covered with dotted or granular-like squamules; context white, without discoloration when bruised; basal mycelium white, without discoloration when bruised. Taste and odor unknown.

Basidia 22–36 × 8–11 μm, clavate, hyaline to yellowish in KOH, 4-spored, sometimes 2-spored. Basidiospores (11) 12–15 × 4–5 μm [Q = 2.5–3.13 (3.33), Q_m = 2.82 ± 0.2], subfusiform and inequilateral in side view with slight suprahilar depression, fusiform to subcylindrical in ventral view, smooth, yellowish to brownish yellow in KOH, yellow to yellow-brown in Melzer's reagent. Hymenophoral trama boletoid composed of 3.5–7 μm wide filamentous hyphae, hyaline to yellowish in KOH, yellowish to yellow in Melzer's reagent. Cheilocystidia broadly clavate to subfusiform, thin-walled, consisting of 2–3 cells, with the terminal cells 25–41 × 5–9 μm. Pleurocystidia 35–57 × 5–10 μm, fusiform to subfusiform or subfusoid-mucronate to ventricose-mucronate, thin-walled. Pileipellis an ixotrichoderm composed of 2.5–4 μm wide gelatinous interwoven hyphae, with clavate to subcylindrical terminal cells 26–73 × 4.5–5.5 μm, pale yellowish to brownish yellow in KOH and yellow to brownish in Melzer's reagent. Pileal trama composed of 3.5–5.5 μm wide interwoven hyphae, colorless to yellowish in KOH and yellowish to pale yellow in Melzer's reagent. Clamp connections absent in all tissues.

Habitat: Scattered on soil in tropical to subtropical forests dominated by plants of the family Fagaceae.

Known distribution: Currently known from central, southeastern, and southern China and Vietnam.

Specimens examined: FUJIAN PROVINCE: Sanming, Sanyuan National Forest Park, alt. 260 m, 26 August 2007, Y.C. Li 1022 (KUN-HKAS 53367); HUNAN PROVINCE: Yizhang County, Mangshan National Forest Park, alt. 880 m, 4 September 2007, Y.C. Li 1087 (KUN-HKAS 53432). HAINAN PROVINCE: Wuzhishan County, Wuzhishan National Nature Reserve, alt. 950 m, 2 August 2009, N.K. Zeng 416 (FHMU 202), the same location, Limu Mountain, alt. 850 m, 3 August 2010, N.K. Zeng 806 (KUN-HKAS 59847). GUANGDONG PROVINCE: Shaoguan, Ruyuan County, Nanling National Nature Reserve, alt. 1100 m, 11 July 2013, T. Guo 761 (KUN-HKAS 81963).

Commentary: *Fistulinella olivaceoalba* was described from Vietnam (Crous *et al.* 2018), based on its morphological characteristics and molecular evidence provided by Wu *et al.* (2014, 2016a). It is characterized by the small strongly gelatinous basidiomata with distinct greenish or olivaceous color in

Fig. 10.1 a *Fistulinella olivaceoalba* T.H.G. Pham, Yan C. Li & O.V. Morozova. Photo by N.K. Zeng (FHMU 202).

Fig. 10.1 b *Fistulinella olivaceoalba* T.H.G. Pham, Yan C. Li & O.V. Morozova. Photo by Y.C. Li (KUN-HKAS 53367).

the pileus, the pallid to pinkish hymenophore, the long cylindrical to clavate septate cheilocystidia and the fusiform pleurocystidia. *Fistulinella olivaceoalba* resembles *F. cinereoalba* Fulgenzi & T.W. Henkel from Guyana, South America. However, *F. cinereoalba* lacks the greenish color in the basidiomata. Moreover, the internal transcribed spacer (ITS) sequence for *F. cinereoalba* deposited in GenBank is significantly different from that of *F. olivaceoalba* (percent identity 93%). Our phylogenetic analysis reveals that *Fistulinella* is polyphyletic. Sequences of *F. olivaceoalba* form an unique clade and occupy the basal position of the subfamily Austroboletoideae, while sequences of *F. salmonea* Yan C. Li & Zhu L. Yang, *F. viscida* (McNabb) Singer and *F. prunicolor* (Cooke & Massee) Watling form another monophyletic clade and occupy the sub-basal position of the subfamily Austroboletoideae. Since the type specimen for the genus *Fistulinella* is conserved in alcohol, no sequences are available for elucidating its phylogenetic position. Further collections from the type locality are needed to clarify the phylogenetic position and morphological characteristics of the genus *Fistulinella*. However, *F. olivaceoalba* matches the morphological characteristics found in *Fistulinella* well and we therefore treat this species as *Fistulinella* for the present.

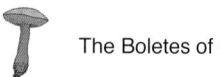

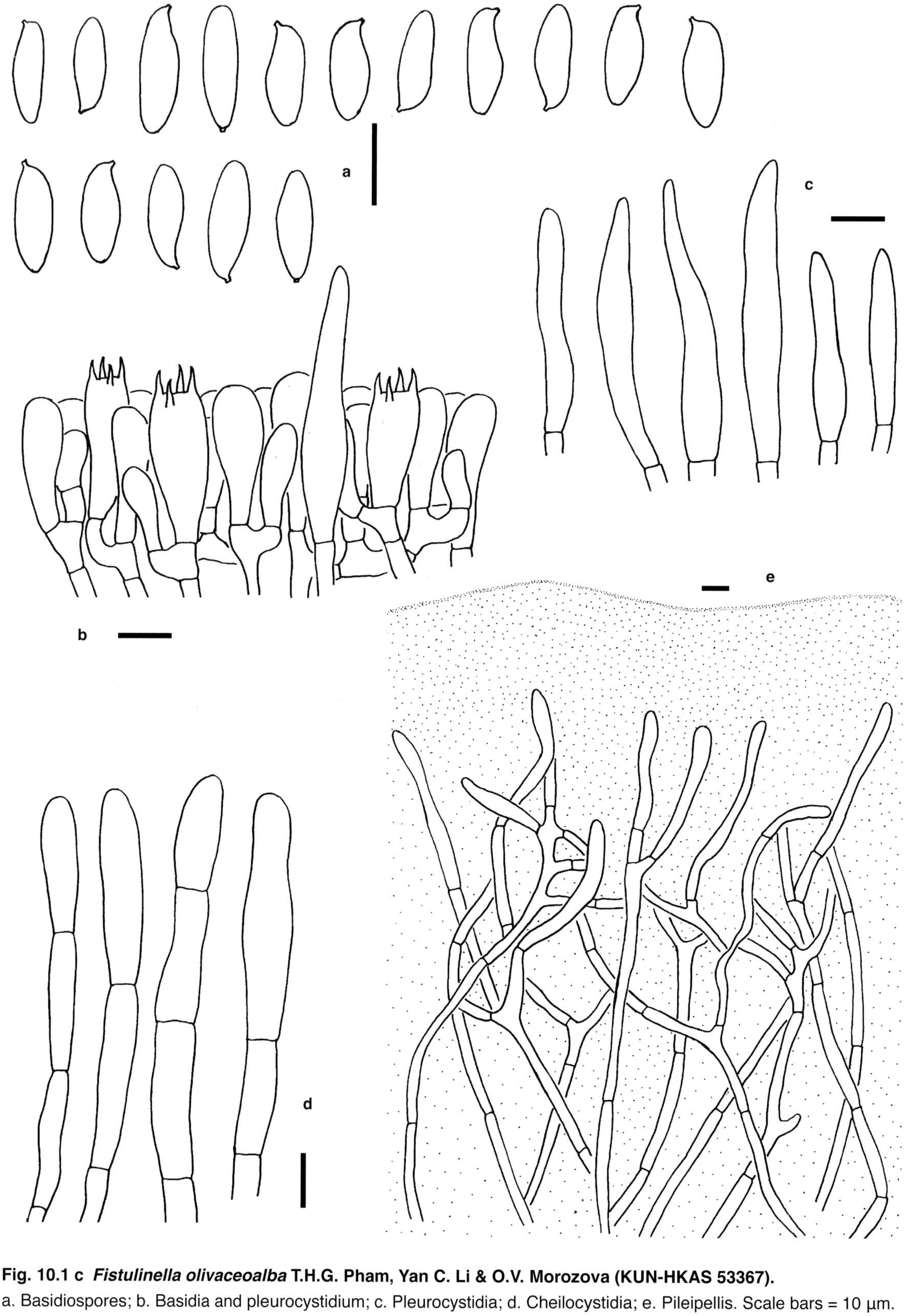

Fig. 10.1 c *Fistulinella olivaceoalba* T.H.G. Pham, Yan C. Li & O.V. Morozova (KUN-HKAS 53367).
a. Basidiospores; b. Basidia and pleurocystidium; c. Pleurocystidia; d. Cheilocystidia; e. Pileipellis. Scale bars = 10 μm.

10.2 *Fistulinella salmonea* Yan C. Li & Zhu L. Yang, The Boletes of China: *Tylopilus* s.l. 107 (2021)

MycoBank: MB 834702

Etymology: The epithet "*salmonea*" refers to the color of the pileus.

Type: CHINA, GUANGDONG PROVINCE: Fengkai County, Heishiding Nature Reserve, alt. 800 m, 1 June 2013, K. Zhao 246 (KUN-HKAS 80671, GenBank Acc. No.: MT154766 for nrLSU, MT110404 for *rpb1*).

Basidioma small to medium-sized. Pileus 3–6 cm in diam., subhemispherical to convex or applanate; surface always covered with white tissues when young, white tissues easily washed away with age, with the background pallid castaneous (7E7–8), brownish salmon (7D8) to dark reddish brown (8C7–8), slightly darker in the center, viscid to gelatinous when wet, shiny when dry, rugose, margin extended; context white but reddish brown (7D8) beneath pileipellis, without discoloration when injured. Hymenophore free to depressed around apex of stipe; surface white to dingy white when young and pinkish (13A2) to brownish pink (10D5–6) when mature, without discoloration when injured; pores angular, 0.3–1 mm wide; tubes 5–10 mm long, concolorous or a little paler than hymenophoral surface, without discoloration when bruised; spore print brown (7D6) to reddish brown (8D6). Stipe clavate, tapering upwards and enlarged downwards, 4–8 × 0.4–1 cm, white to cream when young, pale reddish brown downwards when old;

Fig. 10.2 a *Fistulinella salmonea* Yan C. Li & Zhu L. Yang. Photos by F. Li (KUN-HKAS 106340).

Fig. 10.2 b *Fistulinella salmonea* Yan C. Li & Zhu L. Yang. Photo by K. Zhao (KUN-HKAS 80671, type).

surface always covered with concolorous fibrillose squamules; basal mycelium white, without discoloration when bruised; context white but cream-yellow to reddish brown when bruised or old. Taste and odor mild.

Basidia 29–40 × 8–11 µm, clavate to narrowly clavate, hyaline to yellowish in KOH, 4-spored. Basidiospores [80/4/4] (10) 10.5–12.5 (13.5) × 4–5 µm (Q = 2.33–2.78, Q_m = 2.55 ± 0.13), subfusiform and inequilateral in side view with slight suprahilar depression, fusiform to subcylindrical in ventral view, smooth, yellowish to brownish yellow in KOH, yellow to yellow-brown in Melzer's reagent. Hymenophoral trama boletoid composed of 4.5–10 µm wide filamentous hyphae, hyaline to yellowish in KOH, yellowish to yellow in Melzer's reagent. Cheilo- and pleurocystidia 45–60 × 6–9 µm, narrowly subfusiform to fusoid-ventricose, hyaline to pale yellow in KOH, with 1–2 constrictions in the upper parts, thin-walled. Pileipellis an ixotrichoderm composed of 3–7 µm wide interwoven filamentous hyphae with cylindrical terminal cells 24–71 × 3–5 µm, pale reddish brown to brownish in KOH and reddish brown to dark brown in Melzer's reagent. Pileal trama composed of broad (up to 19 µm wide) interwoven hyphae, yellowish to pale brownish in KOH and yellow to brownish in Melzer's reagent. Clamp connections absent in all tissues.

Habitat: Scattered on soil in subtropical forests dominated by plants of the family Fagaceae.

Known distribution: Currently known from southern China.

Additional specimens examined: CHINA, GUANGDONG PROVINCE: Fengkai County, Heishiding Nature Reserve, alt. 800 m, 25 May 2012, F. Li 289 (KUN-HKAS 106323), the same location, 26 May 2012, F. Li 396 (KUN-HKAS 106449), 13 June 2012, F. Li487 (KUN-HKAS 106340).

Commentary: *Fistulinella salmonea* is characterized by the gelatinous basidiomata, the castaneous to cinnamon brownish or reddish brown pileus, the extended pileal margin and the ixocutis pileipellis. *Fistulinella salmonea* is morphologically similar to *F. campinaranae* Singer, a species described from Brazil (Singer 1978). However, *F. campinaranae* differs from *F. salmonea* in its pallid white pileus always with brown areas, larger basidiospores measuring 11.5–16.5 × 4–6 µm, and South America distributions.

Chapter 10 *Fistulinella* Henn. 93

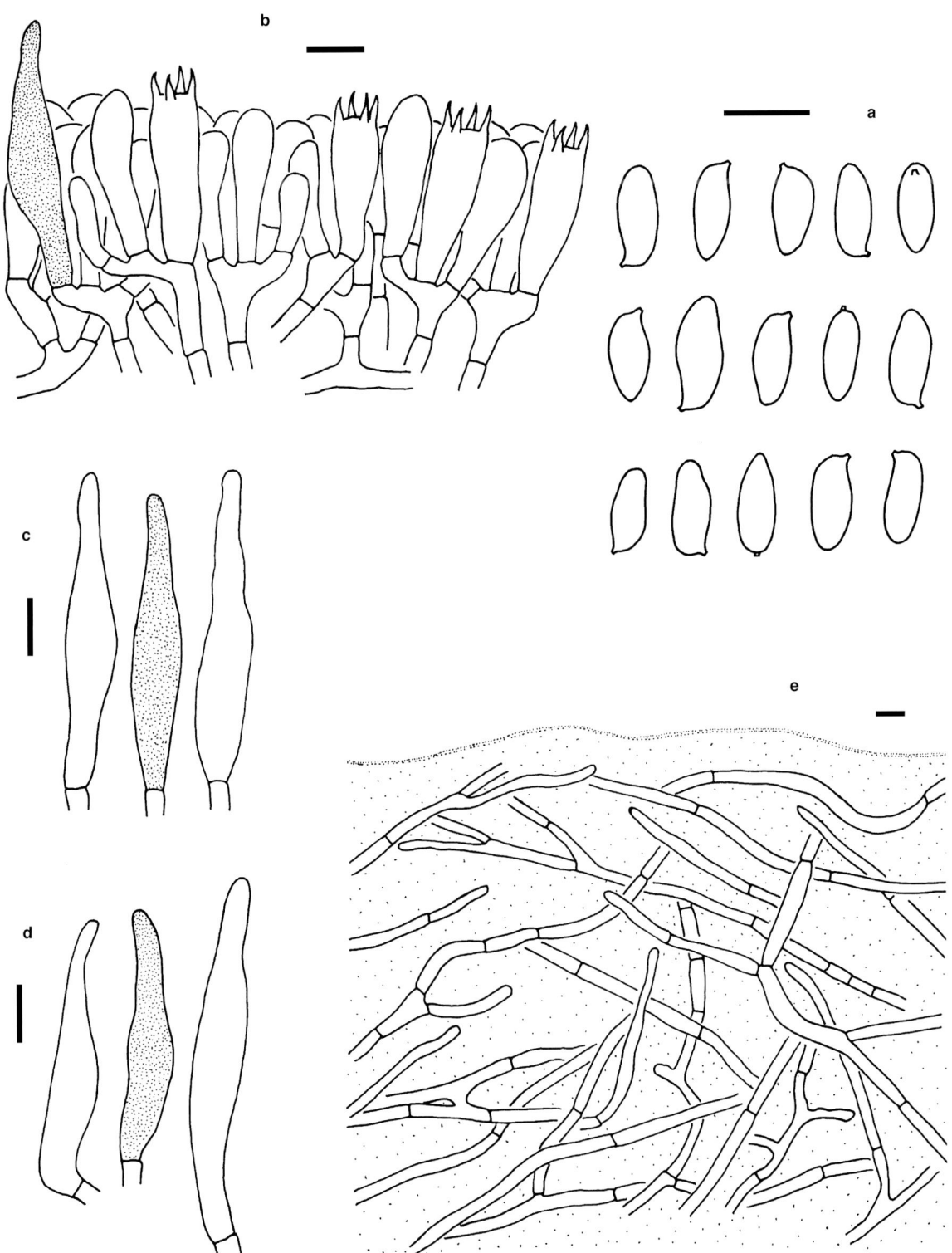

Fig. 10.2 c *Fistulinella salmonea* Yan C. Li & Zhu L. Yang (KUN-HKAS 80671, type).
a. Basidiospores; b. Basidia and cheilocystidium; c. Cheilocystidia; d. Pleurocystidia; e. Pileipellis. Scale bars = 10 μm.

Chapter 11

Harrya Halling, Nuhn & Osmundson

Harrya Halling, Nuhn & Osmundson, Aust Syst Bot 25: 422 (2012)

Type species: *Boletus chromipes* Frost [as "*chromapes*"], Bull Buffalo Soc Nat Sci 2: 105 (1874).

Diagnosis: This genus is different from other genera in the Boletaceae in the combination of the dry and wine-red tinge pileus, the white to pinkish or pink hymenophore, the whitish to cream surface of stipe covered with pink to reddish scabrous squamules, the bright yellow to chrome-yellow base of stipe, the white to cream context in pileus and the yellow to golden yellow context in stipe which is without discoloration when injured, and the trichoderm pileipellis.

Pileus hemispherical when young, subhemispherical or plano-convex or applanate, subtomentose to fibrillose, dry, becoming wine-red to brownish red or brownish pink when bruised; context white to cream but always with purple-red to purplish pink tinge beneath pileipellis, without discoloration when injured. Hymenophore adnate when young and then depressed around apex of stipe; surface initially white and then pinkish to pink when mature, without discoloration when touched; pores subangular or roundish; tubes concolorous or much paler than hymenophoral surface, without discoloration when injured. Spore print pinkish to reddish. Stipe central, white to cream or yellowish on the upper part, but yellowish to yellow downwards and chrome-yellow to bright yellow at base; surface covered with pink to reddish scabrous squamules; basal mycelium golden yellow or chrome-yellow. Basidiospores subfusiform to cylindrical; surface smooth. Hymenial cystidia subfusoid-ventricose or clavate. Pileipellis a cutis, an epithelium, or a trichoderm. Clamp connections absent in all tissues.

Commentary: The genus was originally described by Halling *et al*. (2012b) based on materials from Australia and North America, and harbored two species at that time, namely *Ha. atriceps* Halling, G.M. Muell. & Osmundson and *Ha. chromipes* (Frost) Halling, Nuhn, Osmundson & Manfr. Binder. The initially white to pinkish and then pink hymenophore, the pink to reddish scabrous squamules on the surface of stipe, the chrome-yellow to golden yellow base of stipe, and the smooth basidiospores in *Harrya* are similar to those of the species in *Chiua*, *Hymenoboletus*, *Royoungia*, and *Zangia*. However, they are different in morphology (including the color of context in pileus and stipe, the discoloration when injured, and the structure of the pileipellis) and molecular data (Li *et al*. 2011; Halling *et al*. 2012b; Wu *et al*. 2016a). Currently, five species are known from China.

Key to the species of the genus *Harrya* in China

1. Pileus dark olive to yellowish olive, or olive-gray to dark olive-brown ·· 2
1. Pileus without any olive or green tinge·· 3
2. Pileus olive-gray to dark olive-brown; pileipellis a trichoderm composed of 4–6 μm wide filamentous hyphae·· *Ha. atrogrisea*
2. Pileus olive to dark olive or grayish green to yellowish olive; pileipellis an epithelium composed of 4.5–10 μm wide moniliform cells·· *Ha. moniliformis*
3. Pileus with pink to reddish or purplish pink tinge ·· 4
3. Pileus pale brownish to brownish gray, without pink to reddish or purplish pink tinge·················
··· *Ha. alpina*
4. Pileal surface brownish pink to pink; basidiospores 13–17.5 × 4–5.5 μm; pileipellis a cutis composed of 4.5–6 μm wide subrepent filamentous hyphae ·· *Ha. chromipes*
4. Pileal surface violet-brown to grayish ruby or grayish red; basidiospores relatively short 12.5–14 × 4.5–6 μm; pileipellis a trichoderm to palisadoderm composed of 5–15 μm wide interwoven hyphae ··· *Ha. subalpina*

11.1 *Harrya alpina* Yan C. Li & Zhu L. Yang, Fungal Divers 81: 87 (2016)

Basidioma small. Pileus 3–5 cm in diam., hemispherical and then subhemispherical to applanate, brownish gray to pale brown, a little darker in the center, and paler towards margin, always staining a brownish red to brownish pink tinge when injured; surface slightly viscid when wet, and finely tomentose when dry; context white to cream, but grayish pink to brownish pink beneath pileal surface, without color change when bruised. Hymenophore adnate when young, adnate to depressed around apex of stipe when mature; surface initially white to pallid and then pinkish to pink, without discoloration when injured; pores subangular to roundish, 0.5–1 mm wide; tubes up to 15 mm long, concolorous or much paler than hymenophoral surface, without discoloration when injured. Spore print pink-white to red-white. Stipe 3–7 × 1–1.8 cm, subcylindrical to clavate, enlarged downwards, white to cream or somewhat yellowish, but chrome-yellow or golden yellow at base; surface sparsely covered with pinkish red to red scabrous squamules, without color change when touched; context solid, cream to yellowish at upper part and yellow to bright yellow downwards, but chrome-yellow at base, without color change when injured; basal mycelium golden yellow or chrome-yellow. Taste and odor mild.

Basidia 24–39 × 8–11 µm, clavate, 4-spored, hyaline in KOH. Basidiospores 11.5–15 (16.5) × 4.5–5.5 µm (Q = 2.27–3.22, Q_m = 2.69 ± 0.19), subfusiform to subcylindrical in side view, fusiform to subcylindrical in ventral view, smooth, hyaline to yellowish in KOH, yellow to brownish yellow in Melzer's reagent. Hymenophoral trama boletoid composed of 3–8 µm wide filamentous hyphae, hyaline to yellowish in KOH, yellowish to yellow in Melzer's reagent. Cheilocystidia 36–50 × 5.5–10 µm, fusiform to subfusiform or clavate, usually hyaline to yellowish in

Fig. 11.1 a *Harrya alpina* Yan C. Li & Zhu L. Yang. Photos by B. Feng (KUN-HKAS 99454).

KOH, thin-walled. Pleurocystidia 44–60 × 6.5–9 µm, subfusiform to ventricose or narrowly clavate, hyaline to yellowish in KOH, thin-walled. Caulocystidia forming the scabrous squamules over the surface of stipe, subfusiform, ventricose, clavate or lanceolate with sharp apex. Pileipellis a trichoderm composed of 4.5–11 µm wide interwoven hyphae with subcylindrical terminal cells 29–60 × 5.5–12 µm, yellowish to pale brownish in KOH and yellow to brownish in Melzer's reagent. Pileal trama composed of 5.5–12 µm wide interwoven hyphae, hyaline to yellowish in KOH and yellowish to brownish in Melzer's reagent. Clamp connections absent in all tissues.

Habitat: Scattered on soil in the alpine meadow.

Known distribution: Currently known from southwestern China.

Fig. 11.1 b *Harrya alpina* Yan C. Li & Zhu L. Yang. Photo by B. Feng (KUN-HKAS 68589).

Specimens examined: CHINA, YUNNAN PROVINCE: Deqin County, Benzilan Town, Baima Snow Mountain, alt. 3500 m, 6 September 2007, B. Feng 99 (KUN-HKAS 52820, type), the same location, 18 August 2008, Y.C. Li 1525 (KUN-HKAS 56365); Dali, Cangshan National Forest Park, alt. 3600 m, 12 August 2010, B. Feng 808 (KUN-HKAS 68589). SICHUAN PROVINCE: Kangding, Waze Town, Waze Village, alt. 3450 m, 8 September 2016, B. Feng KD100 (KUN-HKAS 99454).

Commentary: *Harrya alpina* was originally described by Wu *et al.* (2016a) from China, and is characterized by the brownish gray to pale brown pileus staining brownish red to brownish pink when touched, the white to cream pileal context without discoloration when bruised, the initially white and then pinkish hymenophore, the white to cream or somewhat yellowish stipe which is bright yellow or golden yellow at base, the trichoderm pileipellis, and the distribution in alpine areas with high altitudes ranging from 3400 m to 3600 m. In the original description of Wu *et al.* (2016a), line drawing of *Ha. alpina*, labeled Fig. 43, was mistakenly used the line drawing of *Aureoboletus catenarius* G. Wu & Zhu L. Yang. Therefore, we provide the correct line drawing of *Ha. alpina* here.

Morphologically, *Ha. alpina* can be easily confused with *Ha. subalpina* Yan C. Li & Zhu L. Yang due to their similar appearance and distribution in the areas with high altitudes. However, *Ha. subalpina* has a violet-brown to grayish ruby or grayish red pileus, and a distribution in subalpine areas with altitude ranging from 2600 m to 3000 m (Wu *et al.* 2016a).

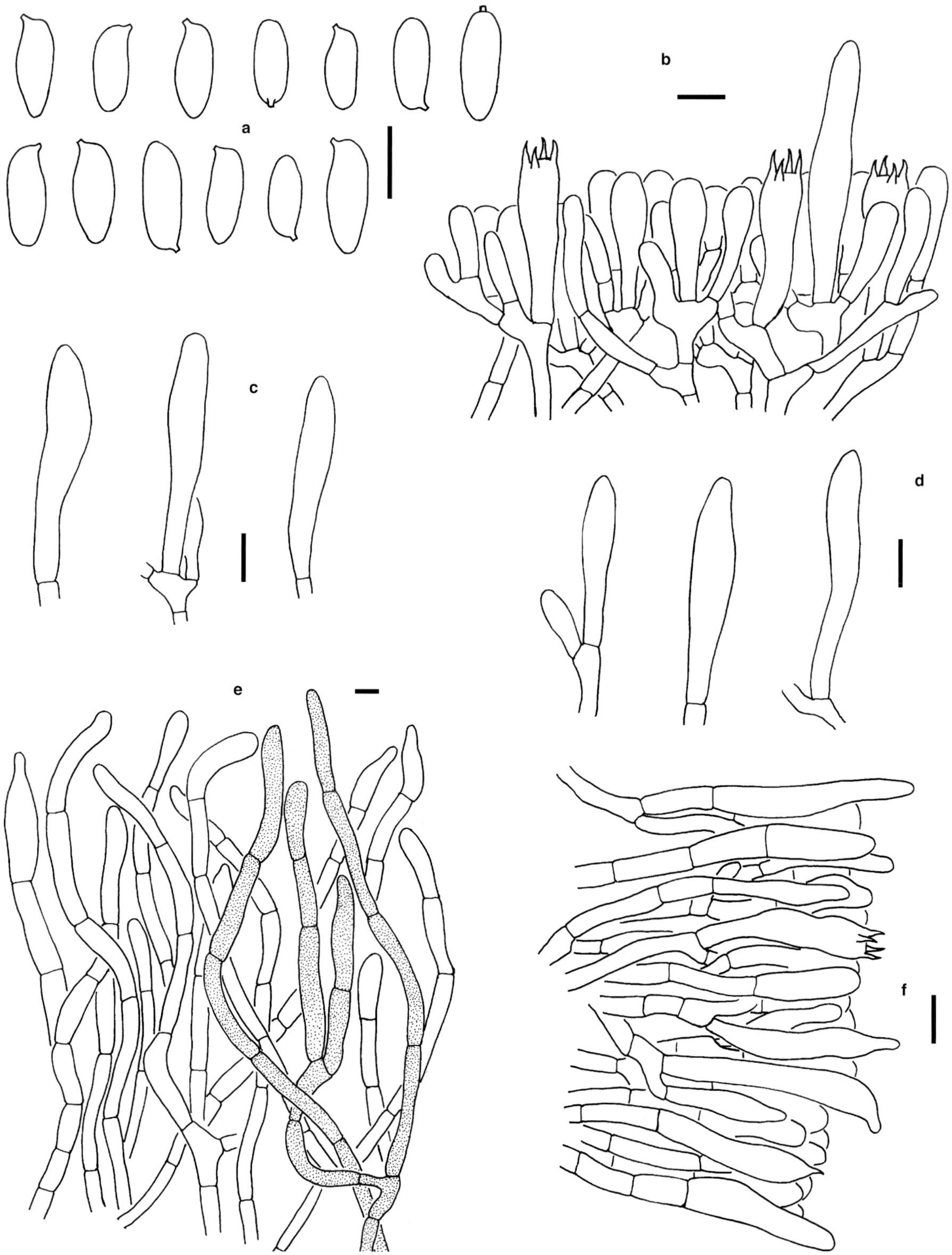

Fig. 11.1 c *Harrya alpina* Yan C. Li & Zhu L. Yang (KUN-HKAS 52820, type).
a. Basidiospores; b. Basidia and pleurocystidium; c. Cheilocystidia; d. Pleurocystidia; e. Pileipellis; f. Stiptipellis. Scale bars = 10 μm.

11.2 *Harrya atrogrisea* Yan C. Li & Zhu L. Yang, Fungal Divers 81: 88 (2016)

Basidioma small to medium-sized. Pileus 3–7 cm in diam., hemispherical when young, subhemispherical to convex or plano-convex when mature, dark olive-brown to olive-gray, staining a grayish red to brownish red tinge when bruised, slightly darker in the center; surface fibrillose to tomentose, dry; context white to cream but pinkish red to grayish red beneath the pileal surface, without discoloration when injured. Hymenophore adnate when young and depressed around apex of stipe when mature; surface white to dingy white when young and becoming pinkish to pink when mature, without discoloration when touched; pores angular to roundish, 0.5–1 mm wide; tubes up to 20 mm long, concolorous or much paler than hymenophoral surface, without discoloration when injured. Spore print pinkish to reddish. Stipe cylindrical, 4–8 × 0.8–1.5 cm, white to cream on the upper part, but yellowish to yellow downwards and golden yellow to chrome-yellow at base; surface covered with red to pinkish red scabrous squamules, without color change when touched; context white to cream in the upper part, cream to yellowish downwards but bright yellow to chrome-yellow at base, without color change when bruised; basal mycelium golden yellow to chrome-yellow. Taste and odor mild.

Basidia 18–37 × 10–15 μm, clavate, 4-spored, hyaline in KOH. Basidiospores 10–15 × 4–5.5 μm [Q = (2.22) 2.27–2.67 (3), Q_m = 2.54 ± 0.2], subfusiform in side view, fusiform to subcylindrical in ventral view, smooth, hyaline to yellowish in KOH, yellow to brownish yellow in Melzer's reagent. Hymenophoral trama boletoid composed of 5–11 μm wide filamentous hyphae, hyaline to yellowish in KOH, yellowish to yellow in Melzer's reagent. Cheilocystidia 25–40 × 5–6 μm, subfusiform or ventricose, hyaline to yellowish in KOH, thin-walled. Pleurocystidia much bigger than cheilocystidia, 55–75 × 10–16 μm, morphologically

Fig. 11.2 a *Harrya atrogrisea* Yan C. Li & Zhu L. Yang. Photo by Y.C. Li (KUN-HKAS 107178).

Fig. 11.2 b *Harrya atrogrisea* Yan C. Li & Zhu L. Yang. Photos by X.T. Zhu (KUN-HKAS 68182).

similar to cheilocystidia, hyaline to yellowish, thin-walled. Caulocystidia forming the scabrous squamules over the surface of stipe, clavate, lanceolate to ventricose or subfusiform with sharp apex. Pileipellis a trichoderm composed of 4–6 μm wide interwoven hyphae with cylindrical terminal cells 40–85 × 4.5–6 μm, yellowish to pale brownish in KOH and yellow to brownish in Melzer's reagent. Pileal trama composed of 6.5–9.5 μm wide interwoven hyphae, hyaline to yellowish in KOH and yellowish to brownish in Melzer's reagent. Clamp connections absent in all tissues.

Habitat: Scattered on soil in subtropical to temperate forests dominated by plants of the family Fagaceae or in the mixed forests dominated by plants of the families Fagaceae and Pinaceae.

Known distribution: Currently known from southwestern China.

Specimens examined: CHINA, YUNNAN PROVINCE: Heqing County, alt. 3000 m, 28 July 2006, Z.L. Yang 4745 (KUN-HKAS 50542, type); Baoshan, Longyang District, Shuizhai Town, Haitang Village, alt. 2350 m, 21 July 2009, Y.C. Li 1726 (KUN-HKAS 59473), the same location, 22 July 2009, Y.C. Li 1757 (KUN-HKAS 59504); Ninglang County, Lugu Lake, alt. 2750 m, 10 July 2010, X.T. Zhu 006 (KUN-HKAS 68182); Jin'an County, alt. 2700 m, 3 August 2011, Y.C. Li 2590 (KUN-HKAS 107178).

Commentary: *Harrya atrogrisea* was originally described from temperate forests in China (Wu *et al*. 2016a). This species is characterized by the dark olive-brown to dark olive-gray pileus staining brownish red to grayish red when injured, the initially white and then pinkish hymenophore, the white to cream or somewhat yellowish stipe with a golden yellow or chrome-yellow base, the red to pinkish red scabrous stipe, and the trichoderm pileipellis. *Harrya atrogrisea* can be confused with *Ha. alpina* as they share a somewhat brownish to grayish pileus. However, *Ha. alpina* has relatively broad hyphae (4.5–11 μm wide) in the pileipellis, and a distribution in alpine areas (Wu *et al*. 2016a; see our description above).

Our multi-locus phylogenetic analysis (Fig. 4.1) indicates that *Ha. atrogrisea* is related to *Ha. moniliformis* Yan C. Li & Zhu L. Yang. However, *Ha. moniliformis* has an olive to dark olive and then grayish green to yellowish olive pileus, relatively slender hymenial cystidia which are not more than 11 μm wide, an epithelium pileipellis composed of 4.5–10 μm wide moniliform cells, and a distribution in subtropical forests (Wu *et al*. 2016a; see our description below).

11.3 *Harrya chromipes* (Frost) Halling, Nuhn, Osmundson & Manfr. Binder [as "*chromapes*", Aust Syst Bot 25: 422 (2012)

Basionym: *Boletus chromipes* Frost [as "*chromapes*"], Bull Buffalo Soc Nat Sci 2: 105 (1874).

Synonyms: *Ceriomyces chromipes* (Frost) Murrill [as "*chromapes*"], Mycologia 1: 145 (1909); *Krombholzia chromipes* (Frost) Singer [as "*chromapes*"], Ann Mycol 40: 34 (1942); *Leccinum chromipes* (Frost) Singer [as "*chromapes*"], Amer Midl Nat 37: 124 (1947); *Tylopilus chromipes* (Frost) A.H. Sm. & Thiers [as "*chromapes*"], Mycologia 60: 949 (1968); *Tylopilus hongoi* Wolfe & Bougher, Aust Syst Bot 6: 190 (1993); *Tylopilus cartagoensis* Wolfe & Bougher, Austral Syst Bot 6: 191 (1993); *Leccinum cartagoense* (Wolfe & Bougher) Halling & G.M. Muell., Kew Bull 54: 747 (1999).

Basidioma small to medium-sized. Pileus 3–8 cm in diam., hemispherical when young, subhemispherical to applanate when mature, reddish pink to brownish pink, slightly darker in the center, staining brownish red to pinkish red when injured; surface dry, nearly glabrous to fibrillose; context solid and then spongy, white to pallid with brownish red to pinkish red tinge beneath pileal surface, without color change when injured. Hymenophore adnate when young and depressed around apex of stipe when mature; surface initially white to pallid and then pinkish or pink, without discoloration when touched; pores angular to roundish, 1–1.5 mm wide; tubes up to 25 mm long, concolorous or a little paler than hymenophoral surface, without discoloration when injured. Spore print pinkish to reddish. Stipe 6–7.5 × 1–1.5 cm, clavate to subcylindrical, cream to yellowish, but bright yellow to chrome-yellow at base; surface covered with pink to pinkish red or red scabrous squamules, without color change when bruised; context cream to yellowish, but chrome-yellow at base, without color change when injured; basal mycelium chrome-yellow or golden yellow. Taste and odor mild.

Fig. 11.3 a *Harrya chromipes* (Frost) Halling, Nuhn, Osmundson & Manfr. Binder. Photos by P.M. Wang (KUN-HKAS 98289).

Fig. 11.3 b *Harrya chromipes* (Frost) Halling, Nuhn, Osmundson & Manfr. Binder. Photo by J.W. Liu (KUN-HKAS 103183).

Basidia 25–40 × 9–12 μm, clavate, 4-spored, hyaline in KOH. Basidiospores 12–19 × 4–6 μm [Q = (2.5) 2.73–3.63 (3.8), Q_m = 3.15 ± 0.28], subfusiform to subcylindrical, smooth, hyaline to yellowish in KOH, yellow to brownish yellow in Melzer's reagent. Hymenophoral trama boletoid composed of 3.5–13 μm wide filamentous hyphae, hyaline to yellowish in KOH, yellowish to yellow in Melzer's reagent. Cheilo- and pleurocystidia 28–64 × 5–11 μm, fusiform to subfusiform or clavate, hyaline to yellowish in KOH, thin-walled. Caulocystidia forming the scabrous squamules over the surface of stipe, ventricose, clavate, lanceolate or subfusiform. Pileipellis a cutis composed of 4.5–6 μm wide filamentous hyphae with subcylindrical terminal cells 38–70 × 4–5.5 μm, yellowish to pale brownish in KOH and yellow to brownish in Melzer's reagent. Pileal trama composed of 6–10 μm wide interwoven hyphae, hyaline to yellowish in KOH and yellowish to brownish in Melzer's reagent. Clamp connections absent in all tissues.

Habitat: Scattered on soil in forests dominated by plants of the families Betulaceae, Fagaceae, and Pinaceae.

Known distribution: Currently known from China, Japan, and North America.

Specimens examined: CHINA, HEILONGJIANG PROVINCE: Fuyuan County, alt. 50 m, 5 August 2004, T. Bau 2552 (HMJAU 2552). CHINA, YUNNAN PROVINCE: Nanjian County, Houqing

Town, Qinshan, alt. 2200 m, 30 August 2003, Z.L. Yang 3909 (KUN-HKAS 25093). CHINA, HUBEI PROVINCE: Xingshan County, Muyu Town, alt. 1700 m, 14 July 2012, Q. Cai 784 (KUN-HKAS 75538). CHINA, XIZANG AUTONOMOUS REGION: Linzhi County, Lulang Town, alt. 3400 m, 6 September 2006, J.F. Liang 518 (KUN-HKAS 51229), the same location, Gongcuo Lake, alt. 3400 m, 7 September 2006, J.F. Liang 562 (KUN-HKAS 51273). CHINA, SICHUAN PROVINCE: Kangding County, alt. 3700 m, 14 August 2005, Z.W. Ge 921 (KUN-HKAS 49416), the same location, 5 September 2016, P.M. Wang KD34 (KUN-HKAS 98289), the same location, 27 August 2017, J.W. Liu 530301MF0672 (KUN-HKAS 103183). JAPAN, TOKYO, National Nature Museum, 11 August 1969, T. Hongo 3954 (TNS-F-172603, type of *Tylopilus hongoi* Wolfe & Bougher).

Commentary: *Harrya chromipes* was originally described by Frost (1874) from Vermont, USA as *Boletus chromipes* Frost. Since then, this species has been placed in *Ceriomyces* Murrill (Murrill 1909), *Krombholzia* P. Karst. (Singer 1942), *Leccinum* (Singer 1947), and *Tylopilus* (Smith and Thiers 1968). Because Murrill's concept of *Ceriomyces* is a mixture of several modern genera, and the name *Krombholzia* was originally used for a member of the plant family Gramineae (Fournier 1876), subsequent placement of *B. chromipes* has been based primarily on either color of the spore deposit or the type of surface ornamentation of the stipe. Thus, Smith and Thiers (1968, 1971), Wolfe and Bougher (1993) and Watling and Li (1999) were inclined to consider the spore color more nearly like that of a *Tylopilus*, whereas Singer (1947, 1986), Snell and Dick (1970), Binder and Besl (2000) judged that the stipe ornamentation was of a scabrous nature as in a *Leccinum*. However, Halling *et al.* (2012b) established the genus *Harrya* to accommodate *B. chromipes* based on molecular and morphological evidence, which was accepted by others (Nuhn *et al.* 2013; Li *et al.* 2014b; Zhao *et al.* 2014; Orihara *et al.* 2016; Wu *et al.* 2016a, 2016b).

Our multi-locus phylogenetic analysis (Fig. 4.1) indicates that *Ha. chromipes* is closely related to *Ha. atriceps* which was originally described from Costa Rica, Central America. However, *Ha. atriceps* differs from *Ha. chromipes* in its black to dark gray or dark grayish brown pileus, relatively small basidiospores measuring 9.1–11.9 × 4.2–6.3 μm, and small hymenial cystidia measuring 35–48 × 4–7 μm (Halling *et al.* 2012b).

11.4 *Harrya moniliformis* Yan C. Li & Zhu L. Yang, Fungal Divers 81: 90 (2016)

Basidioma small to medium-sized. Pileus 3–7 cm in diam., subhemispherical to convex or plano-convex, dark olive to olive when young and then yellowish olive to grayish green when mature, much darker in the center, always with reddish brown tinge when touched; surface dry, fibrillose to tomentose; context solid to spongy, white but pink to purplish pink beneath pileal surface, without discoloration when injured. Hymenophore adnate when young and adnate to depressed around apex of stipe when mature; surface initially white to pallid and then pinkish to pink, without discoloration when touched; pores angular to roundish, 0.5–2 mm wide; tubes up to 15 mm long, concolorous or a little paler than hymenophoral surface, without discoloration when injured. Spore print pink-white to red-white (Fig. 3.4 e). Stipe cylindrical to clavate, 4–9 × 0.4–1.5 cm, white to cream, but bright yellow or chrome-yellow at base; surface covered with pink to red scabrous squamules, without color change when touched; context cream to yellowish in the upper part, but bright yellow or chrome-yellow at base, without color change when bruised; basal mycelium golden yellow to chrome-yellow. Taste and odor mild.

Basidia 28–30 × 11–19 μm, clavate, 4-spored, hyaline in KOH. Basidiospores (10) 12–15 × 4–5.5 μm (Q = 2.32–3.25, Q_m = 2.68 ± 0.16), subfusiform with slight suprahilar depression, fusiform to cylindrical in ventral view, smooth, hyaline to yellowish in KOH, yellow to brownish yellow in Melzer's reagent. Hymenophoral trama boletoid composed of 2.5–8 μm wide filamentous hyphae, hyaline to yellowish in KOH, yellowish to yellow in Melzer's reagent. Cheilocystidia 25–48 × 6–9 μm, fusiform to subfusiform

Fig. 11.4 a *Harrya moniliformis* Yan C. Li & Zhu L. Yang. Photos by T. Guo (KUN-HKAS 81805).

or clavate but with obtuse apex, usually hyaline to yellowish in KOH, thin-walled. Pleurocystidia 45–72 × 7.5–11 μm, narrowly subfusiform, spindly in the upper parts, hyaline to yellowish, thin-walled. Caulocystidia forming the scabrous squamules over the surface of stipe, ventricose, clavate, lanceolate or subfusiform with sharp apex. Pileipellis an epithelium composed of 4.5–10 μm wide moniliform cells, with clavate to cylindrical terminal cells 8–20 × 6–10 μm, yellowish to pale brownish in KOH and yellow to brownish in Melzer's reagent. Pileal trama composed of 5.5–10.5 μm wide interwoven hyphae, hyaline to yellowish in KOH and yellowish to brownish in Melzer's reagent. Clamp connections absent in all tissues.

Fig. 11.4 b *Harrya moniliformis* Yan C. Li & Zhu L. Yang. Photo by Z.L. Yang (KUN-HKAS 52215).

Habitat: Scattered on soil in subtropical forests dominated by plants of the family Fagaceae, or in mixed forests of plants of the families Fagaceae and Pinaceae.

Known distribution: Currently known from southwestern China.

Specimens examined: CHINA, YUNNAN PROVINCE: Kunming, Qiongzhu Temple, alt. 2100 m, 27 October 2005, Y.C. Li 429 (KUN-HKAS 49627), the same location, 14 october 2006, Y.C. Li 744 (KUN-HKAS 51181); Kunming, Miaogao Temple, alt. 2100 m, 6 August 2006, Y.C. Li 699 (KUN-HKAS 51136); Kunming, Qiongzhu Temple, alt. 2100 m, 8 August 2007, Y.C. Li 963 (KUN-HKAS 52650, type), the same location and date, Z.L. Yang 4798 (KUN-HKAS 52215), the same location, 6 September 2012, T. Guo 603 (KUN-HKAS 81805).

Commentary: *Harrya moniliformis* was originally described from southwestern China (Wu *et al.* 2016b). This species is characterized by the dark olive to olive and then yellowish olive to grayish green pileus, the initially white to pallid and then pinkish to pink hymenophore, the cream to yellowish stipe covered with pink to red scabrous squamules, the chrome-yellow to bright yellow base of stipe, the white pileal context without discoloration when injured, and the moniliform pileipellis. *Harrya moniliformis* is morphologically similar to *Zangia olivacea* Yan C. Li & Zhu L. Yang which has an olive to olive-green or yellowish green pileus. However, *Z. olivacea* has a rugose and pulverescent pileus, an ixohyphoepithelium pileipellis, relatively broad basidiospores measuring 12–15.5 × 6–7 μm, and asymmetrically bluish color change when injured (Li *et al.* 2011).

Our multi-locus phylogenetic analysis (Fig. 4.1) indicates that *Ha. moniliformis* is closely related to *Ha. atrogrisea*. However, *Ha. atrogrisea* has an olive-gray to dark olive-brown pileus always staining grayish red to brownish red tinge when touched, relatively narrow cheilocystidia (25–40 × 5–6 μm) and broad pleurocystidia (55–75 × 10–16 μm), a trichoderm pileipellis composed of 4–6 μm wide interwoven hyphae, and a distribution in subtropical to temperate forests (Wu *et al.* 2016a; see our description above).

11.5 *Harrya subalpina* Yan C. Li & Zhu L. Yang, Fungal Divers 81: 92 (2016)

Basidioma small to medium-sized. Pileus 3.5–6 cm in diam., subhemispherical to plano-convex or applanate, violet-brown grayish brown when young and grayish red to grayish ruby when mature or bruised, slightly darker in the center; surface dry, nearly glabrous to fibrillose; context pallid to white, but pink to purplish pink beneath pileipellis, without discoloration when injured. Hymenophore adnate when young, adnate to depressed around apex of stipe when mature; surface white to dingy white when young, pinkish to pink when mature, without discoloration when touched; pores angular to roundish, 0.5–1.5 mm wide; tubes up to 15 mm long, concolorous or a little paler than hymenophoral surface, without discoloration when injured. Spore print pink-white to red-white. Stipe 3.5–7 × 1–1.6 cm, subcylindrical to clavate, always enlarged downwards, white to cream but bright yellow to chrome-yellow at base; surface covered with red to pink scabrous squamules, without color change when injured; context white to yellowish, but chrome-yellow or golden yellow at base, without color change when injured; basal mycelium chrome-yellow to golden yellow. Taste and odor mild.

Basidia 23–48 × 9–15 μm, clavate, 4-spored, hyaline in KOH. Basidiospores 12.5–15 × 4.5–6 μm [Q = 2.1–2.8, Q_m = 2.43 ± 0.12], subfusiform to subcylindrical in side view with slight suprahilar depression, fusiform to cylindrical in ventral view, smooth, hyaline to yellowish in KOH, yellow to brownish yellow in Melzer's reagent. Hymenophoral trama boletoid composed of 4.5–10 μm wide filamentous hyphae, hyaline to yellowish in KOH, yellowish to yellow in Melzer's reagent. Cheilo-and pleurocystidia 49–62 × 7–9 μm, subfusiform to lanceolate, with 1–2 constrictions in the middle parts, hyaline to yellowish in KOH, thin-walled. Caulocystidia forming the scabrous squamules over the surface of stipe, clavate to subfusiform or ventricose. Pileipellis a trichoderm composed of 5–15 μm wide interwoven hyphae, with terminal cells 17–50 × 5.5–10.5 μm, pale yellowish to brownish yellow in KOH and yellow to brown in Melzer's reagent. Pileal trama composed of 5.5–15 μm wide interwoven hyphae, hyaline to yellowish in KOH and yellowish to brownish in Melzer's reagent. Clamp connections absent in all tissues.

Fig. 11.5 a *Harrya subalpina* Yan C. Li & Zhu L. Yang. Photo by Y.Y. Cui (KUN-HKAS 79751).

Fig. 11.5 b *Harrya subalpina* Yan C. Li & Zhu L. Yang. Photos by Q. Cai (KUN-HKAS 58820).

Habitat: Scattered on soil in subalpine forests dominated by Pinaceae plants, or in mixed forests dominated by plants of the families Fagaceae and Pinaceae.

Known distribution: Currently known from southwestern China.

Specimens examined: CHINA, YUNNAN PROVINCE: Yulong County, Shitou Town, Liju Village, alt. 2650 m, 23 August 2007, B. Feng 88 (KUN-HKAS 52809); Weixi County, Tacheng Town, Bazhu Village, alt. 2600 m, 8 September 2008, X.F. Tian 405 (KUN-HKAS 90194, type); Yulong County, Lijiang Alpine Botanic Garden, alt. 3000 m, 27 August 2009, G. Wu 152 (KUN-HKAS 57684) and Q. Cai 155 (KUN-HKAS 58820), the same location, 29 July 2006, Z.L. Yang 4749 (KUN-HKAS 50546), the same location, 15 August 2013, Y.Y. Cui 75 (KUN-HKAS 79751).

Commentary: *Harrya subalpina* was originally described based on materials from southwestern China (Wu *et al.* 2016a). This species is characterized by the initially violet-brown to grayish brown and then grayish red to grayish ruby pileus, the white context with pink to purplish pink beneath pileipellis and without color change when injured, the white to dingy white then pinkish to pink hymenophore, the white to cream stipe with a bright yellow to chrome-yellow base, the trichoderm pileipellis, and the distribution in subalpine areas with altitudes ranging from 2600 m to 3000 m.

Harrya subalpina is phylogenetically related and morphologically similar to *Ha. atrogrisea* and *Ha. moniliformis* (Fig. 4.1). However, *Ha. atrogrisea* differs from *Ha. subalpina* in its olive-gray to dark olive-brown pileus, relatively broad pleurocystidia measuring 55–75 × 10–16 μm, trichoderm pileipellis composed of relatively narrow hyphae which are 4–6 μm wide, and distribution in subtropical to temperate forests (Wu *et al.* 2016a; see our description above). *Harrya moniliformis* differs in its olive to dark olive then grayish green to yellowish olive pileus, epithelium pileipellis composed of 4.5–10 μm wide moniliform cells, and distribution in subtropical forests (Wu *et al.* 2016a; see our description above).

Chapter 12
Hymenoboletus Yan C. Li & Zhu L. Yang

Hymenoboletus Yan C. Li & Zhu L. Yang, Fungal Divers 81: 100 (2016)

Type species: *Hymenoboletus luteopurpureus* Yan C. Li & Zhu L. Yang, Fungal Divers 81: 101 (2016), Figs. 53 e–f, 55.

Diagnosis: This genus can be distinguished from other genera in the Boletaceae by its combination of the purplish red stipe which is yellowish to yellow at apex and golden yellow or chrome-yellow at base, the white to pallid context in pileus without discoloration when injured, the yellow to chrome-yellow context in stipe without discoloration when injured, the white to pinkish or pink hymenophore, and the smooth basidiospores.

Pileus subhemispherical or convex; surface dry, fibrillose to finely subtomentose; context white to cream, without color change when bruised. Hymenophore adnate to depressed around apex of stipe; surface white to pinkish when young and pinkish or pink when mature, without discoloration when touched; pores subangular or roundish; tubes concolorous or a little paler than hymenophoral surface, without discoloration when injured. Spore print pinkish to pink. Stipe central, purple to purple-pink or purple-red, but yellowish to yellow at apex and bright yellow to chrome-yellow at base; surface covered with concolorous scabrous squamules; basal mycelium chrome-yellow. Basidiospores smooth, fusiform to cylindrical. Hymenial cystidia subfusiform to subfusoid-ventricose. Pileipellis an intricate trichoderm composed of broad interwoven hyphae; or a hymeniderm composed of a layer of cells which looks like a hymenium; or a hyphoepithelium with an outer layer composed of filamentous hyphae and an inner layer composed of vertically arranged moniliform hyphae. Clamp connections absent in all tissues.

Commentary: This genus can be easily confused with *Chiua*, *Harrya*, *Royoungia*, and *Zangia*, but differs in the color of the stipe, the structure of the pileipellis, the discoloration of the context, and the molecular data. Currently, four species are known in this genus, including three new species.

Key to the species of the genus *Hymenoboletus* in China

1. Pileus brownish to reddish brown or dark purplish red ·· 2
1. Pileus dark green to olive-green, or yellowish green to grayish green, without brownish to reddish brown or purplish red tinge ·· 3
2. Pileus dark purplish red to black-red; pileipellis a hymeniderm composed of a layer of cells which looks like a hymenium ·· *Hy. luteopurpureus*
2. Pileus brownish to reddish brown; pileipellis a hyphoepithelium, with an outer layer composed of filamentous hyphae and an inner layer composed of vertically arranged moniliform hyphae ·· *Hy. jiangxiensis*
3. Pileus dark green to olive-green; pileipellis an intricate trichoderm composed of interwoven filamentous hyphae ·· *Hy. filiformis*
3. Pileus yellowish green to grayish green; pileipellis a hyphoepithelium, with an outer layer composed of filamentous hyphae and an inner layer composed of erect moniliform hyphae ·· *Hy. griseoviridis*

12.1 *Hymenoboletus filiformis* Yan C. Li & Zhu L. Yang, The Boletes of China: *Tylopilus* s.l. 130 (2021)

MycoBank: MB 834705

Etymology: The epithet "*filiformis*" refers to the structure of pileipellis.

Type: CHINA, YUNNAN PROVINCE: Kunming, Qiongzhu Temple, alt. 2000 m, 28 July 2013, Y.J. Hao 1020 (KUN-HKAS 82811, GenBank Acc. No.: MT154777 for nrLSU, MT110371 for *tef1-α*, MT110446 for *rpb2*).

Basidioma small. Pileus 3–4 cm in diam., subhemispherical to nearly applanate, dry, subtomentose to fibrillose, dark green (28E8) to yellowish green (29C8) or grayish green (30E6–7), slightly darker in the center; context whitish to pallid, without color change when injured. Hymenophore adnate when young, depressed around apex of stipe when mature; surface pinkish (13A2) when young and pink (12A3–4) when mature, without discoloration when touched; pores angular to roundish, 0.3–0.5 mm wide; tubes up to 9 mm long, concolorous or a little paler than hymenophoral surface, without discoloration when injured. Spore print pinkish to pink. Stipe clavate, 4–6 × 0.5–0.8 cm, pink to purple (11B3–5), but cream to yellowish (2A4–5) at apex and golden yellow to chrome-yellow at base; surface covered with concolorous scabrous squamules, without color change when touched; context white at apex, yellowish to yellow downwards and golden yellow or chrome-yellow at base, without color change when bruised; basal mycelium bright yellow or golden yellow. Taste and odor mild.

Basidia 25–35 × 8–12 μm, clavate, 4-spored, hyaline in KOH. Basidiospores [20/1/1] (10.5) 11.5–12 (13) × 4.5–5.5 μm (Q = 2.09–2.78, Q_m = 2.46 ± 0.21), subfusiform to subcylindrical and inequilateral in side view with slight suprahilar depression, fusiform to cylindrical in ventral view, smooth, hyaline to yellowish in KOH, yellow to brownish yellow in Melzer's reagent. Hymenophoral trama boletoid composed of 4–8 μm wide filamentous hyphae, hyaline to yellowish in KOH, yellowish to yellow in Melzer's reagent. Cheilo- and pleurocystidia 32–46 × 5–9 μm, subfusiform to ventricose or broadly lanceolate, always with 1–2 constrictions in the middle parts and usually hyaline to yellowish in KOH, thin-walled. Caulocystidia forming the scabrous squamules over the surface of stipe, ventricose, clavate, lanceolate or subfusiform with sharp apex. Pileipellis an intricate trichoderm composed of 4–11 μm wide interwoven hyphae with clavate to narrowly pyriform or broadly lanceolate terminal cells 29–62 × 5–7 μm, yellowish to pale brownish in KOH and yellow to brownish in Melzer's reagent. Pileal trama composed of 4.5–8 μm wide interwoven hyphae, hyaline to yellowish in KOH and yellowish to brownish in Melzer's reagent. Clamp connections absent in all tissues.

Habitat: Scattered on soil in subtropical forests dominated by plants of the family Fagaceae.

Known distribution: Currently known from southwestern China.

Commentary: *Hymenoboletus filiformis* has a green to yellowish green or grayish green pileus and an intricate trichoderm pileipellis. These traits are similar to those of the species in *Chiua*. However, species in *Chiua* have a yellow to chrome-yellow context in the pileus and yellow or pinkish scabrous squamules over the stipe surface. *Hymenoboletus filiformis* also shares somewhat same colored pileus with *Hy. griseoviridis*

Chapter 12 *Hymenoboletus* Yan C. Li & Zhu L. Yang

Fig. 12.1 a *Hymenoboletus filiformis* Yan C. Li & Zhu L. Yang. Photos by Y.J. Hao (KUN-HKAS 82811, type).

Yan C. Li & Zhu L. Yang. However, *Hy. griseoviridis* has a rugose pileus and a hyphoepithelium pileipellis. Phylogenetically, *Hy. filiformis* clusters together with *Hy. luteopurpureus* Yan C. Li & Zhu L. Yang. However, these two species differ in the color of the pileus and the structure of the pileipellis.

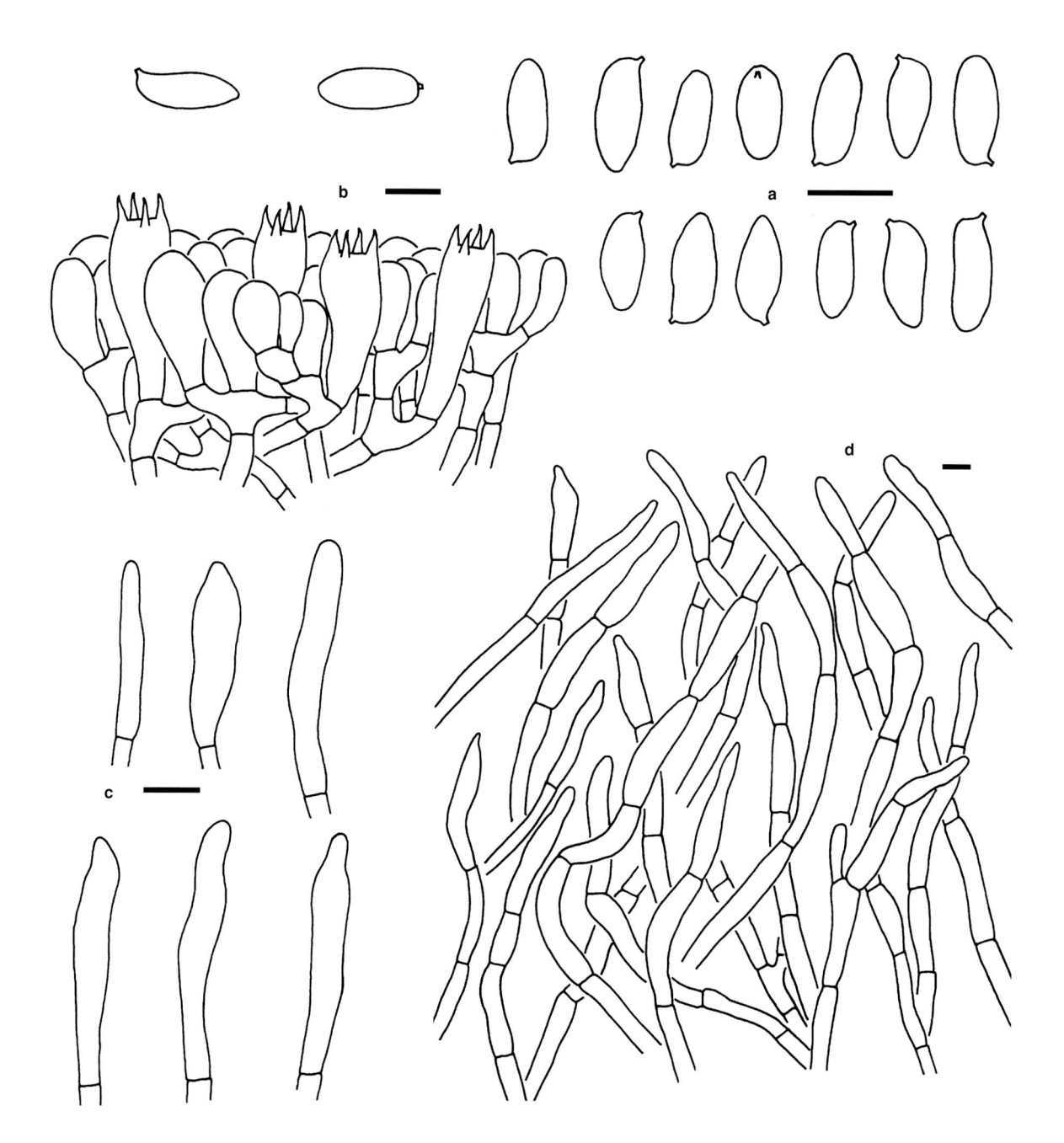

Fig. 12.1 b *Hymenoboletus filiformis* **Yan C. Li & Zhu L. Yang (KUN-HKAS 82811, type).**
a. Basidiospores; b. Basidia; c. Cheilo- and pleurocystidia; d. Pileipellis. Scale bars = 10 μm.

12.2 *Hymenoboletus griseoviridis* Yan C. Li & Zhu L. Yang, The Boletes of China: *Tylopilus* s.l. 133 (2021)

MycoBank: MB 834704

Etymology: The epithet "*griseoviridis*" refers to the color of the pileus.

Type: CHINA, HUBEI PROVINCE: Xingshan County, Muyu Town, Lishu Village, alt. 1800 m, 14 July 2012, X.B. Liu 76 (KUN-HKAS 75687, GenBank Acc. No.: MT154781 for nrLSU, MW165269 for *tef1-α*).

Basidioma small to medium-sized. Pileus 2–6 cm in diam., subhemispherical to hemispherical, dark green (28E8) to olive-green (30E7–8) or grayish green (1D5–6); surface always rugose, fibrillose to tomentose, dry; context white to cream, without discoloration when injured. Hymenophore adnate when young and depressed around apex of stipe when mature; surface white when young and becoming pinkish (13A2) or pink (12A3–4) when mature, without discoloration when touched; pores angular to roundish, 0.5–1 mm wide; tubes up to 6 mm long, concolorous or a little paler than hymenophoral surface. Spore print pink-white (13A2) to pink (12A3–4). Stipe clavate, 4–6 × 0.5–0.8 cm, pink to purple (11B3–5) in middle part, yellow (2A7–8) to yellowish (2A4–5) at apex and golden yellow or chrome-yellow at base; surface covered with concolorous scabrous squamules, without color change when touched; context cream to yellowish in the upper part, but yellow to bright yellow downwards and chrome-yellow or golden yellow at base, without color change when bruised; basal mycelium chrome-yellow or golden yellow. Taste and odor mild.

Basidia 28–45 × 10–14 μm, clavate to broadly clavate, hyaline to yellowish in KOH, 4-spored. Basidiospores [60/3/3] (9) 10–13 × 4–5.5 μm [Q = (2.09) 2.2–2.56 (2.63), Q_m = 2.39 ± 0.14], subfusiform to subcylindrical and inequilateral in side view with slight suprahilar depression, fusiform to cylindrical in ventral view, smooth, hyaline to yellowish in KOH, yellow to brownish yellow in Melzer's reagent. Hymenophoral trama boletoid composed of 3.5–9 μm wide filamentous hyphae, hyaline to yellowish in KOH, yellowish to yellow in Melzer's reagent. Cheilo- and pleurocystidia 30–58 ×

Fig. 12.2 a *Hymenoboletus griseoviridis* Yan C. Li & Zhu L. Yang. Photo by Q. Zhao (KUN-HKAS 78816).

Fig. 12.2 b *Hymenoboletus griseoviridis* Yan C. Li & Zhu L. Yang. Photos by X.B. Liu (KUN-HKAS 75687, type).

6–10 μm, narrowly clavate, always with obtuse apex, usually hyaline to yellowish in KOH, thin-walled. Caulocystidia forming the scabrous squamules over the surface of stipe, ventricose, clavate, lanceolate or subfusiform with sharp apex. Pileipellis a hyphoepithelium: outer layer composed of 6–12 μm wide interwoven hyphae with terminal cells 28–78 × 6–10 μm, yellowish to pale brownish in KOH and yellow to brownish in Melzer's reagent; inner layer composed of 18–30 μm wide concatenated cells arising from 4–10 μm wide radially arranged filamentous hyphae. Pileal trama composed of broad (up to 15 μm wide) interwoven hyphae, hyaline to yellowish in KOH and yellowish to brownish in Melzer's reagent. Clamp connections absent in all tissues.

Habitat: Scattered on soil in subtropical forests dominated by plants of the family Fagaceae.

Known distribution: Currently known from central China.

Additional specimens examined: CHINA, HUBEI PROVINCE: Xingshan County, Muyu Town, Lishu Village, alt. 1800 m, 16 July 2012, Q. Zhao 1557 (KUN-HKAS 78816). HUNAN PROVINCE: Chengbu County, Erbaoding, alt. 1300 m, 9 August 2009, Z.H. Chen 30371 (MHHNU 30371).

Commentary: *Hymenoboletus griseoviridis* is characterized by the dark green to olive-green or grayish green pileus, the rugose pileal surface, the white to cream pileal context, and the hyphoepithelium pileipellis. It shares the color of the pileus with *Hy. filiformis*, but the two species differ in the morphology of the pileus and the structure of the pileipellis. Our phylogenetic analysis indicates that *Hy. griseoviridis* is closely related to *Hy. jiangxiensis* Yan C. Li & Zhu L. Yang. However, these two species differ from each other in the color of pileus and the morphology of hymenial cystidia.

Chapter 12 *Hymenoboletus* Yan C. Li & Zhu L. Yang   115

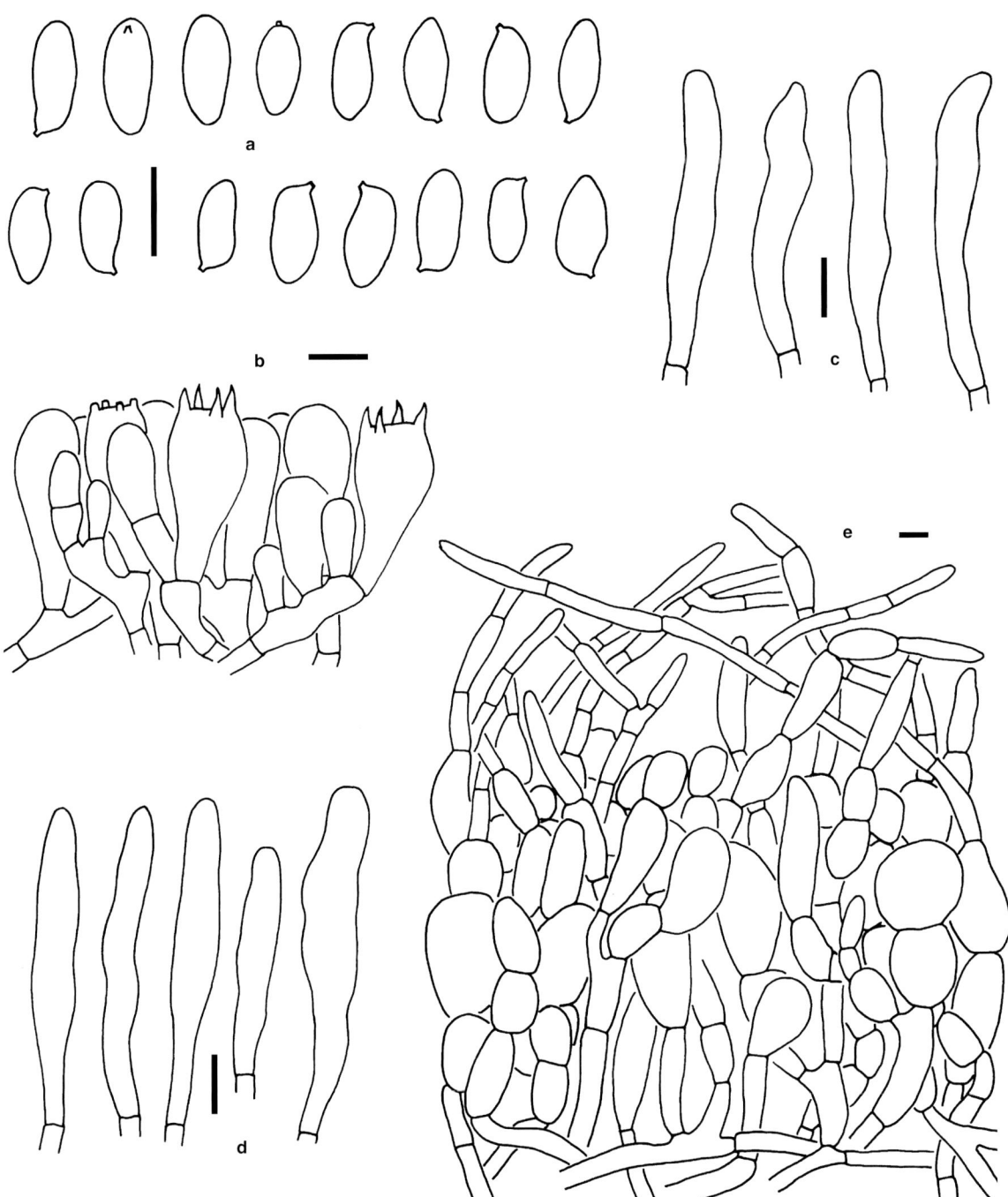

Fig. 12.2 c *Hymenoboletus griseoviridis* **Yan C. Li & Zhu L. Yang (KUN-HKAS 75687, type).**
a. Basidiospores; b. Basidia; c. Cheilocystidia; d. Pleurocystidia; e. Pileipellis. Scale bars = 10 μm.

12.3 *Hymenoboletus jiangxiensis* Yan C. Li & Zhu L. Yang, The Boletes of China: *Tylopilus* s.l. 137 (2021)

MycoBank: MB 834706

Etymology: The epithet "*jiangxiensis*" refers to the type location: Jiangxi Province.

Type: CHINA, JIANGXI PROVINCE: Jinggangshan County, Jinggangshan Nature Reserve, alt. 800 m, 6 August 2012, G. Wu 816 (KUN-HKAS 76988, GenBank Acc. No.: MT154778 for nrLSU, MT110372 for *tef1-α*, MT110447 for *rpb2*).

Basidioma small to medium-sized. Pileus 3–5.5 cm in diam., convex to applanate, dark brown (6E7–8) to reddish brown (5D7–8) when young, much paler in color when mature; surface glabrous, viscid when wet; context white to cream, without color change when injured. Hymenophore adnate when young and adnate to depressed around apex of stipe when mature; surface initially white to pallid and then pinkish (13A2) or pink (12A3–4), without discoloration when touched; pores angular to roundish, 0.5–1 mm wide; tubes up to 4 mm long, concolorous or much paler than hymenophoral surface, without discoloration when injured. Spore print pink-white (13A2) to pink (12A3–4). Stipe 5–7 × 0.4–0.6 cm, clavate to subcylindrical, sometimes obtuse downwards, purplish red to reddish brown, but white to cream or yellowish at apex and chrome-yellow or golden yellow at base; surface covered with concolorous scabrous squamules, without color change when touched; context cream to yellowish in the upper part, yellowish to yellow downwards and chrome-yellow or golden yellow at base, without color change when bruised; basal mycelium chrome-yellow or golden yellow. Taste and odor mild.

Basidia 27–33 × 9–11 μm, clavate, 4-spored, hyaline in KOH. Basidiospores [40/2/2] 10.5–13 (14) × 4.5–5.5 μm [Q = (2.1) 2.18–2.67 (2.78), Q_m = 2.35 ± 0.11], subfusiform and inequilateral in side view, fusiform to cylindrical in ventral view, smooth, hyaline to yellowish in KOH, yellow to brownish yellow in Melzer's reagent. Hymenophoral trama boletoid composed of 3.5–9 μm wide filamentous hyphae, hyaline to yellowish in KOH, yellowish to yellow in Melzer's reagent. Cheilo- and pleurocystidia 35–59 × 8–12 μm, fusiform to subfusiform or subfusoid-ventricose, thin-walled. Caulocystidia forming the scabrous squamules over the surface of stipe, morphologically similar to hymenial cystidia. Pileipellis a hyphoepithelium: outer layer composed of 5–10 μm wide filamentous hyphae with clavate to subcylindrical terminal cells 25–69 × 6–10 μm, pale yellowish to brownish yellow in KOH and yellow to brownish in Melzer's reagent; inner layer made up of 16–30 μm wide subglobose concatenated cells arising from 4–12 μm wide radially arranged filamentous hyphae. Pileal trama composed of broad (up to 15 μm wide) interwoven hyphae, hyaline to yellowish in KOH and yellowish to brownish in Melzer's reagent.

Habitat: Scattered on soil in subtropical forests dominated by plants of the family Fagaceae.

Known distribution: Currently known from southeastern China.

Additional specimen examined: CHINA, JIANGXI PROVINCE: Jinggangshan County, Jinggangshan Nature Reserve, alt. 900 m, 6 August 2012, G. Wu 837 (KUN-HKAS 77009).

Commentary: *Hymenoboletus jiangxiensis* differs from other species in this genus by its convex to

Chapter 12 *Hymenoboletus* Yan C. Li & Zhu L. Yang   117

Fig. 12.3 a *Hymenoboletus jiangxiensis* Yan C. Li & Zhu L. Yang. Photo by G. Wu (KUN-HKAS 76988, type).

Fig. 12.3 b *Hymenoboletus jiangxiensis* Yan C. Li & Zhu L. Yang. Photo by G. Wu (KUN-HKAS 77009).

applanate brown to reddish brown pileus, and hyphoepithelium pileipellis with an outer layer composed of filamentous hyphae and an inner layer composed of subglobose concatenated cells. Phylogenetically, *Hy. jiangxiensis* is closely related to *Hy. griseoviridis*. However, *Hy. griseoviridis* has a subhemispherical to hemispherical dark green to olive-green pileus and narrowly clavate hymenial cystidia measuring 30–58 × 6–10 μm.

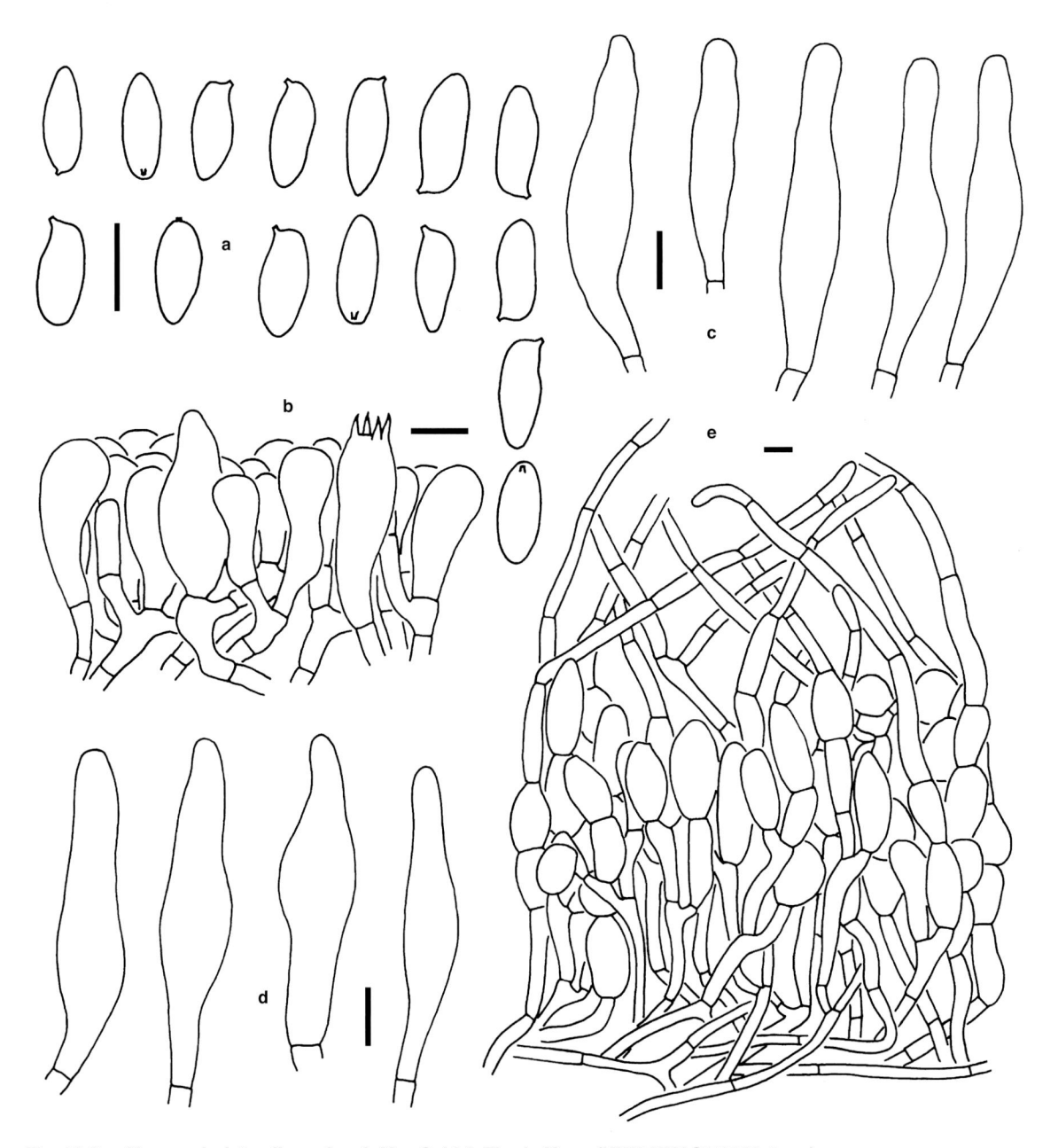

Fig. 12.3 c *Hymenoboletus jiangxiensis* Yan C. Li & Zhu L. Yang (KUN-HKAS 76988, type).
a. Basidiospores; b. Basidia and cheilocystidium; c. Cheilocystidia; d. Pleurocystidia; e. Pileipellis. Scale bars = 10 μm.

12.4 *Hymenoboletus luteopurpureus* Yan C. Li & Zhu L. Yang, Fungal Divers 81: 101 (2016)

Basidioma small. Pileus 2–4 cm in diam., subhemispherical to convex or plano-convex, initially dark red to red or brownish red and then dark purplish red to pinkish red, slightly darker in the center; surface dry, sometimes rugose, fibrillose to tomentose; context pallid to whitish or cream, without color change when injured. Hymenophore adnate when young and adnate to depressed around apex of stipe when mature; surface initially white and then pinkish or pink when mature, without discoloration when touched; pores angular to roundish, 0.5–1 mm wide; tubes up to 5 mm long, concolorous or a little paler than hymenophoral surface, without discoloration when injured. Spore print pink-white to pink. Stipe 3–5 × 0.3–0.5 cm, subcylindrical to clavate, pink to purple, but cream to yellowish at apex and chrome-yellow to bright yellow at base; surface covered with concolorous scabrous squamules, without color change when touched; context cream to yellowish, but chrome-yellow to bright yellow at base, without color when bruised; basal mycelium chrome-yellow or golden yellow. Taste and odor mild.

Basidia 25–30 × 8–10 μm, clavate, 4-spored, hyaline in KOH. Basidiospores 9–12.5 × 4–5 μm (Q = 1.95–2.76, Q_m = 2.32 ± 0.14), subfusiform in side view, fusiform to cylindrical in ventral view, smooth, hyaline to yellowish in KOH, yellow to brownish yellow in Melzer's reagent. Hymenophoral trama

Fig. 12.4 a *Hymenoboletus luteopurpureus* Yan C. Li & Zhu L. Yang. Photo by Z.L. Yang (KUN-HKAS 41694, type).

Fig. 12.4 b *Hymenoboletus luteopurpureus* Yan C. Li & Zhu L. Yang. Photo by J.F. Liang (KUN-HKAS 55828).

boletoid composed of 3–8 μm wide filamentous hyphae, hyaline to yellowish in KOH, yellowish to yellow in Melzer's reagent. Cheilocystidia 30–52 × 6–8.5 μm, clavate to fusiform or subfusiform always with obtuse apex, usually hyaline to yellowish in KOH, thin-walled. Pleurocystidia 43–69 × 6.5–10 μm, lanceolate to subfusiform or narrowly clavate, with 1–2 constrictions in the middle parts, hyaline to yellowish, thin-walled. Caulocystidia forming the scabrous squamules over the surface of stipe, lanceolate, subfusiform, ventricose, or clavate. Pileipellis a hymeniform composed of a layer of cystidia-like cells 28–60 × 4–11 μm, yellowish brown to brownish in KOH and yellow-brown to dark brown in Melzer's reagent. Pileal trama composed of 5.5–7.5 μm wide interwoven hyphae, hyaline to yellowish in KOH and yellowish to brownish in Melzer's reagent. Clamp connections absent in all tissues.

Habitat: Scattered on soil in southern subtropical forests dominated by plants of the families Fagaceae and Pinaceae.

Known distribution: Currently known from southwestern China.

Specimens examined: CHINA, YUNNAN PROVINCE: Longling County, alt. 1600 m, 11 September 2002, Z.L. Yang 3574 (KUN-HKAS 41694, type); Jinghong County, Dadugang Town, alt. 1350 m, 23 June 2008, J.F. Liang 816 (KUN-HKAS 55828, as KUN-HKAS 46334 in Wu *et al*. 2016a).

Commentary: *Hymenoboletus luteopurpureus* originally described from southwestern China by Wu *et al*. (2016a) is characterized by the initially dark red to red or brownish red and then dark purplish red to pinkish red pileus, the initially white and then pinkish or pink hymenophore, the pink to purple stipe, the whitish to pallid or cream pileal context, the hymeniform pileipellis, and the distribution in southern subtropical forests. Phylogenetically, *Hy. luteopurpureus* clusters together with *Hy. filiformis* with strong support. However, *Hy. filiformis* can be distinguished from *Hy. luteopurpureus* by its green to yellowish green or grayish green pileus, intricate trichoderm pileipellis which are 4–11 μm wide, and distribution in subtropical forests.

Chapter 13

Indoporus A. Parihar, K. Das, Hembrom & Vizzini

Indoporus A. Parihar, K. Das, Hembrom & Vizzini, Cryptog Mycol 39(4): 453 (2018)

Type species: *Indoporus shoreae* A. Parihar, K. Das, Hembrom & Vizzini, Cryptog Mycol 39(4): 453 (2018).

Diagnosis: The genus is different from other genera in the Boletaceae in its extended pileal margin, gray or grayish white to grayish pink hymenophore, big hymenophoral pores (up to 4 mm wide), at first reddish then blackish discoloration in the context when injured, and smooth basidiospores.

Basidioma stipitate-pileate with tubular hymenophore. Pileus subhemispherical, dark gray to blackish gray; surface tomentose, dry, distinctly extended at margin; context spongy, white to cream, becoming at first red and then black when injured. Hymenophore adnate when young and adnate to depressed around apex of stipe when mature; surface gray to grayish pink, staining at first red and then black when touched; pores angular to subangular, 2–4 mm wide; tubes concolorous with hymenophoral surface, staining at first red and then black when injured. Stipe central, stubby, concolorous with pileal surface, becoming at first red and then black when injured; context spongy, white to cream, becoming at first red and then black when bruised; basal mycelium cream. Basidiospores subfusiform, smooth, yellow to brownish yellow in KOH. Hymenial cystidia fusiform to subfusoid-ventricose or narrowly clavate. Pileipellis a trichoderm composed of inflated subcylindrical cells. Clamp connections absent in all tissues.

Commentary: *Indoporus* was originally described based on a species from India (Parihar *et al.* 2018). It is notable that the characteristic of initially red and then black discoloration is shared by species in the genera *Abtylopilus* and *Anthracoporus*. However, these genera differ in the morphology of the pileal margin, the color of the hymenophore, the size of the hymenophoral pores, and the structure of the pileipellis. Phylogenetically, *Indoporus* nests in the subfamily Boletoideae, although its relationships to other genera remain unresolved in the former and present studies (Crous *et al.* 2018). Currently, one species, newly described here, is known from China.

13.1 *Indoporus squamulosus* Yan C. Li & Zhu L. Yang, The Boletes of China: *Tylopilus* s.l. 144 (2021)

MycoBank: MB 834707

Etymology: The epithet "*squamulosus*" refers to the morphology of the pileal surface.

Type: CHINA, SICHUAN PROVINCE: Miyi County, Wanqiuyizu Town, alt. 1300 m, 27 July 2012, Y.J. Hao 641 (KUN-HKAS 76299, GenBank Acc. No.: MT154708 for nrLSU, MT110334 for *tef1-α*, MT110375 for *rpb1*).

Basidioma small to medium-sized. Pileus 4–8 cm in diam., subhemispherical to convex when young, applanate when mature, black (4F3) to blackish (4E2) in the center and gray (3D1) to pallid gray (2D1) towards margin, densely covered with concolorous granular-like squamules; surface rimose, always cracked into small squamules on a grayish (4B1) to whitish (2A1) background, margin extended; context white (1A1) to grayish (4B1), quickly staining reddish then slowly blackish when injured. Hymenophore adnate to slightly depressed around apex of stipe; surface white to grayish white (2D1) or gray (8C1), becoming reddish then blackish when injured; pores angular to nearly round, 2–4 mm wide; tubes up to 20 mm long, becoming reddish then blackish when touched. Stipe 3.5–7 × 0.8–1.5 cm, solid, central, subcylindrical to clavate, grayish (4B1) to whitish (2A1) when young, dark gray to black when aged or touched; context concolorous with that of pileus, becoming reddish then blackish when cut; basal mycelium whitish to cream, staining reddish then slowly blackish.

Basidia 22–30 × 6–9 µm, clavate, 4-spored, yellow to brownish yellow in KOH. Basidiospores [60/3/3] 7.5–9.5 (10.5) × 3.5–4.5 µm (Q = 1.88–2.43, Q_m = 2.21 ± 0.17), elongated to subcylindrical and

Fig. 13.1 a *Indoporus squamulosus* Yan C. Li & Zhu L. Yang. Photos by Y.J. Hao (KUN-HKAS 76299, type).

Fig. 13.1 b *Indoporus squamulosus* Yan C. Li & Zhu L. Yang. Photos by L.H. Han (KUN-HKAS 84835).

inequilateral in side view with indistinct suprahilar depression, elongated to cylindrical in ventral view (Fig. 3.3 e), smooth, yellow to brownish yellow in KOH, brown to dark brownish in Melzer's reagent. Hymenophoral trama boletoid; hyphae subcylindrical to cylindrical, 3.5–8 μm wide, hyaline to yellowish in KOH. Cheilocystidia 64–100 × 6–14 μm, fusiform to subfusiform or subfusoid-mucronate, often with 2-3 constrictions in the middle parts, thin-walled, yellow to brownish yellow in KOH. Pleurocystidia morphologically somewhat similar to cheilocystidia, but much bigger 85–127 × 11–18 μm, fusiform to subfusiform always with a long apex and pedicel, thin-walled, yellow to brownish yellow in KOH. Pileipellis a palisadoderm composed of 6–16 μm wide vertically arranged hyphae with subcylindrical to clavate or sometimes subfusiform terminal cells 17–75 × 8–16 μm, pale yellowish brown to pale brownish in KOH and yellowish brown to brownish in Melzer's reagent. Pileal trama composed of 6.5–12 μm wide interwoven hyphae, yellowish to pale brownish in KOH and yellow to brownish in Melzer's reagent. Clamp connections absent in all tissues.

Habitat: Scattered on soil in tropical and southern subtropical forests dominated by plants of the family Fagaceae, sometimes mixed with plants of the family Pinaceae.

Known distribution: Currently known from southern and southwestern China.

Additional specimens examined: CHINA, HAINAN PROVINCE: Wuzhishan County, Wuzhishan National Nature Reserve, alt. 1200 m, 31 July 2010, N.K. Zeng 784 (KUN-HKAS 107153). YUNNAN PROVINCE: Guangnan County, Zhetu Town, Jiulong Mountain, alt. 1580 m, 8 August 2014, L.H. Han 539 (KUN-HKAS 84835).

Commentary: *Indoporus squamulosus* is characterized by its black to blackish or gray pileus, extended pileal margin, white to grayish white or gray hymenophore, and initially red and then black discoloration when injured. These traits are similar to those of *Strobilomyces*. However, species in *Strobilomyces* have distinctly ornamented basidiospores. Phylogenetically *I. squamulosus* forms a distinct lineage within the Boletaceae although its relationships to other genera remain unresolved in present and other former studies (Crous *et al.* 2018).

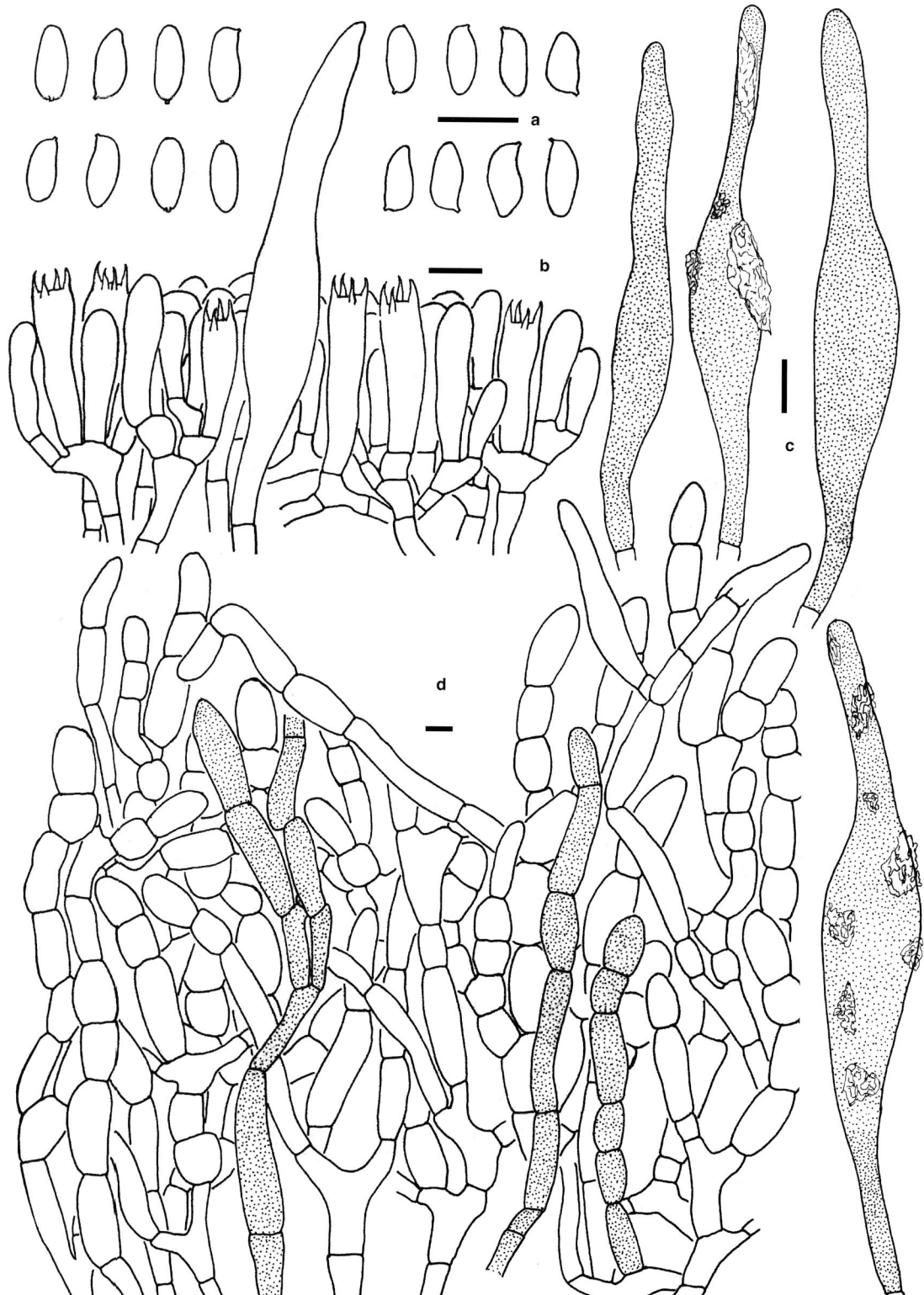

Fig. 13.1 c *Indoporus squamulosus* Yan C. Li & Zhu L. Yang (KUN-HKAS 76299, type).
a. Basidiospores; b. Basidia and cheilocystidium; c. Cheilo- and pleurocystidia; d. Pileipellis. Scale bars = 10 μm.

Chapter 14

Leccinellum Bresinsky & Manfr. Binder

Leccinellum Bresinsky & Manfr. Binder, Regensb Mykol Schr 11: 231 (2003)

Type species: *Leccinellum nigrescens* (Singer) Bresinsky & Manfr. Binder, Regensb Mykol Schr 11: 232 (2003).

Diagnosis: The genus can be distinguished from other genera in the Boletaceae by its whitish or yellow hymenophore without color change or staining brownish to ferruginous or at first reddish then blackish when injured, brown to blackish scabrous squamules over the surface of stipe, epithelium pileipellis, and smooth basidiospores.

Basidioma stipitate-pileate with tubular hymenophore, sometimes secotioid. When stipitate-pileate, pileus subhemispherical, convex or applanate, glabrous to fibrillose, dry, sometimes viscid when wet; context white to cream, without color change or staining brownish to ferruginous, or at first red and then slowly black when bruised. Hymenophore adnate when young and adnate to depressed around apex of stipe when mature; surface white to yellowish or yellow, without color change or staining brownish to ferruginous, or initially red and then slowly black when injured; tubes concolorous or much paler than hymenophoral surface, without color change or staining brownish to ferruginous, or initially red and then slowly black when injured. Spore print dull red to light brownish. Stipe whitish to dingy white, covered with brown to blackish dotted squamules; context white to cream, without color change or staining brownish to ferruginous, or at first red and then slowly back when injured; basal mycelium whitish to white. Basidiospores fusiform to cylindrical, smooth. Hymenial cystidia subfusiform to subfusoid-ventricose. Pileipellis an epithelium. When secotioid, pileus subhemispherical, nearly glabrous to fibrillose; gleba partially enclosed, dingy white to grayish white, staining light brown when bruised, irregularly to angularly loculate; columella present; stipe clavate to subcylindrical, always attenuate downwards, concolorous with pileus; surface covered with concolorous fibrillose squamules, staining a brownish tinge when bruised; basal mycelium cream; basidiospores smooth, subfusiform; hymenial cystidia absent; pileipellis an epithelium. Clamp connections absent in all tissues.

Commentary: *Leccinellum* Bresinsky & Manfr. Binder was erected to accommodate *Leccinum* section *Luteoscabra*, including species with white or yellow hymenophores and context, with *Leccinellum nigrescens* (Singer) Bresinsky & Manfr. Binder as the type species (Bresinsky and Besl 2003). The genus *Leccinum* has been defined and recognized differently by different mycologists (Gray 1821; Snell 1942; Smith *et al*. 1966, 1967, 1968; Smith and Thiers 1971; Singer 1986; Šutara 1989; Lannoy and Estades 1995). In a early study, Kuo and Ortiz-Santana (2020) proposed a broad circumscription of *Leccinum* that includes several sequestrate genera and *Leccinellum* based on their morphological and phylogenetic studies. However, considering the morphological differences and our multi-locus phylogenetic analysis, we follow Binder and Besl (2000) and Bresinsky and Besl (2003) and treat *Leccinum* and *Leccinellum* as independent genera.

The coarsely verrucose stipe in *Leccinellum* can be observed in species of *Chiua*, *Harrya*, *Hymenoboletus*, *Leccinum*, *Zangia*, *Royoungia* and *Sutorius*. However, *Chiua*, *Harrya*, *Hymenoboletus*, *Zangia*, and *Royoungia* all have pink to pinkish hymenophores when mature, pink to purplish pink scabrous squamules on the stipe which do not darken with age, and yellow to chrome-yellow bases to the stipe (den Bakker and Noordeloos 2005, Li *et al.* 2011, Halling *et al.* 2012b; Wu *et al.* 2016a). *Sutorius* is characterized by the purplish red to chocolate-brown hymenophore, the pallid to lilaceous context, and the intricate trichoderm pileipellis (den Bakker and Noordeloos 2005, Halling *et al.* 2012a). In this book, Six species from China, which are easily misidentified as *Tylopilus* based on the pallid to whitish or yellowish hymenophores and the white to whitish or yellowish context without discoloration or staining brownish to ferruginous when bruised, are documented and illustrated.

Key to the species of the genus *Leccinellum* in China

1. Basidiomata stipitate-pileate ··· 2
1. Basidiomata secotioid, stipitate ·· *L. cremeum*
2. Basidiomata bright in color ·· 3
2. Basidiomata umber in color ··· 4
3. Basidiomata yellow to chrome-yellow; hymenophore yellow to yellowish when young, cream to pale ferruginous when mature ··· *L. citrinum*
3. Basidiomata red to orange-red; hymenophore yellowish red to orange-red when young, yellow to orange-yellow when mature ·· *L. sinoaurantiacum*
4. Pileus without brown to castaneous tinge ··· 5
4. Pileus brown to castaneous ·· *L. castaneum*
5. Pileus dark gray to gray; stipe concolorous with pileal surface but yellow to orange-yellow at base ·· *L. griseopileatum*
5. Pileus brown to grayish brown; stipe concolorous with pileal surface but without any yellow to orange-yellow tinge at base ·· *L. onychinum*

14.1 *Leccinellum castaneum* Yan C. Li & Zhu L. Yang, The Boletes of China: *Tylopilus* s.l. 151 (2021)

MycoBank: MB 834708

Etymology: The epithet "*castaneum*" refers to the color of the basidiomata.

Type: CHINA, YUNNAN PROVINCE: Yulong County, Shitou Town, alt. 2600 m, 26 July 2008, Q. Zhao 8184 (KUN-HKAS 55179, GenBank Acc. No.: MT154744 for nrLSU).

Basidioma small to medium-sized. Pileus 3–6 cm wide, subhemispherical to convex or applanate, reddish brown (12E8) or chestnut-brown (8C7–8) to pale brownish (7D7–8); surface viscid when wet, always cracked into small squamules on a white to cream background when mature; context white to pallid, without color change when bruised. Hymenophore adnate when young and adnate to slightly depressed around apex of stipe when mature; surface initially white to pallid and then grayish white (11C1) to dingy white (9B1), without color change when touched; pores angular, up to 1 mm wide; tubes up to 10 mm long, concolorous or much paler than hymenophoral surface, without discoloration when injured. Spore print dull red (10C3) to brownish (7D6) (Fig. 3.4 f). Stipe clavate to subcylindrical, 3–5 × 0.5–1.5 cm, white to

Fig. 14.1 a *Leccinellum castaneum* Yan C. Li & Zhu L. Yang. Photo by G. Wu (KUN-HKAS 57592).

grayish white, covered with black scabrous squamules; context soft, white to pallid, without color change when injured; basal mycelium white. Taste and odor mild.

Basidia 33–41 × 10–13 μm, 4-spored, clavate, hyaline to yellowish in KOH. Basidiospores [40/2/2] (14.5) 15.5–18.5 × 5–6 μm [Q = (2.64) 2.82–3.5 (3.6), Q_m = 3.05 ± 0.24], subfusiform in side view with slight suprahilar depression, fusiform to cylindrical in ventral view (Fig. 3.3 f), smooth, somewhat thick-walled (up to 1 μm thick), yellow-brown to ocherous in KOH, brown to dark brownish in Melzer's reagent. Hymenophoral trama boletoid; hyphae cylindrical 3.5–10 μm wide. Cheilo- and pleurocystidia 36–60 × 6–11 μm, fusiform to subfusiform or subfusoid-ventricose, thin-walled. Caulocystidia similar to hymenial cystidia, brown to dark brownish yellow in KOH. Pileipellis an epithelium composed of subglobose to globose concatenated cells with subglobose to ellipsoid or broadly clavate terminal cells 9–32 × 6–31 μm, yellowish to pale brownish in KOH and yellow to brownish in Melzer's reagent. Pileal trama composed of 4–9 μm wide more or less radially arranged filamentous hyphae, hyaline to yellowish in KOH and yellowish to brownish in Melzer's reagent. Clamp connections absent in all tissues.

Habitat: Scattered on the ground in temperate forests dominated by plants of the families Fagaceae and Pinaceae.

Known distribution: Currently known from southwestern China.

Fig. 14.1 b *Leccinellum castaneum* Yan C. Li & Zhu L. Yang. Photo by Q. Zhao (KUN-HKAS 55179, type).

Chapter 14 *Leccinellum* Bresinsky & Manfr. Binder 129

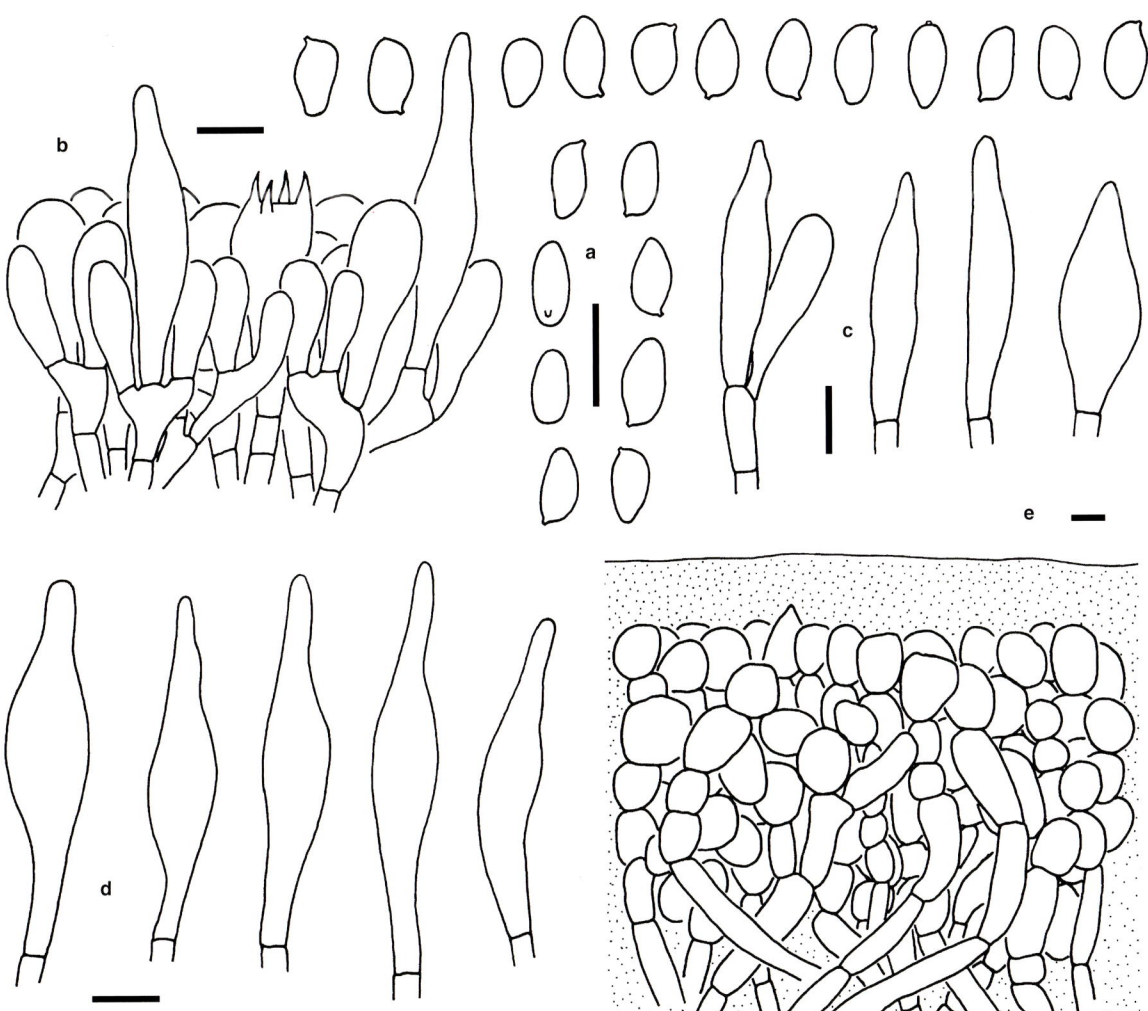

Fig. 14.1 c *Leccinellum castaneum* Yan C. Li & Zhu L. Yang (KUN-HKAS 55179, type).
a. Basidiospores; b. Basidia and pleurocystidia; c. Cheilocystidia; d. Pleurocystidia; e. Pileipellis. Scale bars = 10 μm.

Additional specimen examined: CHINA, YUNNAN PROVINCE: Yongping County, alt. 2100 m, 31 July 2009, G. Wu 60 (KUN-HKAS 57592).

Commentary: *Leccinellum castaneum* is well characterized by the chestnut-brown to pale brownish pileus, the white to pallid context without discoloration when bruised, the white to pallid or grayish white hymenophore, and the epithelium pileipellis composed of subglobose to globose concatenated cells. The chestnut-brown to pale brownish pileus and the epithelium pileipellis of *L. castaneum* are reminiscent of *Porphyrellus castaneus* Yan C. Li & Zhu L. Yang, but *P. castaneus* has a grayish white to grayish pink or brownish pink hymenophore, and white to pallid context becoming asymmetrically bluish when bruised (Wu *et al*. 2016a). In our multi-locus phylogenetic analysis (Fig. 4.1), *L. castaneum* forms an independent lineage. Its phylogenetic relationships with other species in this genus remain unresolved.

14.2 *Leccinellum citrinum* Yan C. Li & Zhu L. Yang, The Boletes of China: *Tylopilus* s.l. 154 (2021)

MycoBank: MB 834710

Etymology: The epithet "*citrinum*" refers to the color of the basidiomata.

Type: CHINA, HUNAN PROVINCE: Yizhang County, Mangshan National Forest Park, alt. 1800 m, 2 September 2007, Y.C. Li 1065 (KUN-HKAS 53410, GenBank Acc. No.: KT990585 for nrLSU, KT990937 for *rpb1*, KT990421 for *rpb2*).

Basidioma very small to small. Pileus 2–5 cm wide, subhemispherical to convexo-applanate, yellow (5B4–6) to yellowish (4A3–6) when young, brownish yellow (3A4–6) when mature; surface viscid when wet, glabrous; context white to cream, without discoloration when bruised. Hymenophore adnate when young, adnate to slightly depressed around apex of stipe when mature; surface yellow (5B4–6) when young and becoming yellowish (4A3–6) to rust-brown (4C3–7) when mature, without discoloration when touched; pores subangular to roundish, up to 1 mm wide; tubes up to 8 mm long, concolorous or much paler than hymenophoral surface, without color change when injured. Spore print brownish (7D6) to reddish brown (8D5). Stipe clavate, 5–8 × 0.5–1 cm, concolorous with pileal surface, densely covered with concolorous scabrous squamules; context white to pallid, without color change when bruised; basal mycelium yellow to chrome-yellow. Taste and odor mild.

Basidia 25–39 × 10–14 µm, clavate to broadly clavate, 4-spored, hyaline in KOH. Basidiospores [60/3/3] (12.5) 13–15.5 (16.5) × 4–5 µm [Q = (2.6) 2.7–3.44 (3.88), Q_m = 3.2 ± 0.31], subcylinderical to subfusiform in side view with slight suprahilar depression, cylinderical to fusiform in ventral view, smooth, somewhat thick-walled (up to 1 µm thick), yellowish to brownish yellow in KOH, yellow to yellow-brown in Melzer's reagent. Hymenophoral trama boletoid; hyphae cylindrical 3.5–8 µm wide. Cheilo- and pleurocystidia 36–60 ×

Fig. 14.2 a *Leccinellum citrinum* Yan C. Li & Zhu L. Yang. Photo by Y.C. Li (KUN-HKAS 53410, type).

Fig. 14.2 b *Leccinellum citrinum* Yan C. Li & Zhu L. Yang. Photos by G. Wu (KUN-HKAS 99924).

8–16 μm, subfusiform to subfusoid-ventricose, usually with a sharp apex and long pedicel, thin-walled. Caulocystidia similar to hymenial cystidia, yellow to brownish yellow in KOH. Pileipellis an epithelium composed of inflated concatenated cells with subglobose to ellipsoid or broadly clavate terminal cells 10–45 × 6–28 μm, colorless to yellowish in KOH and yellowish to pale yellow in Melzer's reagent. Pileal trama composed of 6–11 μm wide more or less radially arranged filamentous hyphae, colorless to yellowish in KOH and yellowish to pale yellow in Melzer's reagent. Clamp connections absent in all tissues.

Habitat: Scattered on soil in subtropical forests dominated by plants of the family Fagaceae.

Known distribution: Currently known from central China.

Additional specimens examined: CHINA, HUNAN PROVINCE: Yizhang County, Mangshan National Forest Park, alt. 1800 m, 4 September 2007, Y.C. Li 1082 (KUN-HKAS 53427), the same location, 13 September 2016, G. Wu 1805 (KUN-HKAS 99924).

Commentary: *Leccinellum citrinum* is similar to *Fistulinella lutea* Redeuilh & Soop in that they share the yellow to yellowish basidioma and the viscid pileus. However, *F. lutea* has a white to pinkish hymenophore and an ixotrichoderm pileipellis (Redeuilh and Soop 2006).

In our multi-locus phylogenetic analysis (Fig. 4.1), *L. citrinum* is related to *L. sinoaurantiacum* (M.

Zang & R.H. Petersen) Yan C. Li & Zhu L. Yang. However, *L. sinoaurantiacum* has a red to scarlet or orange-red to yellow-red pileus, a yellowish red or orange-red to yellow hymenophore, and relatively big basidiospores 15.5–18 × 5–6.5 μm (Zang *et al*. 2001).

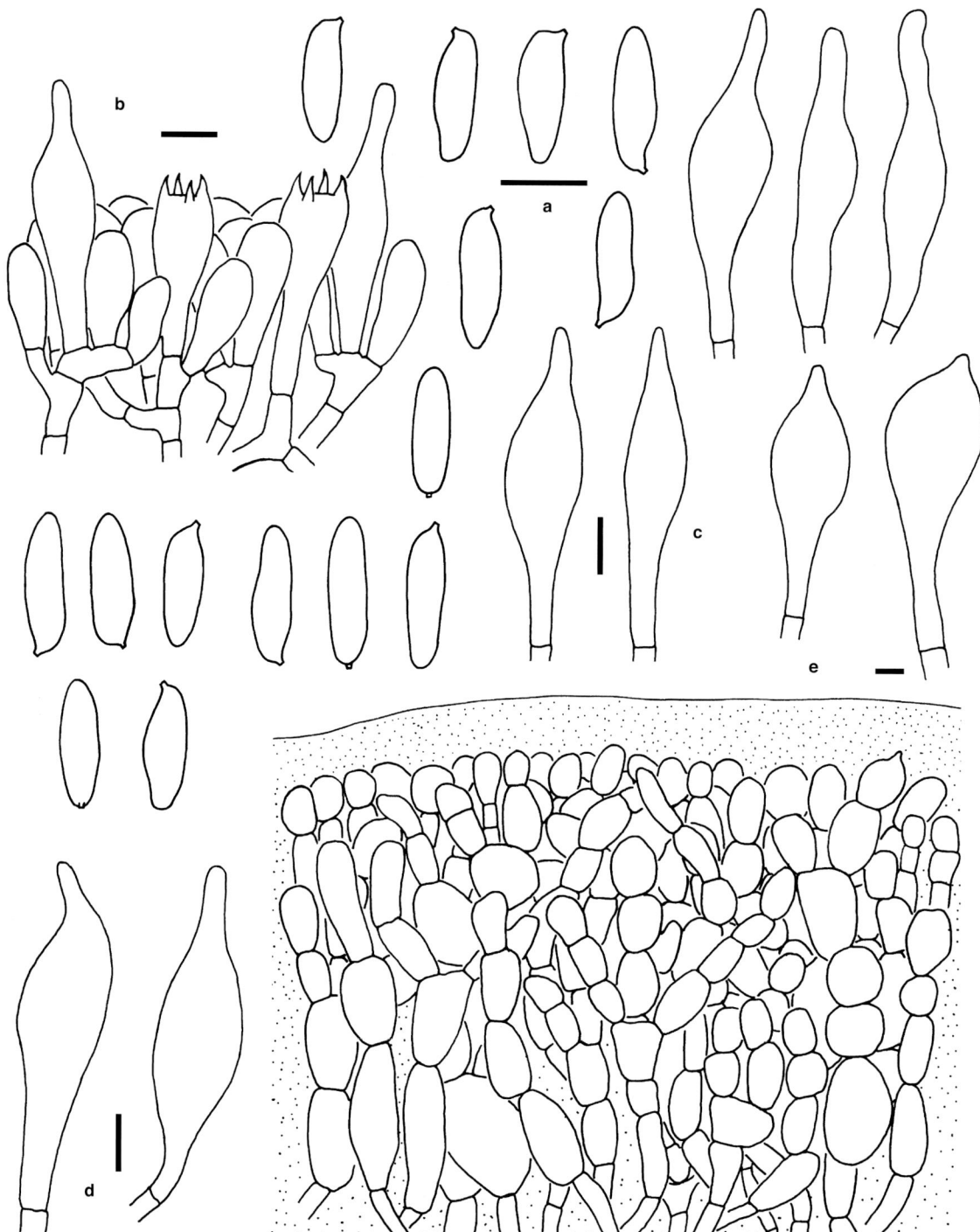

Fig. 14.2 c *Leccinellum citrinum* **Yan C. Li & Zhu L. Yang (KUN-HKAS 53410, type).**
a. Basidiospores; b. Basidia and cheilocystidia; c. Cheilocystidia; d. Pleurocystidia; e. Pileipellis. Scale bars = 10 μm.

14.3 *Leccinellum cremeum* Zhu L. Yang & G. Wu, Fungal Divers 81: 102 (2016)

Basidioma very small to small, secotioid, stipitate. Pileus 2–4 cm wide, subhemispherical; surface glabrous, yellowish to cream, staining brownish when touched. Gleba partially enclosed, pallid to dingy white or grayish white, staining brownish when bruised, irregularly to angularly loculate; loculi 0.5–1.5 mm wide; columella present, grayish white to dingy white. Stipe 2–4 × 0.5–1 cm, clavate to subcylindrical, always attenuate downwards; surface concolorous with pileal surface, covered with concolorous fibrillose squamules, staining a brownish tinge when injured; basal mycelium cream to grayish white.

Basidia 20–38 × 10–15 μm, clavate, 4-spored, hyaline in KOH. Basidiospores 13.5–19 × 5–7.5 μm [Q = (2.27) 2.29–2.92 (3), Q_m = 2.59 ± 0.23], subfusiform to subcylindrical and inequilateral in side view with slight suprahilar depression, fusiform to cylindrical in ventral view, smooth, somewhat thick-walled (up to 1 μm thick), yellow-brown to ocherous in KOH, brown to dark brownish in Melzer's reagent. Cystidia absent. Trama composed of colorless, thin-walled, 3–5 μm wide filamentous hyphae. Pileipellis a hyphoepithelium: outer layer composed of 3–5 μm wide filamentous hyphae, yellowish to pale brownish in KOH and yellow to yellow-brown in Melzer's reagent; inner layer composed of vertically arranged 10–25 μm wide moniliform hyphae, colorless to yellowish in KOH and yellowish to pale yellow in Melzer's reagent. Clamp connections absent in all tissues.

Habitat: Scattered on soil in temperate forests dominated by plants of the genera *Picea* and *Quercus*.

Known distribution: Currently known from southwestern China.

Specimen examined: CHINA, YUNNAN PROVINCE: Shangri-La County, Luoji Town, Bitahai Nature Reserve, alt. 3200 m, 26 July 2013, Q. Zhao T22183 (KUN-HKAS 90639, type).

Commentary: *Leccinellum cremeum* was originally described and illustrated by Wu *et al.* (2016a) as a new species of *Leccinellum*. It is characterized by the yellowish to cream basidiomata, the pallid to whitish gleba staining brownish when injured, and the hyphoepithelium pileipellis with an outer layer composed of filamentous hyphae and an inner layer composed of subglobose to globose cells. In our multi-locus phylogenetic analysis, *L. cremeum* forms an independent lineage within *Leccinellum* (Fig. 4.1). Its phylogenetic relationships with other species in this genus remain unresolved.

Fig. 14.3 *Leccinellum cremeum* Zhu L. Yang & G. Wu. Photo by Q. Zhao (KUN-HKAS 90639, type).

14.4 *Leccinellum griseopileatum* Yan C. Li & Zhu L. Yang, The Boletes of China: *Tylopilus* s.l. 159 (2021)

MycoBank: MB 834709

Etymology: The epithet "*griseopileatum*" refers to the gray to grayish basidiomata.

Type: CHINA, JIANGXI PROVINCE: Longnan County, Jiulianshan National Nature Reserve, alt. 500 m, 13 June 2012, G. Wu 888 (KUN-HKAS 77060, GenBank Acc. No.: MT154747 for nrLSU, MT110357 for *tef1-α*, MT110429 for *rpb2*).

Basidioma small. Pileus 3.5 cm wide, subhemispherical to convexo-applanate, viscid when wet, glabrous, gray (1C1–2) to grayish (1B1) or brownish gray (3C2–3); context white to grayish white or cream, without discoloration when injured. Hymenophore adnate; surface dingy white (2B1) to grayish

Fig. 14.4 a *Leccinellum griseopileatum* Yan C. Li & Zhu L. Yang. Photos by G. Wu (KUN-HKAS 77060, type).

Chapter 14 *Leccinellum* Bresinsky & Manfr. Binder 135

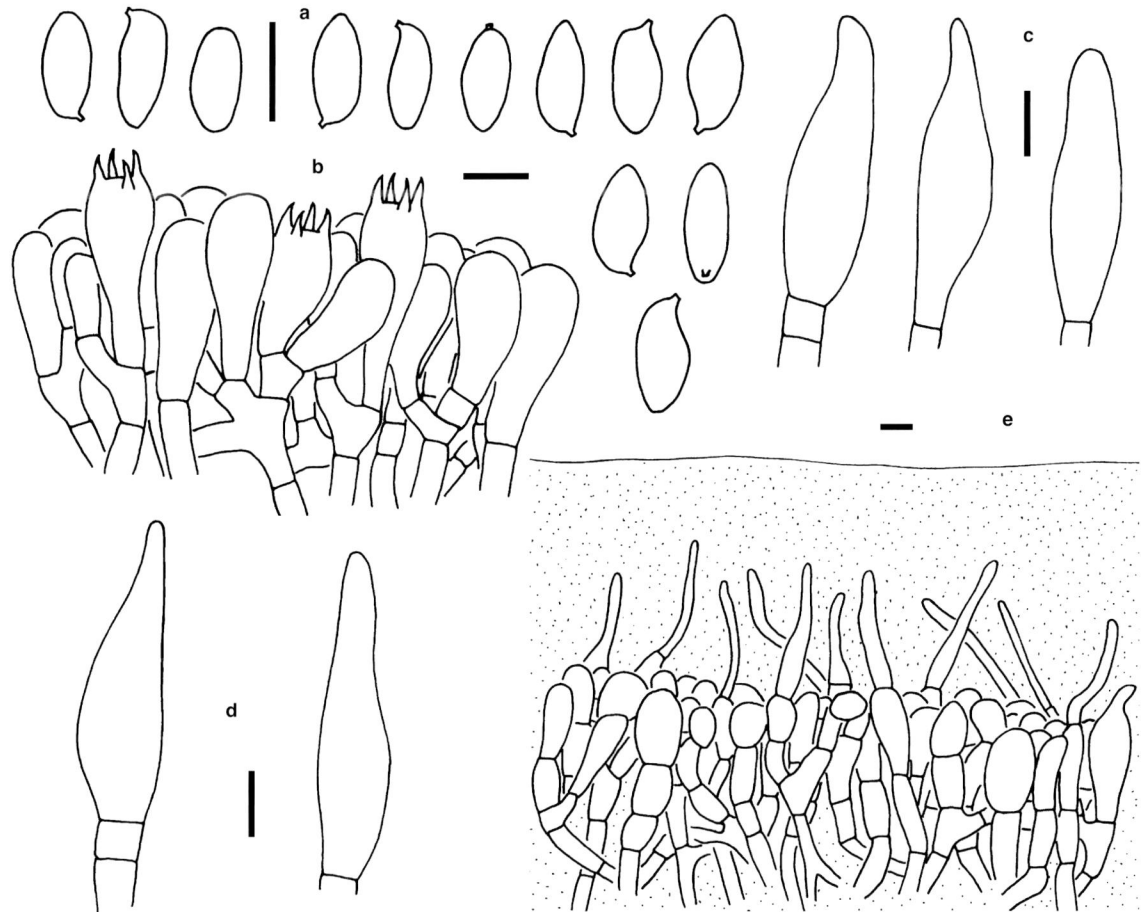

Fig. 14.4 b *Leccinellum griseopileatum* Yan C. Li & Zhu L. Yang (KUN-HKAS 77060, type).
a. Basidiospores; b. Basidia; c. pleurocystidia; d. Cheilocystidia; e. Pileipellis. Scale bars = 10 μm.

white (1A1), staining brownish (4B2–3) to ferruginous or rufescent (5C6–7) when injured; pores roundish, 0.3–1 mm wide; tubes 3–8 mm long. Spore print brownish (7D6) to reddish brown (8D5). Stipe 5 × 0.5–0.8 cm, clavate always enlarged downwards, concolorous with pileal surface, but white (4A2) to cream (2B2–3) at the apical part, and grayish yellow (4B2–4) to brownish yellow (5C7–8) at base; surface densely covered with concolorous scabrous squamules; context pale yellowish (4A2–3), without color change when injured; mycelium on stipe base yellowish (4A2–4) to yellow (4A5–8) or orange-yellow (5B2–5), without color change when injured. Taste and odor mild.

Basidia 19–32 × 8–11 μm, clavate to broadly clavate, 4-spored, hyaline in KOH. Basidiospores (10) 10.5–12 × 4.5–5.5 μm (Q = 2.1–2.56, Q_m = 2.31 ± 0.14) subcylindrical in side view with slight suprahilar depression, cylindrical in ventral view, smooth, somewhat thick-walled (up to 0.5 μm thick), yellow to brownish yellow in KOH, brown to dark brownish in Melzer's reagent. Hymenophoral trama boletoid composed of 2.5–8 μm wide filamentous hyphae, hyaline to yellowish in KOH, yellowish to yellow in Melzer's reagent. Cheilo- and pleurocystidia 30–46 × 10–13 μm, clavate to subfusiform or subfusoid-ventricose, yellowish to yellow in KOH, thin-walled. Caulocystidia forming the squamules over the surface of stipe, clavate, lanceolate, ventricose, or subfusiform. Pileipellis an epithelium composed of inflated concatenated cells with terminal cells subglobose to subfusoid-mucronate or ventricose-mucronate sometimes flagelliform, pale yellowish to brownish yellow in KOH and yellow to brownish in Melzer's reagent. Pileal trama composed of 4–8 μm wide interwoven hyphae, colorless to yellowish in KOH and yellow to yellowish brown in Melzer's reagent. Clamp connections absent in all tissues.

Habitat: Solitary on soil in subtropical forests dominated by plants of the family Fagaceae.

Known distribution: Currently known from southeastern China.

Commentary: *Leccinellum griseopileatum* is characterized by the gray to grayish or brownish gray pileus, the white to grayish white or cream context staining brownish to ferruginous when injured, the grayish yellow to brownish yellow stipe base, and the epithelium pileipellis. Sequences of *L. griseopileatum* form a distinct lineage in our phylogenetic study, however, the relationships between *L. griseopileatum* and the other taxa in this genus are not well resolved in our present phylogenetic analysis (Fig. 4.1).

14.5 *Leccinellum onychinum* Fang Li, Kuan Zhao & Q.L. Deng [as "onyx"], Mycol Progr 15: 1275 (2016)

Basidioma small to medium-sized. Pileus 4–8 cm in diam., hemispherical to subconvex or convex to nearly applanate; surface uneven, dry, gray to dark brown when young, brownish gray when mature; context grayish white to cream, without discoloration when injured. Hymenophore adnate; surface initially cream to yellowish and then yellow to brownish yellow when mature, without discoloration when injured; pores angular to roundish, 0.5–1 mm wide; tubes 4–6 mm long, concolorous or much paler than hymenophoral surface, without discoloration when injured. Stipe 5–10 × 0.6–1.5 cm, clavate to subcylindrical, tapering towards the base, concolorous with pileal surface but much paler at apex, densely covered with concolorous scabrous squamules, without discoloration when injured; context grayish white to cream, without discoloration when bruised; basal mycelium white to dirty-white. Taste and odor mild.

Basidia 23–35 × 9–15 μm, clavate, 4-spored, hyaline to yellowish in KOH. Basidiospores 8–12 × 5–6 μm [Q = (1.64) 1.72–2.19 (2.3), Q_m = 1.95 ± 0.11], elongated to subcylindrical in side view with slight suprahilar depression, elongated to cylindrical in ventral view, smooth, somewhat thick-walled (up to 1 μm thick), yellow-brown to ocherous in KOH, brown to dark brownish in Melzer's reagent. Hymenophoral trama boletoid; hyphae cylindrical 3.5–7 μm wide. Cheilo- and pleurocystidia 28–40 × 8–12 μm, fusiform to subfusiform, thin-walled. Caulocystidia similar to hymenial cystidia, dark brownish yellow in KOH. Pileipellis a hyphoepithelium: outer layer composed of 4–8 μm wide filamentous hyphae, yellowish to brownish yellow in KOH and yellow to brown in Melzer's reagent; inner layer composed of vertically arranged 6–15 μm wide moniliform hyphae, yellowish to pale brownish in KOH and yellow to brownish in Melzer's reagent. Pileal trama composed of 3.5–8 μm wide interwoven hyphae, hyaline to yellowish in KOH and yellowish to brownish in Melzer's reagent. Clamp connections absent in all tissues.

Habitat: Gregarious on soil in subtropical forests dominated by plants of the family Fagaceae.

Known distribution: Currently known from southern China.

Specimens examined: CHINA, GUANGDONG PROVINCE: Fengkai County, Heishiding Nature Reserve, alt. 500 m, 20 July 2012, F. Li 700 (KUN-HKAS 92188, type), the same location, 19 August 2013, F. Li 1497 (KUN-HKAS 82191).

Commentary: *Leccinellum onychinum* was originally described from southern China (Li *et al.* 2016) and is characterized by the gray or dark brown to brownish gray pileus, the cream to yellowish or brownish yellow hymenophore, the white to cream context without discoloration when bruised, and the hyphoepithelium pileipellis with an outer layer composed of filamentous hyphae and an inner layer composed of inflated cells. Morphologically, it shares the color of the pileus and hymenophore with *L. griseopileatum*. However, *L. griseopileatum* differs from *L. onychinum* in its viscid pileus, yellow to orange-yellow base of stipe, and epithelium pileipellis. In our phylogenetic analysis (Fig. 4.1), *L. onychinum* forms an independent lineage. Its phylogenetic relationships with other species in this genus remain unresolved.

138 The Boletes of China: *Tylopilus* s.l.

Fig. 14.5 *Leccinellum onychinum* Fang Li, Kuan Zhao & Qing Li Deng. Photos by F. Li (KUN-HKAS 92188, type).

14.6 *Leccinellum sinoaurantiacum* (M. Zang & R.H. Petersen) Yan C. Li & Zhu L. Yang, The Boletes of China: *Tylopilus* s.l. 164 (2021)

MycoBank: MB 834711

Basionym: *Boletus sinoaurantiacus* M. Zang & R.H. Petersen, Mycotaxon 80: 481 (2001).

Basidioma very small to small. Pileus 2–5 cm in diam., subhemispherical to convex or convexo-applanate; surface gelatinous, red to scarlet when young, orange-red to yellowish red when old, smooth, margin extended; context whitish to cream but orange-red to red beneath pileipellis, without color change when bruised. Hymenophore adnate when young and adnate to depressed around apex of stipe when mature; surface red to yellowish red when young and yellow to yellowish when mature, without discoloration when touched; pores angular, 0.3–1 mm wide; tubes 4–8 mm long, yellow to yellowish, without discoloration when injured. Spore print brownish (7D6) to reddish brown (8D5). Stipe 3–6 × 0.6–1 cm, clavate to subcylindrical, often enlarged downwards, concolorous with pileal surface but much paler at apex, densely covered with concolorous coarsely verrucose squamules, without discoloration when touched; context whitish to cream but orange-red to red beneath stipitipellis, without discoloration when bruised; basal mycelium cream to yellowish. Taste and odor mild.

Basidia 18–27 × 14–16 μm, clavate to broadly clavate, 4-spored, hyaline to yellowish in KOH. Basidiospores (13.5) 15.5–18 (19.5) × 5–6.5 μm [Q = (2.27) 2.29–2.92 (3), Q_m = 2.59 ± 0.23], subfusiform to subcylindrical in side view with slight suprahilar depression, fusiform to cylindrical in ventral view, smooth, somewhat thick-walled (≤1 μm thick), yellow-brown to ocherous in KOH, brown to dark brownish in Melzer's reagent. Hymenophoral trama boletoid; hyphae cylindrical 4–8 μm wide. Cheilo- and pleurocystidia 38–56 × 18–28 μm, fusiform to subfusiform or subfusoid-ventricose, thin-walled. Caulocystidia similar to hymenial cystidia, yellow to brownish yellow in KOH. Pileipellis an ixohyphoepithelium: outer layer consisting of 3.5–6 μm wide gelatinous filamentous hyphae with subcylindrical terminal cells 12–41 × 3.5–5 μm, yellowish to pale brownish in KOH and yellow to

Fig. 14.6 a *Leccinellum sinoaurantiacum* (M. Zang & R.H. Petersen) Yan C. Li & Zhu L. Yang. Photos by Y.C. Li (KUN-HKAS 59439).

Fig. 14.6 b *Leccinellum sinoaurantiacum* (M. Zang & R.H. Petersen) Yan C. Li & Zhu L. Yang. Photos by Y.C. Li (KUN-HKAS 89413).

brownish in Melzer's reagent; inner layer composed of inflated (up to 25 μm wide) concatenated cells, colorless to yellowish in KOH and yellowish to pale yellow in Melzer's reagent. Pileal trama composed of 3–9 μm wide more or less radially arranged filamentous hyphae, colorless to yellowish in KOH and yellowish to pale yellow in Melzer's reagent. Clamp connections absent in all tissues.

Habitat: Scattered on soil in tropical to subtropical forests dominated by plants of the family Fagaceae.

Known distribution: Currently known from southwestern China.

Specimens examined: CHINA, YUNNAN PROVINCE: Simao County, Laiyanghe National Forest Park, alt. 1500 m, 20 June 2000, M. Zang 13418 (KUN-HKAS 36112), the same location, alt. 1680 m, 23 June 2000, M. Zang 13486 (KUN-HKAS 36065, type), the same location, 28 June 2000, M. Zang 13489 (KUN-HKAS 36294), the same location, alt. 1570 m, 2 July 2000, M. Zang 13549 (KUN-HKAS 36437); Yingjiang County, Tongbiguan Town, alt. 1450 m, 14 July 2003, L. Wang 104 (KUN-HKAS 43200), the same location, 19 July 2009, Y.C. Li 1692 (KUN-HKAS 59439); Yingjiang County, Xima Town, Huanglianhe Village, alt. 1400 m, 17 July 2003, L. Wang 158 (KUN-HKAS 43253); Longling County, Xiaoheishan Nature Reserve, alt. 2100 m, 27 August 2002, Z.L. Yang 3268 (KUN-HKAS 41581); Nanjian County, Houqing Town, Qinshan, alt. 2300 m, 30 August 2003, Z.L. Yang 3928 (KUN-HKAS 25092); Pu'er, near the Radar Station on National Road 213#, altitude unknown, 7 September 2008, B. Feng 371 and 372 (KUN-HKAS 55662 and KUN-HKAS 55663, respectively); Pu'er, Jinggu County, Yongping Town, Chahe Village, alt. 1380 m, 30 July 2008, L.P. Tang 526 (KUN-HKAS 54757); Tengchong County, alt. 1700 m, 19 July 2009, Y.C. Li 1692 (KUN-HKAS 59439); Tengchong County, on the way from Tengchong County to Longling County, alt. 1700 m, 11 August 2011, G. Wu 622 (KUN-HKAS 74936); Pingbian County, Xinxian Town, Malongdi Village, alt. 1600 m, 24 September 2011, Y.C. Li 2770 (KUN-HKAS 89413).

Commentary: *Leccinellum sinoaurantiacum* was originally described as *Boletus sinoaurantiacus* M. Zang & R.H. Petersen (2001). This species is characterized by the gelatinous and red to scarlet or orange-red basidiomata, the reddish yellow to yellow hymenophoral surface, the coarsely verrucose stipe, and the ixohyphoepithelium pileipellis. These traits differ from those in *Boletus* typified by *Boletus edulis* Bull., and fit well with the morphological features of the genus *Leccinellum*. Moreover, sequences from the type and other specimens of *L. sinoaurantiacum* nest into the genus *Leccinellum* in our phylogenetic analysis (Fig. 4.1). Thus, a new combination is proposed.

Our multi-locus phylogenetic analysis (Fig. 4.1) indicates that *L. sinoaurantiacum* is related to *L. citrinum*. However, *L. citrinum* has a yellow to bright yellow pileus, a yellow to yellowish but never red hymenophore, and relatively small basidiospores measuring 12.5–16.5 × 4–5 μm.

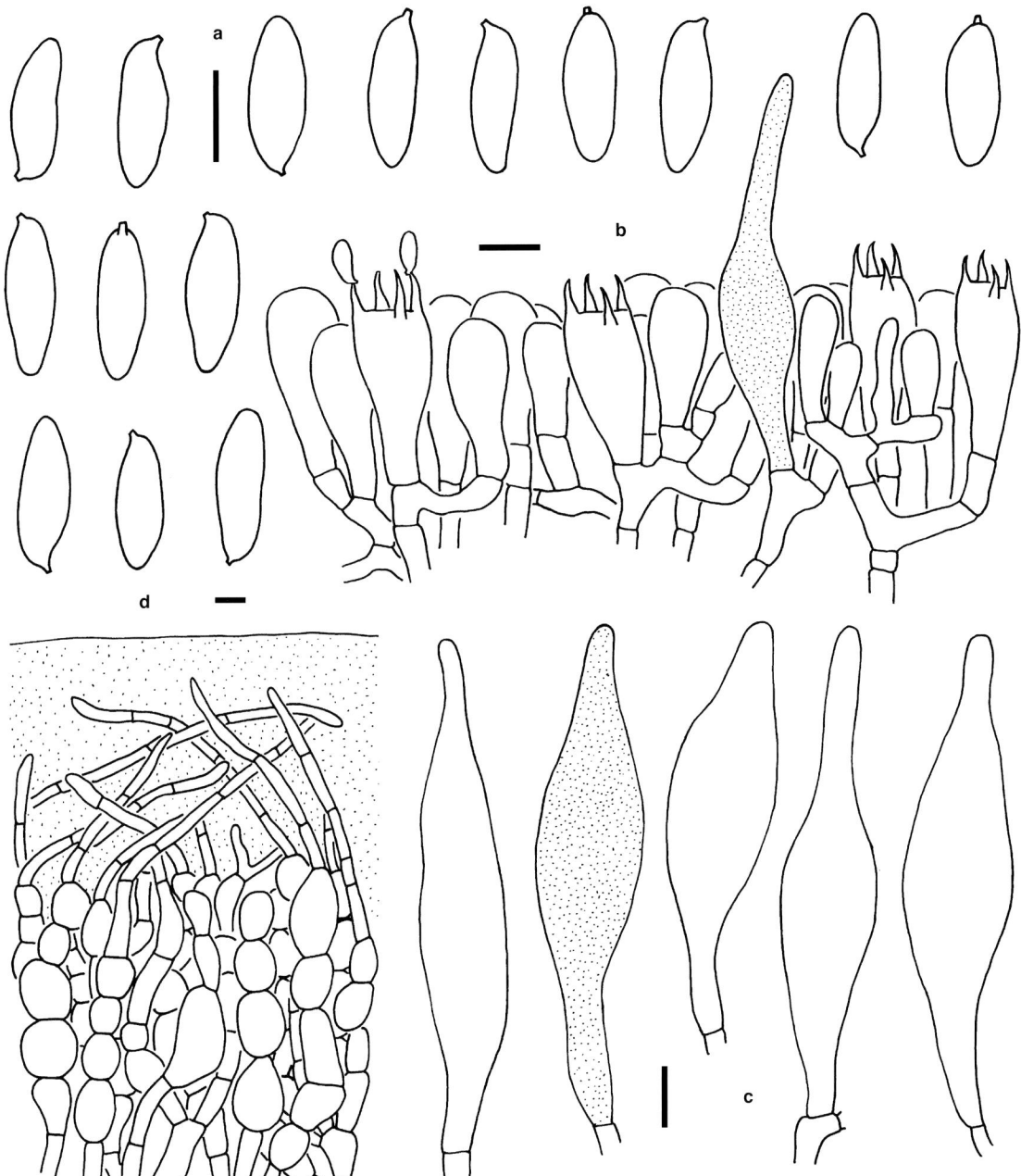

Fig. 14.6 c *Leccinellum sinoaurantiacum* **(M. Zang & R.H. Petersen) Yan C. Li & Zhu L. Yang (KUN-HKAS 89413).** a. Basidiospores; b. Basidia and cheilocystidium; c. Cheilo- and pleurocystidia; d. Pileipellis. Scale bars = 10 μm.

Chapter 15

Mucilopilus Wolfe

Mucilopilus Wolfe, Mycotaxon 10(1): 117 (1979)

Type species: *Porphyrellus viscidus* McNabb, New Zealand J Bot 5: 543. 1967.

Diagnosis: This genus differs from the other genera in the Boletaceae in its viscid pileus and dry stipe, white to pallid context of pileus and stipe without discoloration in pileus but staining yellow to cream yellow in the base of stipe when injured, white to pinkish or pink hymenophore, pink to grayish pink spore print, ixotrichoderm pileipellis, and smooth basidiospores.

Pileus subhemispherical to plano-convex or applanate; surface glabrous, viscid to gelatinous when wet and shiny when dry, margin slightly extended; context white to sordid-white, without discoloration when injured. Hymenophore adnate when young and adnate to slightly depressed around apex of stipe when mature; surface white when young and pinkish to grayish pink or pallid violet when mature, without discoloration when touched; pores angular to roundish; tubes concolorous or a little paler than hymenophoral surface, without discoloration when injured. Spore print pink-white to red-white. Stipe clavate or tapering upwards, glabrous to subglabrous or sometimes with longitudinal striations; context white but yellowish to cream-yellow at base when injured; basal mycelium white, without discoloration when injured. Basidiospores smooth, fusiform, elongated, or cylindrical. Hymenial cystidia present, clavate or fusoid-ventricose to subfusiform. Pileipellis an ixotrichoderm composed of strongly gelatinous filamentous hyphae. Clamp connections absent in all tissues.

Commentary: This genus shares the color of the spore print and pileal context with some species in *Tylopilus* P. Karst. However, *Tylopilus* differs from *Mucilopilus* in that its pileus is non-gelatinous, pileal margin is not extended, and base of stipe without yellow to cream-yellow stains when injured (Wolfe 1979b, 1982). *Mucilopilus* is also similar to *Fistulinella* Henn. in their viscid to gelatinous pileus. However, *Fistulinella* has a strongly gelatinous stipe, brown to reddish brown spore print, and white context in the base of stipe without discoloration when injured. Currently, three species, newly described here are known from China.

15.1 *Mucilopilus cinnamomeus* Yan C. Li & Zhu L. Yang, The Boletes of China: *Tylopilus* s.l. 170 (2021)

MycoBank: MB 834712

Etymology: The epithet "*cinnamomeus*" refers to the color of the basidiomata.

Type: CHINA, YUNNAN PROVINCE: Changning County, alt. 1800 m, 25 July 2009, Y.C. Li 1824 (KUN-HKAS 59572, GenBank Acc. No.: MT154761 for nrLSU, MT110401 for *rpb1*, MT110435 for *rpb2*).

Basidioma small to medium-sized. Pileus 4–8 cm in diam., subhemispherical to applanate, pale orange (5A2–3) to brownish orange (6C7–8) or brownish red (9C6) to grayish red (9C5); margin slightly extended, viscid to gelatinous when wet, shiny when dry, sometimes rugose; context whitish to grayish white, without discoloration when injured. Hymenophore adnate when young and adnate to slightly depressed around apex of stipe when mature; surface white to pinkish (13A2) or grayish pink (14C2), without discoloration when touched; pores angular, 0.5–2 mm wide; tubes up to 10 mm long, concolorous or a little paler than hymenophoral surface, without discoloration when bruised. Spore print reddish (8A2) to pink (8A3). Stipe clavate to subcylindrical, 5–7 × 0.8–1.5 cm, white but always with cream-yellow to yellow stains at base on injury; surface always covered with concolorous pulverous squamules; basal mycelium white, without discoloration when bruised; context white but with cream-yellow to yellow stains at base when bruised. Taste and odor mild.

Basidia 26–37 × 9.5–12 μm, clavate, hyaline to yellowish in KOH, 4-spored. Basidiospores [120/6/6] (8.5) 9–11.5 (12) × 4.5–5.5 (6) μm [Q = (1.82) 1.89–2.22 (2.33), Q_m = 2.04 ± 0.12], elongated to subcylindrical and inequilateral in side view with slight suprahilar depression, elongated to cylindrical in ventral view, smooth under light microscope, finely pitted under SEM (Fig. 3.3 h), yellowish to brownish yellow in KOH, yellow to yellow-brown in Melzer's reagent. Hymenophoral trama boletoid composed of 3.5–8 μm wide filamentous hyphae, hyaline to yellowish in KOH, yellowish to yellow in Melzer's reagent. Cheilocystidia 33–63 × 5.5–7 μm, clavate to narrowly clavate or ventricose, thin-walled. Pleurocystidia 31–56 × 9–17 μm, fusiform to subfusiform or subfusoid-mucronate to ventricose-mucronate with a short pedicel, thin-walled. Pileipellis an ixotrichoderm composed of 3–11 μm wide gelatinous interwoven hyphae, yellowish to pale brownish in KOH and yellowish brown to brownish in Melzer's reagent; terminal cells 9–48 × 4–8 μm, cystidia-like to narrowly clavate, always with long flagelliform ends 28–75 × 2.5–3.5 μm. Pileal trama composed of 4.5–15 μm wide interwoven hyphae, hyaline to yellowish in KOH and yellowish to brownish in Melzer's reagent.

Fig. 15.1 a *Mucilopilus cinnamomeus* Yan C. Li & Zhu L. Yang. Photo by Y.C. Li (KUN-HKAS 59572, type).

Fig. 15.1 b *Mucilopilus cinnamomeus* Yan C. Li & Zhu L. Yang. Photo by J.W. Liu (KUN-HKAS 97728).

Clamp connections absent in all tissues.

Habitat: Scattered on soil in southern subtropical forests dominated by plants of the family Fagaceae.

Known distribution: Currently known from southeastern and southwestern China.

Additional specimens examined: CHINA, YUNNAN PROVINCE: Jinghong County, Dadugang Town, alt. 1300 m, 7 July 2006, Y.C. Li 475 (KUN-HKAS 50229), the same location, 23 July 2007, Y.C. Li 945 (KUN-HKAS 52632), the same location, 10 July 2014, G. Wu 1304 and 1305 (KUN-HKAS 89041 and KUN-HKAS 89042, respectively); Lancang County, Huimin Town, alt. 1300 m, 26 September 2016, J.W. Liu 530828MF224 (KUN-HKAS 97728). FUJIAN PROVINCE: Zhangping County, Xinqiao Town, Chengkou Village, alt. 360 m, 1 September 2009, N.K. Zeng 646 (KUN-HKAS 107160).

Commentary: *Mucilopilus cinnamomeus* is characterized by the pale orange to brownish orange or brownish red and viscid to gelatinous pileus, the slightly extended pileal margin, the whitish to grayish white pileal context without discoloration when injured, the white to pinkish or grayish pink hymenophore, the white stipe with cream-yellow to yellow stains at base on injury, the finely pitted basidiospores under SEM, the ixotrichoderm pileipellis with terminal cells clavate to narrowly clavate or flagelliform, and the distribution in southern subtropical forests.

Mucilopilus cinnamomeus is the only species with ornamented basidiospores in this genus, however the other characteristics, including the viscid pileus, the white context in the stipe with yellowish to cream-yellow stains when injured, and the ixotrichoderm pileipellis fit well the circumscription of *Mucilopilus*. Moreover, in our multi-locus analysis (Fig. 4.1), *M. cinnamomeus* clusters, with moderate support (MLB = 80%), with the *M. castaneiceps* group, which consists of *M.* cf. *castaneiceps*, *M. paracastaneiceps*, and *M. ruber*.

Fig. 15.1 c *Mucilopilus cinnamomeus* Yan C. Li & Zhu L. Yang. Photos by G. Wu (KUN-HKAS 89041).

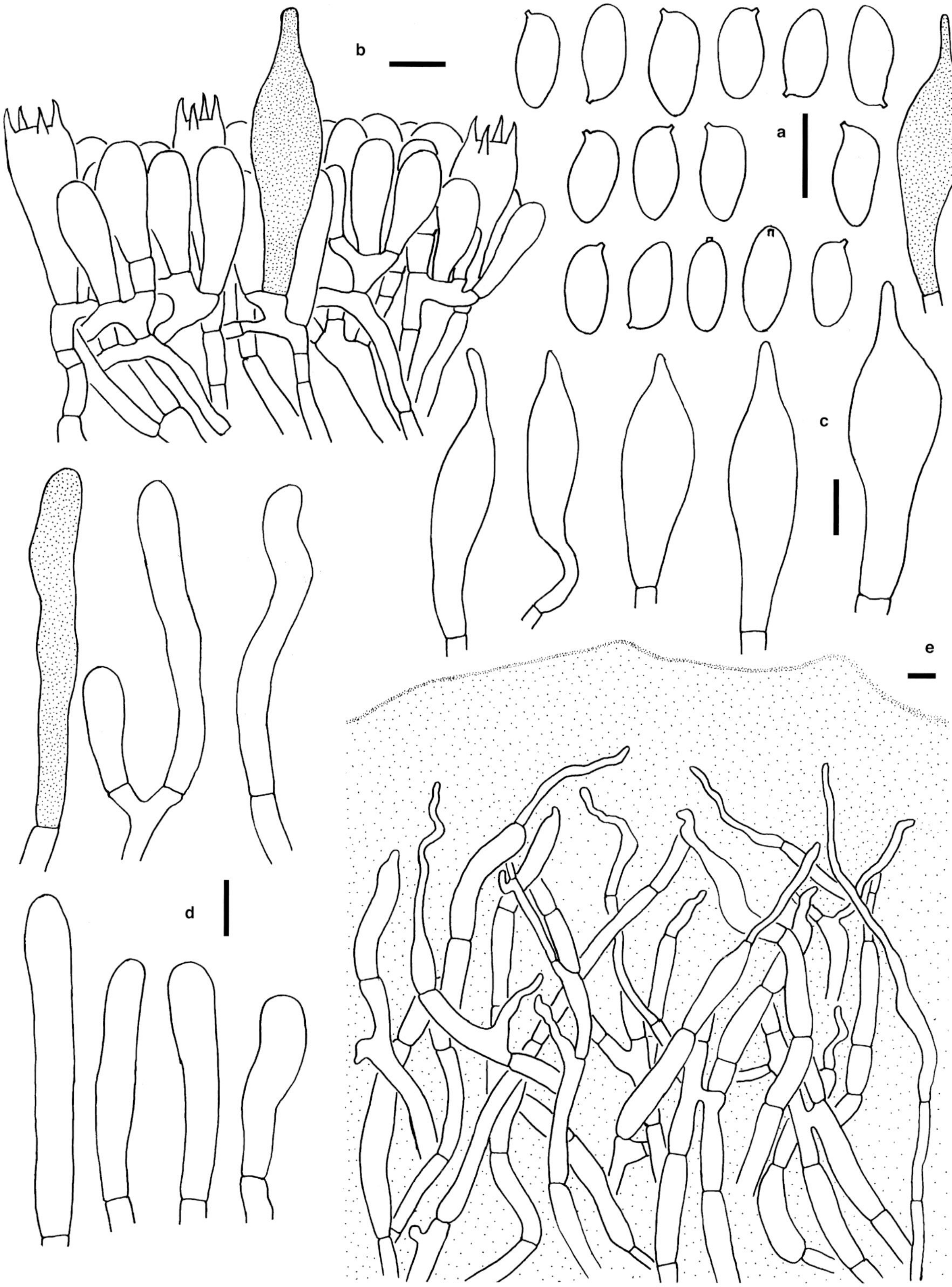

Fig. 15.1 d *Mucilopilus cinnamomeus* Yan C. Li & Zhu L. Yang (KUN-HKAS 59572, type).
a. Basidiospores; b. Basidia and pleurocystidium; c. Pleurocystidia; d. Cheilocystidia; e. Pileipellis. Scale bars = 10 μm.

15.2 *Mucilopilus paracastaneiceps* Yan C. Li & Zhu L. Yang, The Boletes of China: *Tylopilus* s.l. 175 (2021)

MycoBank: MB 834739

Etymology: The epithet "*paracastaneiceps*" refers to its similarity to *M. castaneiceps*.

Type: CHINA, YUNNAN PROVINCE: Jingdong County, Ailao Mountain, alt. 2450 m, 19 July 2006, Y.C. Li 584 (KUN-HKAS 50338, GenBank Acc. No.: KT990555 for nrLSU, KT990755 for *tef1-α*, KT990922 for *rpb1*, KT990391 for *rpb2*).

Basidioma small to medium-sized. Pileus 3–6.5 cm wide, subhemispherical to convexo-applanate, grayish ruby (12D7) to ruby (12D8) when young, brownish red (10D7–8) to grayish red (9C5–6) or pallid red (9A4–5) when mature, margin sometimes slightly extended, viscid when wet, rugose; context whitish, without discoloration when injured. Hymenophore adnate when young, adnate to slightly depressed around apex of stipe; surface white when young and becoming pinkish (13A2) to brownish pink (10D5–6) when mature, without color change when touched; pores angular, 0.3–1 mm wide; tubes 5–8 mm long, concolorous or a little paler than hymenophoral surface, without discoloration when bruised. Spore print pink-white (8A2) to red-white (8A3) (Fig. 3.4 g). Stipe clavate, tapering upwards and enlarged downwards, 3.5–8 × 0.3–0.8 cm, white but always with yellow to cream-yellow stains downwards on injury; surface always covered with concolorous longitudinally arranged reticulum; basal mycelium white without discoloration when injured; context white but cream-yellow to yellowish at stipe base when injured. Taste and odor mild.

Fig. 15.2 a *Mucilopilus paracastaneiceps* Yan C. Li & Zhu L. Yang. Photos by S.P. Jian (KUN-HKAS 107159).

148 The Boletes of China: *Tylopilus* s.l.

Fig. 15.2 b *Mucilopilus paracastaneiceps* Yan C. Li & Zhu L. Yang. Photo by Y.C. Li (KUN-HKAS 50338, type).

Basidia 27–42 × 8–12 μm, clavate to narrowly clavate, 4-spored, sometimes 2-spored, hyaline in KOH. Basidiospores [60/3/3] 12–13.5 × 4.5–5.5 (6.5) μm (Q = 2.14–2.6, Q_m = 2.4 ± 0.15), subfusiform to subcylindrical and inequilateral in side view with slight suprahilar depression, fusiform to cylindrical in ventral view (Fig. 3.3 d), smooth, yellowish to brownish yellow in KOH, yellow to yellow-brown in Melzer's reagent. Hymenophoral trama boletoid composed of 3.5–8 μm wide filamentous hyphae, hyaline to yellowish in KOH, yellowish to yellow in Melzer's reagent. Cheilo- and pleurocystidia 52–66 × 6–8.5 μm, narrowly subfusiform to fusoid-ventricose, yellowish brown to pinkish brown in KOH, thin-walled. Pileipellis an ixotrichoderm composed of 2.5–5 μm wide interwoven hyphae with clavate to subcylindrical terminal cells 18–43 × 2–5 μm, yellowish to pale brownish in KOH and yellowish brown to brownish in Melzer's reagent. Pileal trama composed of 5–11 μm wide interwoven hyphae, hyaline to yellowish in KOH and yellowish to brownish in Melzer's reagent. Clamp connections absent in all tissues.

Habitat: Scattered on soil in subtropical forests dominated by plants of the family Fagaceae.

Known distribution: Currently known from southwestern China.

Specimens examined: CHINA, YUNNAN PROVINCE: Chuxiong, Zixi Mountain, alt. 1500 m, 11 July 2007, Y.C. Li 800 (KUN-HKAS 52487); Tengchong County, Mile Temple, alt. 1980 m, 21 July 2018, S.P. Jian 122 (KUN-HKAS 107159).

Commentary: *Mucilopilus paracastaneiceps* is characterized by the viscid pileus which is grayish ruby to ruby when young and then brownish red to grayish red when mature, the slightly extended pileal margin, the whitish pileal context without discoloration when injured, the white to pinkish or brownish pink hymenophore, the pink-white to red-white spore print, the white stipe with yellow to cream-yellow tinge at base when touched, the ixotrichoderm pileipellis, and the distribution in subtropical forests. In China, this taxon was thought to be *M. castaneiceps* (Hongo) Hid. Takah., originally described from Japan (Li and Yang 2011; Wu *et al.* 2016a). However, *M. castaneiceps* is characterized by the brown (7D6–7, 7E6–7E8) pileus, the smaller and clavate pleurocystidia 25–40 × 7.5–12 μm and the clavate cheilocystidia 40–80 (100) × 6.5–10 μm with 1–2 septa. A single ITS sequence labeled as "*Mucilopilus castaneiceps*" (strain: Tsukuba227, GenBank Acc. No. AB289669) from Japan was released in GenBank. Following blasting the ITS sequences from the Chinese samples against the Japanese strain (Tsukuba227), the percentage identity was found to be only 95%. Thus, the Chinese samples represent a new species and are described as such here. For detailed descriptions and line drawings of *Mucilopilus castaneiceps* and *M. paracastaneiceps* see Li and Yang (2011) and Wu *et al.* (2016a). Phylogenetically, *M. paracastaneiceps* clusters together with a sample (HKAS 71039) we collected from Japan but without statistical support. Moreover, the sample collected from Japan is morphologically different from *M. castaneiceps*. As no sequences were successfully generated from the type specimen, further collections from the type locality are needed to understand the phylogeny and taxonomy of *M. castaneiceps*.

15.3 *Mucilopilus ruber* Yan C. Li & Zhu L. Yang, The Boletes of China: *Tylopilus* s.l. 177 (2021)

MycoBank: MB 834744

Etymology: The epithet "*ruber*" refers to the color of the pileus.

Type: CHINA, YUNNAN PROVINCE: Longling County, Daxue Mountain, alt. 2000 m, 14 June 2014, L.H. Han 259 (KUN-HKAS 84555, GenBank Acc. No.: MT154762 for nrLSU, MT110364 for *tef1-α*, MT110436 for *rpb2*).

Basidioma small to medium-sized. Pileus 4–6 cm in diam., hemispherical to convexo-applanate, yellowish red (8B7–8) to reddish orange (7B7–8) or pale reddish orange (7A6–7) in the center, pinkish or reddish white (8A2) towards margin, margin slightly extended; surface glabrous, viscid to gelatinous when wet, shiny when dry; context whitish to cream, without discoloration when injured. Hymenophore adnate when young and adnate to slightly depressed around apex of stipe when mature; surface white to pinkish (8A2), without discoloration when touched; pores angular, 0.5–1 mm wide; tubes up to 10 mm long, concolorous or a little paler than hymenophoral surface, without discoloration when bruised. Spore print reddish (8A2) to pink (8A3). Stipe subcylindrical to clavate, tapering upwards and enlarged downwards, 7–10 × 0.7–1.2 cm, white but always staining yellow on injury at base; surface always covered with concolorous pulverous squamules; basal mycelium white, without discoloration when bruised; context white but staining cream-yellow to yellowish at stipe base when bruised. Taste and odor mild.

Basidia 33–45 × 11–14 μm, clavate to narrowly clavate, 4-spored, sometimes 2-spored, hyaline in KOH.

Fig. 15.3 a *Mucilopilus ruber* Yan C. Li & Zhu L. Yang. Photos by L.H. Han (KUN-HKAS 84555, type).

Chapter 15 *Mucilopilus* Wolfe   151

Fig. 15.3 b *Mucilopilus ruber* Yan C. Li & Zhu L. Yang. Photos by L.H. Han (KUN-HKAS 84632).

Basidiospores [60/3/3] 10–13 × 4–5.5 (6) μm (Q = 2–2.88, Q_m = 2.22 ± 0.12), elongated to subcylindrical and inequilateral in side view with slight suprahilar depression, elongated to cylindrical in ventral view, smooth, yellowish to brownish yellow in KOH, yellow to yellow-brown in Melzer's reagent. Hymenophoral trama boletoid composed of 3–9 μm wide filamentous hyphae, hyaline to yellowish in KOH, yellowish to yellow in Melzer's reagent. Cheilocystidia not abundant, 25–37 × 12–14 μm, narrowly subfusiform to fusoid-ventricose. Pleurocystidia rare, 31–64 × 8–12 μm, narrowly subfusiform to fusoid-ventricose, hyaline to yellowish in KOH, yellow to yellowish brown in Melzer's reagent, thin-walled. Pileipellis an ixotrichoderm composed of 3–5 μm wide interwoven hyphae with subfusiform to subcylindrical terminal cells 12–25 × 4–5 μm, yellowish to pale brownish in KOH and yellowish brown to brownish in Melzer's reagent. Pileal trama composed of 2–20 μm wide interwoven hyphae, hyaline to yellowish in KOH, yellowish brown to brownish in Melzer's reagent. Clamp connections absent in all tissues.

Habitat: Scattered on soil in subtropical forests dominated by plants of the family Fagaceae.

Known distribution: Currently known from southwestern China.

Additional specimens examined: CHINA, YUNNAN PROVINCE: Longling County, Gaoligongshan National Forest Park, alt. 2200 m, 18 June 2014, X.B. Liu 340 (KUN-HKAS 86992), the same location, 19 June 2014, L.H. Han 336 (KUN-HKAS 84632).

Commentary: *Mucilopilus ruber* is characterized by the yellowish red to reddish orange or pale reddish orange to reddish white and gelatinous pileus, the slightly extended pileal margin, the whitish to cream context without discoloration when injured, the white to pinkish hymenophore, the white stipe with cream-yellow to yellowish tinge at base when injured, the ixotrichoderm pileipellis, and the distribution in subtropical forests.

Mucilopilus ruber is phylogenetically related and morphologically similar to *M. paracastaneiceps*. However, *M. ruber* differs from *M. paracastaneiceps* in its yellowish red to reddish orange or pale reddish orange to reddish white pileus, short and broad cheilocystidia measuring 25–37 × 12–14 μm, broad pleurocystidia measuring 31–64 × 8–12 μm.

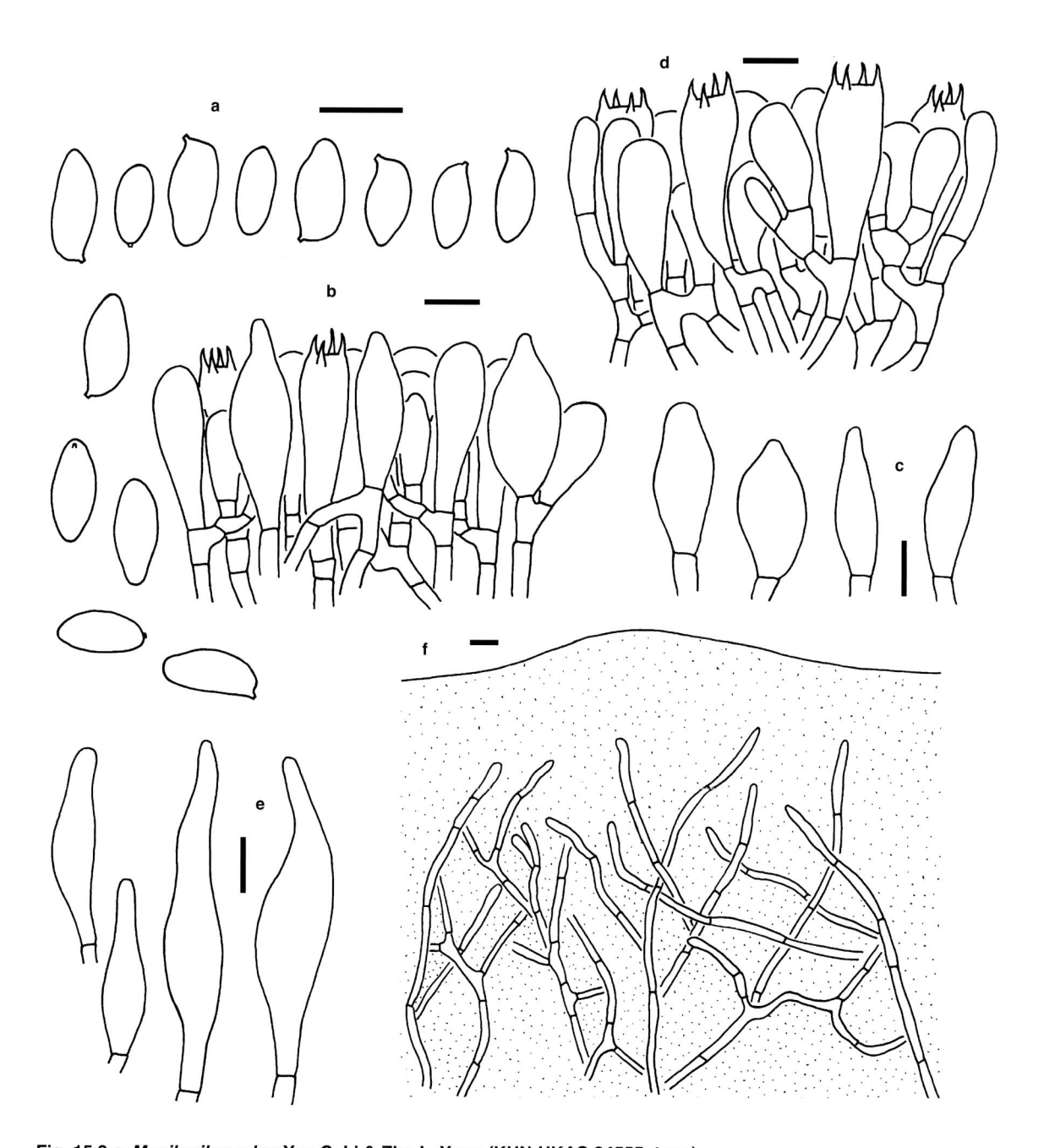

Fig. 15.3 c *Mucilopilus ruber* Yan C. Li & Zhu L. Yang (KUN-HKAS 84555, type).
a. Basidiospores; b. Basidia and cheilocystidia; c. Cheilocystidia; d. Basidia; e. Pleurocystidia; f. Pileipellis. Scale bars = 10 μm.

Chapter 16

Porphyrellus E.-J. Gilbert

Porphyrellus E.–J. Gilbert, Les Livres du Mycologue Tome I–IV, Tom. III: Les Bolets. 75: 99 (1931)

Type species: *Boletus porphyrosporus* Fr. & Hök, Boleti, Fungorum Generis, Illustration. 13 (1835).

Diagnosis: The genus is different from other genera in the Boletaceae in its umber basidiomata, pallid to white or grayish context without discoloration or becoming asymmetrically blue or at first bluish then reddish when injured, white to pinkish or grayish pink hymenophore without discoloration or staining blue or at first staining asymmetrically blue then reddish when injured, palisadoderm or epithelium pileipellis, and smooth basidiospores.

Pileus hemispherical to subhemispherical or applanate; surface, dry, nearly glabrous, or finely tomentose to fibrillose, or scaly, sometimes rimose; context white, without discoloration or becoming asymmetrically blue or at first asymmetrically blue then rufescent when injured. Hymenophore adnate when young, adnate to depressed around apex of stipe when mature; surface white to grayish white when young, grayish pink to blackish pink when mature, without color change or staining blue or at first bluish then reddish when injured; pores subangular to roundish; tubes concolorous or a little paler than hymenophoral surface, without color change or staining blue or initially bluish and then reddish when injured. Spore print orange-red, brownish red, brown or brownish. Stipe always concolorous with pileal surface; surface glabrous or tomentose; basal mycelium white, without discoloration when injured. Basidiospores smooth, subfusiform to oblong. Hymenial cystidia abundant, subfusiform to subfusoid-ventricose. Pileipellis an epithelium composed of moniliform cells or a palisadoderm composed of vertically arranged broad hyphae. Clamp connections absent in all tissues.

Commentary: *Porphyrellus* was proposed as a genus by Gilbert (1931) based on its smooth red-brown to purple-brown basidiospores. The genus was once treated as a subgenus of *Tylopilus* (Smith and Thiers 1971; Wolfe 1979a). However, *Porphyrellus* is easily distinguished from *Tylopilus* by its umber basidioma, red-brown to purple-brown basidiospores, and mild taste of its context staining blue or initially bluish and then reddish brown when injured. Currently, seven species are known from China, including three new species and one new record to China.

Key to the species of the genus *Porphyrellus* in China

1. Pileus dark brown to brown; pileipellis a palisadoderm composed of broad vertically arranged hyphae ·· 2
1. Pileus lighter in color, pileipellis a palisadoderm composed of broad vertically arranged hyphae; or pileus with dark brown o brown tinge, but pileipellis an epithelium composed of inflated concatenated cells ·· 3
2. Pileus large, up to 10 cm wide; pileal surface nearly glabrous or tomentose; hymenophoral surface blackish to blackish pink when young; context white, asymmetrically bluish then reddish when injured; basidiospores 12–18 × 5.5–6.5 μm ·· *P. porphyrosporus*
2. Pileus small, up to 3 cm wide; pileal surface scrobiculate; hymenophoral surface whitish when young; context white, without discoloration when injured; basidiospores 8.5–11 × 5–6 μm ········· ··· *P. scrobiculatus*
3. Species with a distribution in temperate forests; basidiospores 10.5–15.5 × 4–6.5 μm; pileipellis a palisadoderm ··· *P. cyaneotinctus*
3. Species with a distribution in subtropical or tropical forests; basidiospores relatively short and less than 12 μm long; pileipellis a palisadoderm, or an epithelium ·· 4
4. Surface of pileus and stipe dark in color but without grayish brown to pale brown or orange-brown tinge; pileipellis an epithelium ·· 5
4. Surface of pileus and stipe grayish brown to pale brown or orange-brown; pileipellis a palisadoderm ··· *P. griseus*
5. Pileus without distinct reddish brown to chestnut-brown tinge; basidiospores 4.5–5.5 μm wide; pleurocystidia morphologically similar to cheilocystidia and more than 14 μm wide ·················· 6
5. Pileus with distinct reddish brown to chestnut-brown tinge; basidiospores relatively slender 3.5–5 μm wide; cheilocystidia 10–18 μm wide; pleurocystidia relatively slender 6–9 μm wide ··· ·· *P. castaneus*
6. Pileus brown to reddish brown or brownish; apex of stipe usually with a bluish ring ···················· ··· *P. orientifumosipes*
6. Pileus black-brown to coffee-brown and then dark brown to brown; apex of stipe without bluish ring ··· *P. pseudofumosipes*

16.1 *Porphyrellus castaneus* Yan C. Li & Zhu L. Yang, Fungal Divers 81: 108 (2016)

Basidioma small to medium-sized. Pileus 3–6 cm wide, subhemispherical to convex or subconvex, initially reddish brown or violet-brown and then brownish red to chestnut-brown or pale brownish, slightly darker in the center; surface dry, tomentose, sometimes minutely cracked with age; context white to pallid, asymmetrically bluish when injured. Hymenophore adnate when young, adnate to depressed around apex of stipe; surface grayish white when young and grayish pink to brownish pink when mature, staining blue when touched; pores angular to roundish, up to 2 mm wide; tubes up to 15 mm long, concolorous or a little paler than hymenophoral surface, becoming bluish when injured. Spore print brown to brownish (Fig. 3.4 h). Stipe clavate to cylindrical, 4–7 × 0.6–1.2 cm, concolorous with pileal surface, fibrillose; context grayish white to white or pallid, without color change or sometimes asymmetrically bluish at apex when bruised; basal mycelium whitish to pallid, without discoloration when bruised. Taste and odor mild.

Basidia 25–38 × 9–13 µm, clavate, 4-spored, hyaline in KOH. Basidiospores 8–12 × 3.5–5 µm [Q = (1.78) 2–2.63 (2.86), Q_m = 2.3 ± 0.21], elongated to subfusiform and inequilateral in side view, elongated to fusiform in ventral view, smooth, yellowish to brownish yellow in KOH, yellow to yellow-brown in

Fig. 16.1 *Porphyrellus castaneus* Yan C. Li & Zhu L. Yang. Photos by Y.C. Li (KUN-HKAS 52554, type).

Melzer's reagent. Hymenophoral trama boletoid composed of 3.5–10 μm wide filamentous hyphae, hyaline to yellowish in KOH, yellowish to yellow in Melzer's reagent. Cheilocystidia abundant, 36–68 × 10–18 μm, clavate to subfusiform, yellowish to yellow-brown in KOH, thin-walled. Pleurocystidia abundant, 35–56 × 6–9 μm, similar to cheilocystidia, yellowish to yellow-brown in KOH, thin-walled. Pileipellis an epithelium composed of inflated concatenated cells 12–21 μm wide, yellowish to pale brownish in KOH and yellowish brown to brownish in Melzer's reagent; terminal cells 33–72 × 10–19 μm, pyriform to subfusiform. Pileal trama composed of 4.5–12 μm wide interwoven hyphae, hyaline to yellowish in KOH and yellowish to brownish in Melzer's reagent. Clamp connections absent in all tissues.

Habitat: Scattered on soil in southern subtropical forests dominated by plants of the family Fagaceae.

Known distribution: Currently known from southern and southwestern China.

Specimens examined: CHINA, YUNNAN PROVINCE: Jingdong County, Ailao Mountain, alt. 2400 m, 17 July 2007, Y.C. Li 869 (KUN-HKAS 52554, type); Tengchong County, Gaoligong Mountain, alt. 2100 m, 11 August 2010, L.P. Tang 1256 (HKAS 63076); Simao County, Caiyanghe National Forest Park, alt. 1300 m, 11 July 2014, L.H. Han 444 (KUN-HKAS 84740). GUANGDONG PROVINCE: Fengkai County, Heishiding Nature Reserve, alt. 250 m, 3 July 2012, F. Li 555 (KUN-HKAS 106345).

Commentary: *Porphyrellus castaneus* was originally described based on materials from southwestern China (Wu *et al.* 2016a). This species is characterized by the initially reddish brown or violet-brown and then brownish red to chestnut-brown or pale brownish pileus, the grayish white to grayish pink or brownish pink hymenophore, the white to pallid context without color change or asymmetrically bluish when bruised, the epithelium pileipellis composed of subglobose to globose concatenated cells, and the distribution in southern subtropical forests.

Porphyrellus castaneus is phylogenetically related and morphologically similar to *P. cyaneotinctus* (A.H. Sm. & Thiers) Singer, *P. griseus* Yan C. Li & Zhu L. Yang, *P. orientifumosipes* Yan C. Li & Zhu L. Yang and *P. pseudofumosipes* Yan C. Li & Zhu L. Yang. However, *P. cyaneotinctus* and *P. griseus* both have brown to grayish brown or gray to smoky-gray pileus and palisadoderm pileipellis composed of broad vertically arranged hyphae. *Porphyrellus orientifumosipes* differs from *P. castaneus* in its bluish ring at apex of stipe and big pleurocystidia measuring 58–74 × 15–19 μm (Wu *et al.* 2016a). *Porphyrellus pseudofumosipes* differs from *P. castaneus* in its black-brown to coffee-brown and then dark brown to brown pileus, relatively wide pleurocystidia which are more than 14 μm wide, epithelium pileipellis composed of relatively narrow concatenated cells which are 8–16 μm wide.

16.2 *Porphyrellus cyaneotinctus* (A.H. Sm. & Thiers) Singer, Beih Nova Hedwigia 102: 64 (1991)

Basionym: *Tylopilus cyaneotinctus* A.H. Sm. & Thiers, Mycologia 60(4): 952 (1968).

Basidioma small to medium-sized. Pileus 4–8 cm wide, subhemispherical to convex or convexo-applanate, reddish brown to gray-brown or smoky-gray, slightly darker in the center; surface dry, always cracked into small squamules on a grayish to whitish background; context grayish white to white or pallid, asymmetrically bluish when cut. Hymenophore adnate when young, depressed around apex of stipe when mature; surface pinkish to brownish pink, staining blue when injured; pores subangular to roundish, up to 3 mm wide; tubes up to 18 mm long, concolorous or much paler than hymenophoral surface, staining blue when injured. Spore print orange-red to brownish red. Stipe clavate, 4–7 × 0.8–1.5 cm, concolorous with pileal surface; context grayish white to white or pallid, asymmetrically bluish at apex when hurt; basal mycelium whitish to grayish white, without discoloration when bruised. Taste and odor mild.

Basidia 28–40 × 9–15 μm, clavate, 4-spored, hyaline in KOH. Basidiospores 10.5–12.5 (15.5) × 4–5.5 (6.5) μm [Q = 2.09–2.5 (2.75), Q_m = 2.36 ± 0.15], subfusiform to subcylindrical in side view, fusiform to cylindrical in ventral view, smooth, yellowish to brownish yellow in KOH, yellow to yellow-brown in Melzer's reagent. Hymenophoral trama boletoid composed of 4–8 μm wide filamentous hyphae, hyaline to yellowish in KOH, yellowish to yellow in Melzer's reagent. Cheilo- and pleurocystidia 42–82 × 11–20 μm, abundant, fusiform to subfusiform or subfusoid-ventricose, hyaline to yellowish in KOH, thin-walled. Pileipellis a palisadoderm composed of 6–17 μm wide vertically arranged hyphae, yellowish brown to brownish in KOH and yellow-brown to dark brown in Melzer's reagent; terminal cells 28–66 × 9–14 μm, subfusiform to cylindrical or sometimes subfusoid-ventricose. Pileal trama composed of broad (up to 21 μm wide) interwoven hyphae, yellowish to pale brownish in KOH and yellowish brown to brownish in Melzer's reagent. Clamp connections absent in all tissues.

Habitat: Scattered or solitary on soil in temperate mixed forests dominated by plants of the families Fagaceae and Pinaceae.

Fig. 16.2 a *Porphyrellus cyaneotinctus* (A.H. Sm. & Thiers) Singer. Photos by Y.J. Hao (KUN-HKAS 80183).

158 The Boletes of China: *Tylopilus* s.l.

Known distribution: Currently known from central China and North America.

Additional specimens examined: CHINA, HENAN PROVINCE: Shangcheng County, Huangbai Mountain, alt. 200 m, 29 June 2013, Y.J. Hao 903 (KUN-HKAS 80183). HUBEI PROVINCE: Shiyan, Yingtaogou Village, alt. 200 m, 1 July 2013, Y.J. Hao 912 (KUN-HKAS 80192).

Commentary: *Porphyrellus cyaneotinctus* was originally described from North America as *Tylopilus cyaneotinctus* Smith & Thiers (Smith and Thiers 1968). This species is characterized by the reddish brown to gray-brown or smoky-gray pileus which is always cracked into small squamules on a grayish to whitish background, the white to pallid context which is asymmetrically bluish when injured, the pinkish to brownish pink hymenophore staining bluish when injured, the palisadoderm pileipellis, and the distribution in temperate forests. Singer *et al*. (1991) transferred it to *Porphyrellus* based on the color of its basidiomata and spore print. Our molecular phylogenetic analysis (Fig. 4.1) indicates that it indeed nests into the clade *Porphyrellus* and is closely related to *P. griseus*. However, *P. griseus* differs from *P. cyaneotinctus* in its small basidiospores measuring 9.5–11.5 × 4.5–5 μm, small hymenial cystidia measuring 34–58 × 8–12 μm.

Fig. 16.2 b *Porphyrellus cyaneotinctus* (A.H. Sm. & Thiers) Singer. Photos by Y.J. Hao (KUN-HKAS 80192).

Chapter 16 *Porphyrellus* E.-J. Gilbert 159

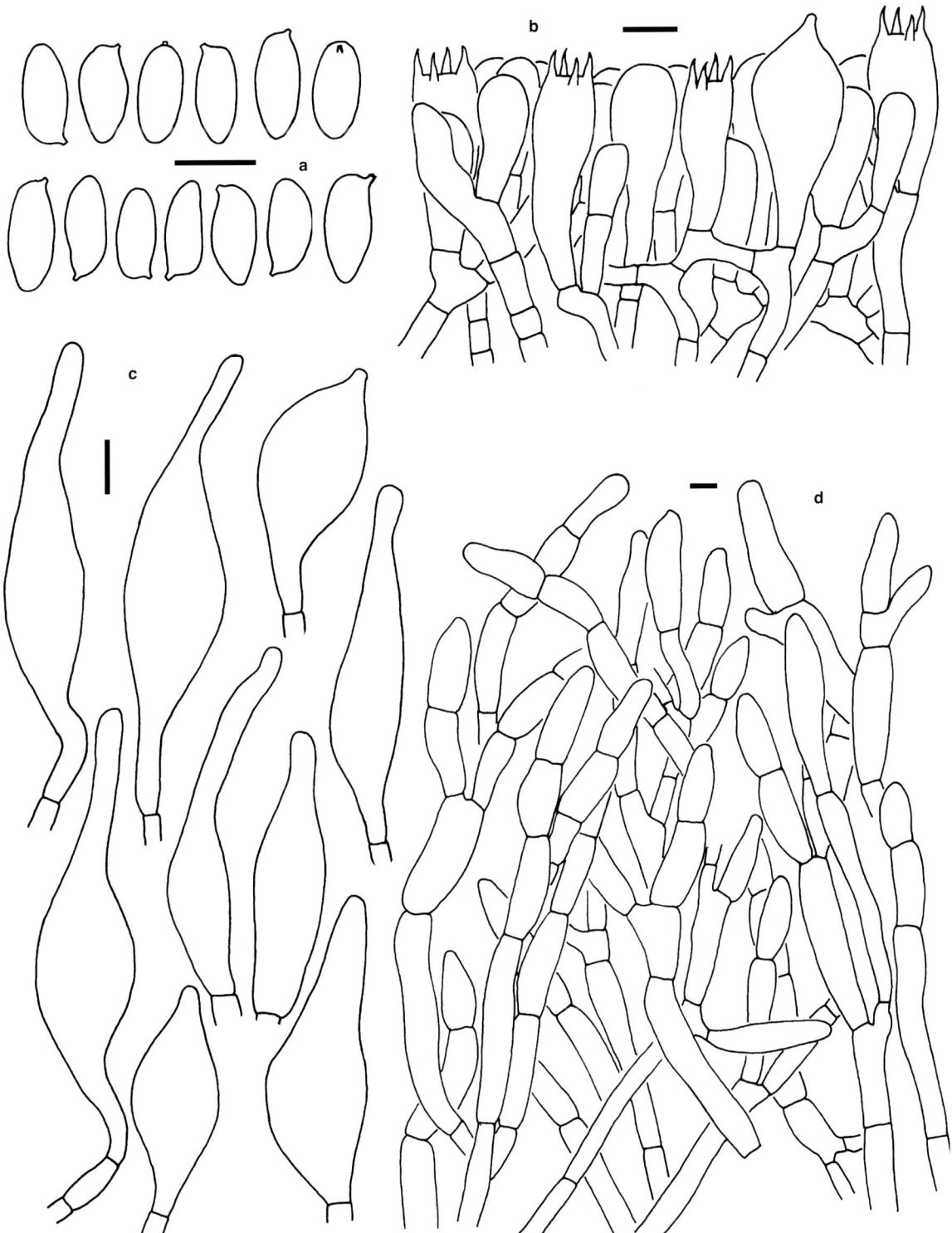

Fig. 16.2 c *Porphyrellus cyaneotinctus* (A.H. Sm. & Thiers) Singer (KUN-HKAS 80183).
a. Basidiospores; b. Basidia and cheilocystidium; c. Cheilo- and pleurocystidia; d. Pileipellis. Scale bars = 10 μm.

16.3 *Porphyrellus griseus* Yan C. Li & Zhu L. Yang, The Boletes of China: *Tylopilus* s.l. 188 (2021)

MycoBank: MB 834713

Etymology: The epithet "*griseus*" refers to the gray basidiomata.

Type: CHINA, CHONGQING MUNICIPALITY: Jiangjin, Bailin Town, Baiguo Village, alt. 500 m, 15 June 2014, Y.J. Hao 1058 (KUN-HKAS 82849, GenBank Acc. No.: MT154716 for nrLSU, MT110414 for *rpb2*).

Basidioma medium-sized. Pileus 5–7.5 cm wide, subhemispherical when young, subconvex to applanate, margin always incurved, grayish brown (6E3–4) to pale brown (6D4–5) or orange-brown (5C3–4); surface dry, rimose; context grayish white to pallid, asymmetrically bluish when injured. Hymenophore adnate to slightly depressed around apex of stipe; surface pinkish (13A2) to brownish pink (10D5–6), becoming bluish when injured; pores angular to roundish, up to 1 mm wide; tubes up to 20 mm long, concolorous or a little paler than hymenophoral surface, becoming bluish when injured. Spore print brown (7D6) to brownish red (8D5–6). Stipe cylindrical, 4–6.5 × 2–3.5 cm, concolorous with pileal surface; context grayish white to pallid, always brownish downwards, asymmetrically bluish at apex when injured; basal mycelium grayish white. Taste and odor mild.

Basidia 24–50 × 7–11 μm, clavate to narrowly clavate, 4-spored, hyaline to yellowish in KOH. Basidiospores [40/2/1] 9.5–11.5 × 4.5–5 μm [Q = 2–2.33, Q_m = 2.18 ± 0.1], elongated to subfusiform in side view, elongated to fusiform in ventral view, smooth, yellowish to brownish olivaceous in KOH, brown to yellow-brown or olivaceous in Melzer's reagent. Hymenophoral trama boletoid composed of 4.5–8 μm wide filamentous hyphae, hyaline to yellowish in KOH, yellowish to yellow in Melzer's reagent. Cheilo- and pleurocystidia 34–58 × 8–12 μm, abundant, fusiform to subfusiform or ventricose, usually hyaline to yellowish in KOH, thin-walled. Pileipellis a palisadoderm composed of 6–18 μm wide vertically arranged hyphae, yellowish to pale brownish in KOH and yellowish brown to brownish in Melzer's reagent; terminal cells 21–45 × 9–18 μm, broadly clavate to cystidioid or pyriform. Pileal trama composed of 7–12 μm wide interwoven hyphae, yellowish to pale brownish in KOH and yellow to brownish in Melzer's reagent. Clamp connections absent in all tissues.

Habitat: Scattered on soil in subtropical forests dominated by plants of the family Fagaceae.

Known distribution: Currently known from central China.

Commentary: *Porphyrellus griseus* is characterized by the grayish brown to pale brown or orange-brown pileus, the grayish white to pallid context, the pinkish to brownish pink hymenophore, the bluish color change when injured, the palisadoderm pileipellis, and the distribution in subtropical forests. Morphologically, it is similar to *P. fumosipes* (Peck) Snell. and *P. orientifumosipes* Yan C. Li & Zhu L. Yang. However, *P. fumosipes* has an olive-brown pileus, a yellowish brown hymenophore when mature, big basidiospores 12–14 (18) × 4–5 (7) μm, and an epithelium pileipellis (Peck 1898; Smith and Thiers 1971; Wolfe and Petersen 1978). *Porphyrellus orientifumosipes* differs from *P. griseus* in its initially reddish brown to purplish olivaceous and then smoky-brown to brown pileus, bluish ring at apex of stipe, and an epithelium pileipellis composed of 16–21 μm wide concatenated cells.

In our molecular phylogenetic analysis (Fig. 4.1), *P. griseus* is closely related to *P. cyaneotinctus*, but *P. cyaneotinctus* differs from *P. griseus* in its reddish brown to gray-brown or smoky-gray pileus, relatively large basidiospores measuring 10.5–15.5 × 4–6.5 μm, relatively large hymenial cystidia measuring 42–82 × 11–20 μm, and distribution in temperate forests.

Fig. 16.3 a *Porphyrellus griseus* Yan C. Li & Zhu L. Yang. Photos by Y.J. Hao (KUN-HKAS 82849, type).

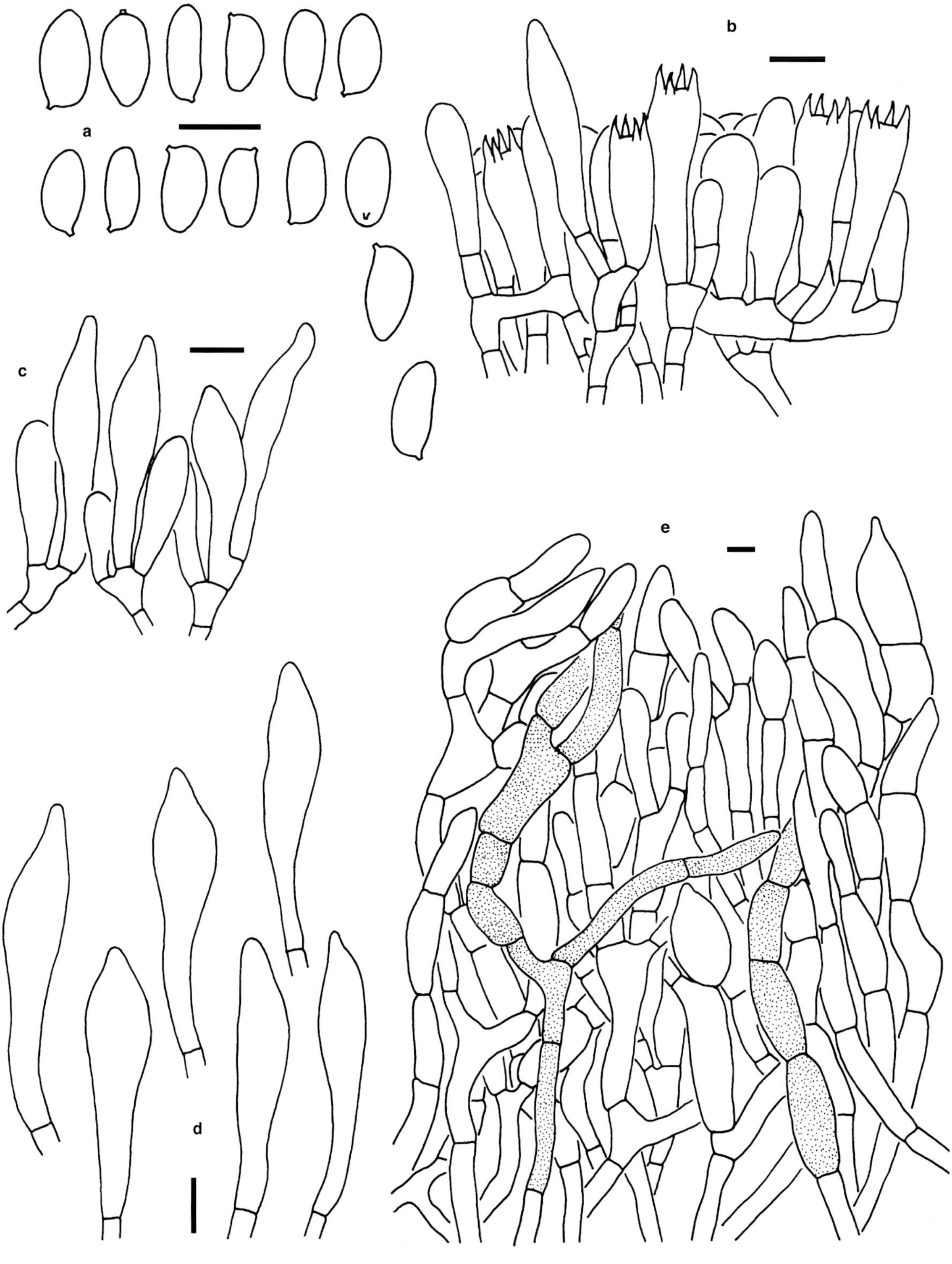

162 The Boletes of China: *Tylopilus* s.l.

Fig. 16.3 b *Porphyrellus griseus* Yan C. Li & Zhu L. Yang (KUN-HKAS 82849, type).
a. Basidiospores; b. Basidia and cheilocystidium; c. Cheilocystidia; d. Pleurocystidia; e. Pileipellis. Scale bars = 10 μm.

16.4 *Porphyrellus orientifumosipes* Yan C. Li & Zhu L. Yang, Fungal Divers 81: 109 (2016)

Basidioma very small to medium-sized. Pileus 2–8 cm wide, hemispherical when young, subhemispherical to applanate or plano-convex when mature, brown to reddish brown or brownish, slightly darker in the center; surface dry, always cracked into large brown to reddish brown or brownish squamules on a whitish to cream background; context grayish white to pallid, always asymmetrically blue when injured. Hymenophore adnate when young, depressed around apex of stipe when mature; surface initially grayish white and then grayish pink to brownish pink when mature, staining bluish when touched; pores subangular to roundish, up to 2 mm wide; tubes up to 20 mm long, concolorous or much paler than hymenophoral surface, staining blue when injured. Spore print brown to brownish red. Stipe cylindrical to clavate, 2–8 × 0.2–1.5 cm, concolorous with pileal surface, always with a bluish ring at apex when mature or aged; context white to pallid, asymmetrically bluish at apex when hurt, white to brownish downwards; basal mycelium grayish white. Taste and odor mild.

Basidia 28–42 × 8–12 μm, clavate, 4-spored, hyaline in KOH. Basidiospores 9–11 × 4.5–5.5 μm [Q = (1.8) 1.9–2.33 (2.44), Q_m = 2.11 ± 0.18], elongated to subfusiform in side view, elongated to fusiform

Fig. 16.4 *Porphyrellus orientifumosipes* Yan C. Li & Zhu L. Yang. Photos by Y.C. Li (KUN-HKAS 53372, type).

in ventral view, smooth, yellowish to brownish yellow in KOH, yellow to yellow-brown in Melzer's reagent. Hymenophoral trama boletoid composed of 4.5–10 μm wide filamentous hyphae, hyaline to yellowish in KOH, yellowish to yellow in Melzer's reagent. Cheilo- and pleurocystidia 55–70 × 14–19 μm, abundant, subfusiform to fusiform, usually hyaline to yellowish in KOH, thin-walled. Pileipellis an epithelium composed of 12–21 μm wide concatenated cells, terminal cells 22–57 × 16–21 μm, pyriform to subfusiform, yellowish to pale brownish in KOH and yellow to brownish in Melzer's reagent. Pileal trama composed of 5–15 μm wide interwoven hyphae, yellowish to pale brownish in KOH and yellowish brown to brownish in Melzer's reagent. Clamp connections absent in all tissues.

Habitat: Scattered on soil in subtropical to tropical forests dominated by plants of the family Fagaceae.

Known distribution: Currently known from southeastern and southwestern China.

Specimens examined: CHINA, ZHEJIANG PROVINCE: Lishui, Baiyunshan Forest Park, alt. 200 m, 23 August 2015, X. M. Jiang 69 (RITF 2697). FUJIAN PROVINCE: Sanming, Sanyuan National Forest Park, alt. 260 m, 26 August 2007, Y.C. Li 1027 (KUN-HKAS 53372, type) and Y.C. Li 1035 (KUN-HKAS 53380). YUNNAN PROVINCE: Mengla County, alt. 1200 m, 9 July 2014, L.H. Han 414 (KUN-HKAS 84710); Malipo County, Babu Town, Nabo Village, alt. 660 m, 20 June 2017, G. Wu 1929 (KUN-HKAS 107182).

Commentary: *Porphyrellus orientifumosipes* described from China by Wu *et al.* (2016a) is characterized by the dark brown to brown or reddish brown pileus, the grayish white to pallid context, the initially grayish white to grayish pinkish and then brownish pink hymenophore, the asymmetrically bluish color change when injured, the bluish ring at apex of stipe when mature or aged, the epithelium pileipellis composed of inflated concatenated cells, and the subtropical to tropical distribution. *Porphyrellus orientifumosipes* is phylogenetically related and morphologically similar to *P. cyaneotinctus*, *P. griseus* and *P. pseudofumosipes*. However, these species are varied in the color of basidiomata, the structure of pileipellis, the size of basidiospores and pleurocystidia (see our commentary under *P. castaneus* above).

16.5 *Porphyrellus porphyrosporus* (Fr. & Hök) E.–J. Gilbert, Les Livres du Mycologue Tome I–IV, Tom. III: Les Bolets: 99 (1931)

Basionym: *Boletus porphyrosporus* Fr., Boleti, Fungorum generis, illustration 13 (1835).

Synonyms: *Porphyrellus porphyrosporus* (Fr. & Hök) E.-J. Gilbert, Les Livres du Mycologue Tome I-IV, Tom. III: Les Bolets 99 (1931); *Tylopilus porphyrosporus* (Fr.) A.H. Sm. & Thiers, The Boletes of Michigan 98 (1971).

Basidioma medium-sized to large. Pileus 5–10 cm wide, hemispherical to subhemispherical, dark brown to brown; surface tomentose, dry; context white to pallid, bluish at first and then reddish brown when hurt. Hymenophore adnate when young, adnate to depressed around apex of stipe when mature; surface blackish pink to grayish pink, staining blue first and then reddish when touched; pores angular, 0.3–1 mm wide; tubes up to 20 mm long; concolorous with hymenophoral surface. Spore print brown to brownish red. Stipe clavate to cylindrical, 4–10 × 1–2.5 cm, concolorous with pileal surface; surface almost glabrous to fibrillose; context grayish white to pallid, initially staining blue and then reddish brown when hurt; basal mycelium grayish white, without discoloration when injured. Taste and odor mild.

Basidia 28–50 × 10–14 μm, clavate, 4-spored, hyaline to yellowish in KOH. Basidiospores 12–18 × 5.5–6.5 μm [Q = (1.8) 2–2.75 (3.1), Q_m = 2.35 ± 0.26], elongated to subcylindrical in side view with slight suprahilar depression, elongated to cylindrical in ventral view, smooth, yellowish to brownish yellow in KOH, yellow to yellow-brown in Melzer's reagent. Hymenophoral trama boletoid; hyphae cylindrical

Fig. 16.5 a *Porphyrellus porphyrosporus* (Fr. & Hök) E.-J. Gilbert. Photos by X.T. Zhu (KUN-HKAS 76537).

Fig. 16.5 b *Porphyrellus porphyrosporus* (Fr. & Hök) E.-J. Gilbert. Photos by J.W. Liu (KUN-HKAS 91013).

4–10 μm wide filamentous hyphae, hyaline to yellowish in KOH, yellowish to yellow in Melzer's reagent. Cheilo- and pleurocystidia 36–56 × 10–14.5 μm, subfusiform to fusiform or ventricose, brownish yellow to pale brownish in KOH, brown to yellow-brown in Melzer's reagent, thin-walled. Pileipellis a trichoderm composed of 8–14 μm wide filamentous hyphae, yellowish to pale brownish in KOH and yellow to brownish in Melzer's reagent; terminal cells 35–80 × 8–12 μm, clavate to cylindrical. Pileal trama composed of 5–11 μm wide interwoven hyphae, yellowish to pale brownish in KOH and yellowish brown to brownish in Melzer's reagent. Clamp connections absent in all tissues.

Habitat: Scattered on soil in temperate to subalpine forests dominated by plants of the family Pinaceae.

Known distribution: Currently known from Europe, East Asia, and North America.

Specimens examined: CHINA, GANSU PROVINCE: Diebu County, Hutou Mountain, alt. 3000 m, 12 August 2012, X.T. Zhu 688 (KUN-HKAS 76537). HUBEI PROVINCE: Xingshan County, Shennongjia National Nature Reserve, alt. 2500 m, 17 July 2012, J. Qin 571 (KUN-HKAS 77972). SICHUAN PROVINCE: Shiqu County, Luoxu Town, alt. 4000 m, 30 July 2005, Z.W. Ge 687 (KUN-HKAS 49182); Kangding County, Gongga Mountain, alt. 3800 m, 4 September 2016, J.W. Liu KD-10 (KUN-HKAS 97803). XIZANG AUTONOMOUS REGION: Bomi County, alt. 3800 m, 3 August 2014, J.W. Liu 205 (KUN-HKAS 91013).

Commentary: *Porphyrellus porphyrosporus*, the type species of *Porphyrellus*, was originally described from Sweden (Fries and Hök 1835) and is characterized by the dark brown to brown basidiomata, the blackish pink to grayish pink hymenophore, the initially bluish and then reddish brown color change when injured, and the trichoderm pileipellis. These traits are greatly different from other species in this genus. In our multi-locus phylogenetic analysis (Fig. 4.1), *P. porphyrosporus* forms an independent lineage. Its phylogenetic relationships with other species in this genus remain unresolved.

16.6 *Porphyrellus pseudofumosipes* Yan C. Li & Zhu L. Yang, The Boletes of China: *Tylopilus* s.l. 195 (2021)

MycoBank: MB 837923

Etymology: The epithet "*pseudofumosipes*" refers to its similarity to *P. fumosipes*.

Type: CHINA, YUNNAN PROVINCE: Lancang County, alt. 1000 m, 26 June 2017, J.W. Liu 530828MF0008 (KUN-HKAS 103784, GenBank Acc. No.: MW114845 for nrLSU).

Basidioma small to medium-sized. Pileus 3–7 cm wide, hemispherical to subhemispherical when young, subhemispherical to applanate when mature, black-brown (7F4–5) to coffee-brown (6E6) when young, dark brown (6F7–8) to brown (6E5–6) when mature, darker in color in the center; surface dry, covered with concolorous tomentose squamules, always cracked with white background exposed when mature; context white to pallid, asymmetrically bluish when injured. Hymenophore depressed around apex of stipe; surface grayish white (8A1) to pale grayish (8B1) when young, and then grayish pink (8B2) when mature, staining blue when touched; pores angular to roundish, up to 2 mm wide; tubes up to 10 mm long, concolorous or a little paler than hymenophoral surface, becoming bluish when injured. Spore print brown (8E5–6). Stipe clavate to cylindrical, 5–10 × 0.8–1.2 cm, concolorous or paler than pileal surface, fibrillose; context grayish white to pallid, without color change or sometimes asymmetrically bluish at apex of stipe when bruised; basal mycelium grayish white. Taste and odor mild.

Basidia 25–36 × 10–14 μm, clavate, 4-spored, hyaline in KOH. Basidiospores [80/4/4] 9–11 (12) × 4.5–5.5 μm (Q = 1.9–2.22, Q_m = 2.1 ± 0.11), elongated to subcylindrical with slight suprahilar depression, elongated to cylindrical in ventral view, smooth, yellowish to pale brownish in KOH, yellow to yellow-brown in Melzer's reagent. Hymenophoral trama boletoid composed of 3–8 μm wide filamentous hyphae, hyaline to yellowish in KOH, yellowish to yellow in Melzer's reagent. Cheilo- and pleurocystidia abundant, 42–63 × 14–19 μm, subfusiform to clavate, yellowish brown to yellow-brown in KOH, brown to dark brown in Melzer's reagent, thin-walled. Pileipellis an epithelium composed of inflated (8–16 μm wide) concatenated cells, yellowish to pale brownish in KOH and yellowish brown to brownish in Melzer's reagent; terminal cells 26–57 × 9–16 μm, pyriform to subfusiform. Pileal trama composed of 4.5–12 μm wide interwoven hyphae, hyaline to yellowish in KOH and yellowish to brownish in Melzer's reagent. Clamp connections absent in all tissues.

Habitat: Scattered to solitary on soil in subtropical forests dominated by plants of the family Fagaceae.

Known distribution: Currently known from southwestern China.

Additional specimens examined: CHINA, YUNNAN PROVINCE: Nanhua County, Wild Mushroom Market, altitude unknown, 18 August, 2011, G. Wu 763 (KUN-HKAS 75078); Longling County, Mengliu Village alt. 1200 m, 17 June 2014, L.H. Han 300 (KUN-HKAS 84596); Lancang County, Menglang Town, Zhaluozhai Village, alt. 1300 m, 21 August 2016, J.W. Liu LC62 (KUN-HKAS 97566).

Commentary: *Porphyrellus pseudofumosipes* is characterized by the black-brown to coffee-brown and then dark brown to brown pileus, the white to pallid context staining asymmetrically bluish when injured, the white or pale grayish to grayish pink hymenophore staining bluish when injured, the epithelium pileipellis, and the distribution in subtropical forests. Morphologically, *P. pseudofumosipes* can be easily confused with *P. fumosipes* (Peck) Snell and *P. orientifumosipes*. However, *P. fumosipes* has an olive-brown pileus, a yellowish brown hymenophore when mature, and relatively big basidiospores 12–14 (18) × 4–5 (7) μm (Peck 1898; Smith and Thiers 1971; Wolfe and Petersen 1978). *Porphyrellus orientifumosipes* differs from *P. pseudofumosipes* in its initially reddish brown to purplish olivaceous and then smoky-brown to brown pileus, a bluish ring at apex of stipe in mature or aged basidioma, and relatively broad pileipellis

Fig. 16.6 a *Porphyrellus pseudofumosipes* Yan C. Li & Zhu L. Yang. Photos by J.W. Liu (KUN-HKAS 103784, type).

hyphae (12–21 μm wide) (Wu *et al.* 2016a; see our description above).

In our multi-locus phylogenetic analysis (Fig. 4.1), *P. pseudofumosipes* clusters, with high support, with *P. castaneus*, *P. cyaneotinctus*, *P. griseus*, and *P. orientifumosipes*. However, *P. castaneus* has an initially violet-brown to reddish brown and then brownish red to chestnut-brown or pale brownish pileus and relatively broad basidiospores (4.5–5.5 μm wide) and pleurocystidia (14–19 μm wide) (Wu *et al.* 2016a; see our description above). *Porphyrellus cyaneotinctus* differs from *P. pseudofumosipes* in its reddish brown to gray-brown or smoky-gray pileus, relatively large basidiospores measuring 10.5–15.5 × 4–6.5 μm, palisadoderm pileipellis composed of 6–17 μm wide vertically arranged hyphae, and distribution in temperate forests (see our description above). *Porphyrellus griseus* can be easily distinguished from *P. pseudofumosipes* by its grayish brown to pale brown or orange-brown pileus, and palisadoderm pileipellis composed of 6–18 μm wide vertically arranged hyphae (see our description above).

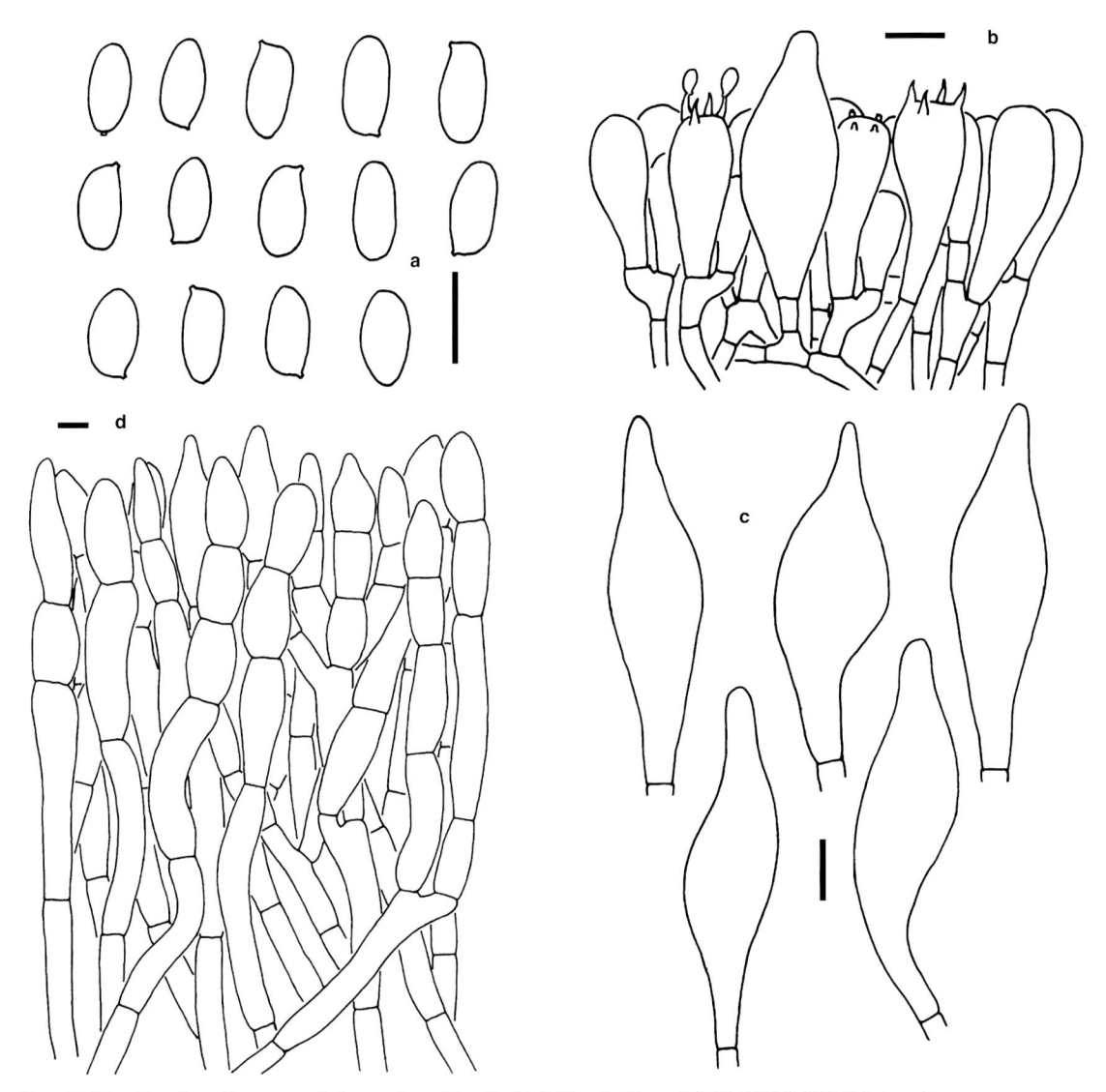

Fig. 16.6 b *Porphyrellus pseudofumosipes* Yan C. Li & Zhu L. Yang (KUN-HKAS 103784, type).
a. Basidiospores; b. Basidia and pleurocystidium; c. Cheilo- and pleurocystidia; d. Pileipellis. Scale bars = 10 μm.

16.7 *Porphyrellus scrobiculatus* Yan C. Li & Zhu L. Yang, The Boletes of China: *Tylopilus* s.l. 199 (2021)

MycoBank: MB 834714

Etymology: The epithet "*scrobiculatus*" refers to the ridges on the pileus.

Type: CHINA, FUJIAN PROVINCE: Sanming, Sanyuan National Forest Park, 26 August 2007, Y.C. Li 1021 (KUN-HKAS 53366, GenBank Acc. No.: KF112480 for nrLSU, KF112241 for *tef1-α*, KF112610 for *rpb1*, KF112716 for *rpb2*).

Basidioma small. Pileus 3–4 cm in diam., subhemispherical, dark brown (9F7–8) to ruby (12E7–8) or maroon (11F6–7), slightly darker in the center; surface wrinkled and pitted, dry, tomentose, margin slightly extended; context whitish to pallid, without discoloration when injured. Hymenophore adnate to slightly depressed around apex of stipe; surface white to pale grayish pink (9B1), without discoloration when injured; pores angular to roundish, up to 1 mm wide; tubes up to 8 mm long, concolorous or much paler than hymenophoral surface, without discoloration when injured. Stipe clavate, 4–6 × 0.6–0.9 cm, concolorous with pileal surface, nearly glabrous or covered with fibrillose squamules; context whitish to pallid, without discoloration when bruised; basal mycelium whitish. Taste and odor mild.

Basidia 30–47 × 8–10 μm, clavate 4-spored, hyaline in KOH. Basidiospores [60/2/1] (8.5) 9–11 × 5–6 μm [Q = (1.5) 1.64–2, Q_m = 1.79 ± 0.12], subellipsoid to elongated and inequilateral in side view with slight suprahilar depression, ellipsoid to elongated in ventral view, smooth, yellowish to brownish yellow in KOH, yellow to yellow-brown in Melzer's reagent. Hymenophoral trama boletoid composed of 3.5–10 μm wide filamentous hyphae, hyaline to yellowish in KOH, yellowish to yellow in Melzer's reagent. Cheilo- and pleurocystidia 43–70 × 11–16 μm, abundant, clavate to subfusiform always with a long pedicel, yellowish to yellow-brown in KOH, thin-walled. Pileipellis a palisadoderm composed of 7–16 μm wide somewhat vertically arranged hyphae with clavate to subfusiform or subfusoid-ventricose terminal cells 28–66 × 7–14 μm, yellowish to pale brownish in KOH and yellowish brown to brownish in Melzer's reagent. Pileal trama composed of 4–9 μm wide interwoven hyphae, pale yellowish to pale brownish in KOH and yellow to yellowish brown in Melzer's reagent. Clamp connections absent in all tissues.

Habitat: Solitary on soil in subtropical forests dominated by plants of the family Fagaceae.

Known distribution: Currently known from southeastern China.

Commentary: *Porphyrellus scrobiculatus* is characterized by the dark brown to ruby or maroon pileus, the wrinkled and pitted pileal surface, the white context without discoloration when injured, and the palisadoderm pileipellis. The wrinkled pileus of *P. scrobiculatus* is similar to that of *Boletus castanopsidis* Hongo and *B. reticuloceps* (M. Zang, M.S. Yuan & M.Q. Gong) Q.B. Wang & Y.J. Yao. However, *B. castanopsidis* differs from *P. scrobiculatus* in its olivaceous pileus, whitish then yellow or sordid brownish hymenophore, reticulate stipe, and relatively large basidiospores measuring 12.5–23 × 4.5–7 μm (Hongo 1973). *Boletus reticuloceps* differs from *P. scrobiculatus* in its ochraceous brown to ochraceous pileus densely covered with brown granular squamules, conspicuously reticulate stipe, white to yellow hymenophore, large basidiospores measuring 15–20 × 5–6.5 μm, and distribution in subalpine coniferous forests (Wang and Yao 2005; Zang 2006). In our multi-locus phylogenetic analysis (Fig. 4.1),

P. scrobiculatus forms an independent lineage. Its phylogenetic relationships with other species in this genus remain unresolved.

Fig. 16.7 a *Porphyrellus scrobiculatus* Yan C. Li & Zhu L. Yang. Photos by Y.C. Li (**KUN-HKAS 53366**, type).

172 The Boletes of China: *Tylopilus* s.l.

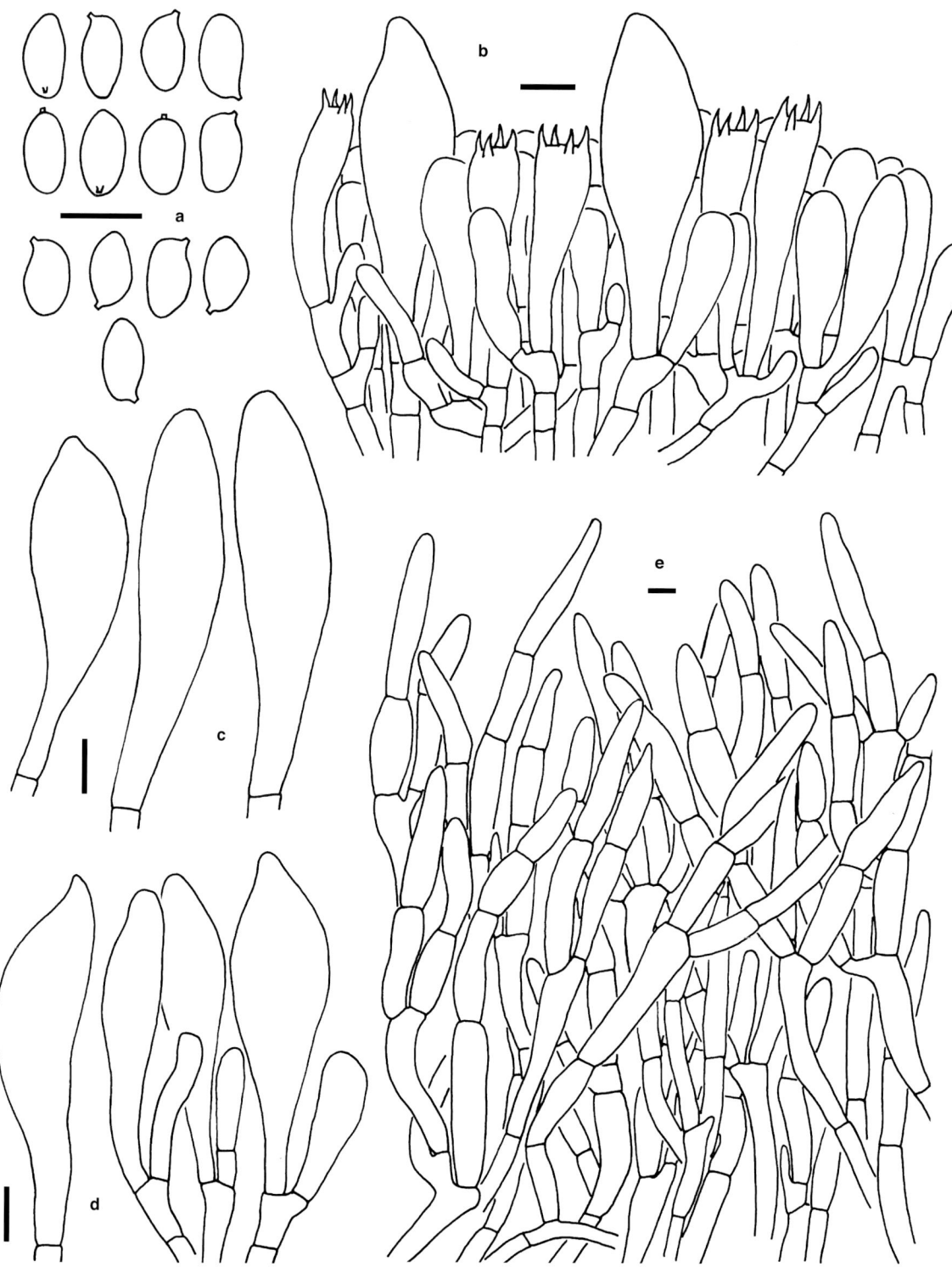

Fig. 16.7 b *Porphyrellus scrobiculatus* Yan C. Li & Zhu L. Yang (KUN-HKAS 53366, type).
a. Basidiospores; b. Basidia and pleurocystidia; c. Pleurocystidia; d. Cheilocystidia; e. Pileipellis. Scale bars = 10 μm.

Chapter 17
Pseudoaustroboletus Yan C. Li & Zhu L. Yang

Pseudoaustroboletus Yan C. Li & Zhu L. Yang, Mycol Progr 13: 1209 (2014)

Type species: *Boletus valens* Corner, *Boletus* in Malaysia. 161 (1972).

Diagnosis: This genus differs from all the other genera in the Boletaceae in the combination of the tomentose to filamentous pileus, the white to pallid context without discoloration in pileus but always with yellow stains in the base of stipe when injured, the initially white then pinkish hymenophore, the distinctly reticulate stipe, the trichoderm pileipellis, and the smooth basidiospores.

Pileus subhemispherical when young and then convex to applanate or plano-convex when mature; surface dry, covered with tomentose to filamentous squamules; context white to pallid, without discoloration when injured. Hymenophore adnate when young and then depressed around apex of stipe when mature; surface initially white to pallid and then pale pinkish, pinkish to pink, without discoloration when injured. Spore print grayish red to pale red. Stipe pallid to white, distinctly reticulate; context white, without color change in the upper part, occasionally staining yellowish at base when bruised; mycelium white. Basidiospores smooth, cylindrical to fusiform. Hymenial cystidia abundant, fusiform to subfusiform or subfusoid-mucronate to ventricose-lanceolate. Pileipellis a trichoderm composed of interwoven filamentous hyphae. Clamp connections absent in all tissues.

Commentary: *Pseudoaustroboletus* generally shares the color of the hymenophore, and the morphology of the stipe with the genus *Austroboletus*. But *Austroboletus* is characterized by the distinctly extended pileal margin, the deep colored hymenophoral and the heavily ornamented basidiospores (Wolfe 1979a; Singer 1986; Fulgenzi *et al.* 2010). These traits differ greatly from those in *Pseudoaustroboletus*. Currently only one species, described here, is known to occur in China.

17.1 *Pseudoaustroboletus valens* (Corner) Yan C. Li & Zhu L. Yang, Mycol Progr 13: 1211 (2014)

Basionym: *Boletus valens* Corner, *Boletus* in Malaysia 161 (1972) [≡ *Boletus albellus* Mass., Kew Bull 206 (1909) (non Peck 1888)].

Synonym: *Tylopilus valens* (Corner) Hongo & Nagas., Rep Tottori Mycol Inst 14: 87 (1976).

Basidioma small to large. Pileus 4–11 cm in diam., hemispherical to subhemispherical when young and subhemispherical to applanate when mature, dark gray, pale fuscous-gray or grayish white, slightly darker in the center; surface dry, inviscid when wet, nearly glabrous, sometimes covered with tomentose to filamentous squamules; context white to pallid, without discoloration when injured. Hymenophore adnate when young and adnate to depressed around apex of stipe when mature; surface pallid to pale pinkish when young and then pinkish to pale pink with age, without discoloration when touched; pores angular to roundish, up to 1 mm wide; tubes up to 15 mm long, concolorous or a little paler than hymenophoral surface, without discoloration when injured. Spore print grayish red to pale red (Fig. 3.4 i). Stipe 6–12 × 1.2–2.5 cm, clavate to subcylindrical, always enlarged downwards, pallid to white; surface entirely covered with concolorous reticulum; context white to whitish, without discoloration when bruised, sometimes occasionally staining yellow to yellowish or

Fig. 17.1 a *Pseudoaustroboletus valens* (Corner) Yan C. Li & Zhu L. Yang. Photos by Y.C. Li (KUN-HKAS 52602).

Fig. 17.1 b *Pseudoaustroboletus valens* (Corner) Yan C. Li & Zhu L. Yang. Photos by L.H. Han (KUN-HKAS 84678).

pinkish yellow at base on injury; basal mycelium white. Taste slight sour, odor indistinct.

Basidia 27–32 × 8.5–12 μm, clavate, hyaline to light yellowish in KOH and yellowish to yellow in Melzer's reagent, thin-walled, 4-spored. Basidiospores (11.5) 12–15 (16) × (4) 4.5–5.5 (6) μm [Q = (2.18) 2.33–3.18 (3.5), Q_m = 2.69 ± 0.21], subcylindrical to subfusiform in side view with slight suprahilar depression, cylindrical to fusiform in ventral view, smooth, somewhat thick-walled (up to 0.5 μm thick), hyaline to yellowish brown in KOH and yellow-brown to brownish in Melzer's reagent. Cheilo- and pleurocystidia 22–62 × 6.5–10 μm, fusiform, subfusiform, subfusoid-mucronate or ventricose-lanceolate, thin-walled, brownish to yellowish brown in KOH and brown to dark brown in Melzer's reagent. Caulocystidia forming the reticulum over the surface of stipe, morphologically similar to hymenial cystidia. Hymenophoral trama bilateral, composed of 5–11.5 μm wide filamentous hyphae, hyaline to yellowish in KOH and yellowish to yellow in Melzer's reagent. Pileipellis a trichoderm composed of 5–10.5 μm wide more or less vertically arranged interwoven filamentous hyphae, yellow to yellowish brown in KOH and yellow-brown to brownish in Melzer's reagent; terminal cells 15–46 × 6–11 μm, subcylindrical to clavate. Pileal trama made up of 3–10.5 μm wide filamentous hyphae, hyaline to yellowish in KOH and yellow to yellowish brown in Melzer's reagent. Clamp connections absent in all tissues.

Habitat: Solitary on the ground in tropical forests dominated by plants of the family Fagaceae.

Known distribution: Currently known from China, Japan, Malaysia, and Singapore.

Specimens examined: CHINA, HAINAN PROVINCE: Qiongzhong County, Limu Mountain, alt. 750 m, 6 May 2009, N.K. Zeng 120 (KUN-HKAS 59835). FUJIAN PROVINCE: Zhangping County, Xinqiao Town, Chengkou Village, alt. 360 m, 3 September 2009, N.K. Zeng 670 (KUN-HKAS 82641). YUNNAN PROVINCE: Jinghong County, Dadugang Town, alt. 1400 m, 21 July 2007, Y.C. Li 915 (KUN-HKAS 52602); Mengla County, alt. 1200 m, 6 July 2014, L.H. Han 382 (KUN-HKAS 84678); Maguan County, Pojiao Town, alt. 1600 m, 14 August 2016, X.H. Wang 4085 (KUN-HKAS 96674). HUNAN PROVINCE: Yizhang County, Mangshan National Forest Park, alt. 1800 m, 2 September 2007, Y.C. Li 1062 (KUN-HKAS 53407). GUANGDONG PROVINCE: Fengkai County, Heishiding Nature Reserve, alt. 200m, 5 June 2012, F. Li 443 (KUN-HKAS 82642), the same location, 14 June 2012, F. Li 507 (KUN-HKAS 82643), 18 July 2012, F. Li 690 (KUN-HKAS 82644) and 15 October 2012, F. Li 1081 (KUN-HKAS 77791). MALAYSIA, Pahang, Fraser's Hill, 16 May 1930, Corner (E 98119); Borneo, Kinabalu, Mesilau, alt. 1800 m, 24 January 1964, Corner RSNB 5082A (E 98120). SINGAPORE, Bukit Timah, 16 August 1939, Corner (E 98118); Reservoir Jungle, 4 September 1939, Corner (E 98116).

Commentary: *Pseudoaustroboletus valens* was originally illegitimately named by Massee (1909) from Singapore as *Boletus albellus* Mass. (non Peck 1888). The legitimate name, *Boletus valens* Corner, provided by Corner (1972), has been used for this species. It was then reported from Japan and China (Corner 1972; Hongo and Nagasawa 1976; Li and Song 2003; Li *et al.* 2014b). This species is characterized by the pale fuscous-gray to grayish pileus, the pallid to pale pinkish then pinkish to pale pink hymenophore, the distinctly reticulate stipe, the smooth basidiospores, and the trichoderm pileipellis.

In our multi-locus phylogenetic analysis (Fig. 4.1), *Ps. valens* forms a monophyletic clade with good support in the subfamily Leccinoideae. Its phylogenetically related genera or species are yet unknown. For detailed descriptions, comparisons with similar species, line drawings of *Ps. valens* see Li *et al.* (2014b).

Chapter 18
Retiboletus Manfr. Binder & Bresinsky

Retiboletus Manfr. Binder & Bresinsky, Feddes Repert. 113: 36 (2002)

Type species: *Boletus ornatipes* Peck, Ann Rep NY St Mus nat Hist 29: 67 (1878).

Diagnosis: The genus is different from other genera in the Boletaceae in its dry and tomentose to fibrillose pileus, white to grayish white or cream to yellowish or pale greenish to olivaceous green context usually with yellowish tinge in the base of stipe when injured, initially pallid to grayish or grayish white to grayish pink and then yellowish or gray-pink hymenophore without discoloration or staining brownish to fuliginous when injured, reticulate stipe, grayish yellow to yellow tinged context in the base of stipe, and trichoderm pileipellis.

Pileus hemispherical, subhemispherical or applanate, dry, subtomentose to fibrillose; context white to grayish white, or cream to yellowish, or grayish olivaceous to olivaceous green without discoloration or staining brownish to fuliginous when injured. Hymenophore adnate when young, adnate or depressed around apex of stipe when mature; surface pallid, grayish white to yellowish or grayish pink, without discoloration or staining brownish to fuliginous when injured; pores angular or roundish; tubes concolorous with hymenophoral surface, without discoloration or staining brownish to fuliginous when injured. Spore print pale brown to brownish. Stipe central, distinctly reticulate, rarely non-reticulate or indistinctly reticulate; basal mycelium white to cream; context white to grayish white, or cream to yellowish, or grayish olivaceous to olivaceous green, but always with yellow to brownish yellow tinge at base, without discoloration or staining brownish to fuliginous when injured. Basidiospores smooth, subfusoid. Hymenial cystidia abundant, subfusiform to fusoid-ventricose. Pileipellis a cutis, or a trichoderm composed of interwoven filamentous hyphae. Clamp connections absent in all tissues.

Commentary: *Retiboletus* has been described in detail by Binder and Bresinsky (2002b) and Zeng *et al.* (2016). Species of this genus have dark gray, gray, mustard-yellow, olive-brown or grayish green pileus, pallid to grayish, grayish pink to gray-pink, or yellowish to yellow hymenophore, reticulate stipe with grayish yellow to yellow tinge at base when hurt or aged; pallid to cream or yellow to vivid yellow context, and smooth basidiospores. Currently nine species, including three species with yellow to vivid yellow hymenophores and six species with pallid to grayish or grayish pink hymenophores, have been reported from China. Here we consider only the later six species, which are easily misidentified as *Tylopilus* s.l. by mycologists based on the pallid to grayish or grayish pink hymenophores. The other three species with yellow to vivid yellow hymenophores are not included as the colors of their hymenophores are greatly different from those in *Tylopilus* s.l. In addition, *Retiboletus brunneolus* Yan C. Li & Zhu L. Yang invalidly published in Wu *et al.* (2016a) is validated below.

Key to the species of the genus *Retiboletus* in China

1. Pileus smoky-gray to gray or black to grayish black, always with an olivaceous tinge; context of pileus white to grayish white with olivaceous tinge, staining brownish to fuliginous when injured; context of stipe dark green to olivaceous when cut, staining brownish to fuliginous when injured ·· 2
1. Pileus without olivaceous tinge; context of pileus white to grayish white without olivaceous tinge, color unchanged when hurt; context of stipe white to grayish white in the upper part and grayish yellow to brownish yellow at base, color without discoloration when injured ················ 3
2. Pileus smoky-gray to gray with olivaceous tinge; hymenophore initially white and then lilac to purplish; basidiospores 9–11 × 4–5 μm ··· *R. zhangfeii*
2. Pileus black to grayish black with olivaceous tinge; hymenophore white to grayish white without lilac to purplish tinge; basidiospores relatively small 8–10.5 × 3.5–4 μm ············ *R. nigrogriseus*
3. Pileus brown to blackish brown or grayish brown to grayish black; stipe entirely reticulate ······· 4
3. Pileus pale brown to grayish brown or black to blackish; stipe without reticulum or with reticulum restricted to the upper part ·· 5
4. Pileus brown to blackish brown; basidiospores 9.5–11 × 4–4.5 μm; pileipellis a palisadoderm composed of vertically arranged broad hyphae up to 13 μm wide; distributed in tropical forests····
   ··· *R. pseudogriseus*
4. Pileus grayish brown to grayish black; basidiospores slightly narrower 9–12 × 3.5–4 μm; pileipellis a trichoderm composed of interwoven filamentous hyphae 4–8 μm wide; distributed in subtropical to temperate forests·· *R. fuscus*
5. Pileus pale brown to grayish brown; stipe without reticulum or with indistinct reticulum on the upper part; basidiospores 10–12.5 × 4.5–5 μm; pleurocystidia 50–80 × 9–14.5 μm; pileipellis a palisadoderm composed of 5–15 μm wide vertically arranged hyphae····················*R. brunneolus*
5. Pileus black to blackish; stipe covered with reticulum over the upper 1/3; basidiospores small 8–10.5 × 3–4.5 μm; pleurocystidia relatively small 26–55 × 6–10 μm; pileipellis a trichoderm composed of 4–8 μm wide interwoven filamentous hyphae·· *R. ater*

18.1 *Retiboletus ater* Yan C. Li & T. Bau, MycoKeys 67: (2020)

Basidioma small. Pileus 3–5 cm in diam., hemispherical to subhemispherical when young, applanate to plano-convex when mature, black to blackish in the center and gray or yellowish gray towards margin; surface dry, tomentose, without discoloration when touched; context pallid gray to cream, without color change when bruised. Hymenophore adnate when young, adnate to slightly depressed around apex of

Fig. 18.1 *Retiboletus ater* Yan C. Li & T. Bau. Photos by Y.C. Li (KUN-HKAS 56069, type).

stipe when mature; surface whitish to cream when young, dingy white to grayish yellow or yellowish when mature, without discoloration when bruised; pores subangular, 0.3–1 mm wide; tubes 5–11mm long, concolorous or a little paler than hymenophoral surface. Spore print pale brown to brownish. Stipe 4–6 × 0.8–1.2 cm, subcylindrical to clavate; surface blackish to gray, distinctly reticulate over the upper 1/3; context solid, white in the upper part, yellowish to cream-yellow downwards, without discoloration when hurt; basal mycelium white. Taste and odor mild.

Basidia 26–38 × 6–10 μm, clavate to narrowly clavate, 4-spored, hyaline in KOH. Basidiospores 7–11 × 3–4.5 (5) μm [Q = (1.89) 2–3.33 (3.67), Q_m = 2.52 ± 0.42], elongated to subcylindrical or subfusiform and inequilateral in side view with slight suprahilar depression, elongated to cylindrical or fusiform in ventral view, smooth, somewhat thick-walled (up to 0.5 μm), hyaline to yellowish in KOH, yellow to brownish yellow in Melzer's reagent. Hymenophoral trama boletoid composed of 3.5–9 μm wide filamentous hyphae. Cheilo- and pleurocystidia 26–55 × 6–10 μm, subfusiform to fusoid-ventricose, usually yellow to brownish yellow in KOH, thin-walled. Pileipellis a palisadoderm composed of 5–15 μm wide interwoven filamentous hyphae, blackish to blackish yellow in KOH and black-yellow to brownish yellow in Melzer's reagent; terminal cells 45–111 × 9–15 μm, clavate to subcylindrical. Pileal trama composed of 5–11 μm wide interwoven hyphae, colorless to yellowish in KOH and yellowish to pale yellow in Melzer's reagent. Clamp connections absent in all tissues.

Habitat: Scattered on soil in tropical forests dominated by plants of the family Fagaceae.

Known distribution: Currently known from southwestern China.

Specimens examined: CHINA, YUNNAN PROVINCE: Jingdong County, Ailao Mountain, alt. 2500 m, 14 July 2008, Y.C. Li 1215 (KUN-HKAS 56069, type, GenBank Acc. No.: MT010611 for nrLSU, MT010621 for *tef1-α*), the same location and date, Y.C. Li 1224 (KUN-HKAS 56078, GenBank Acc. No.: MT010612 for nrLSU, MT010622 for *tef1-α*).

Commentary: *Retiboletus ater* was described by Liu *et al.* (2020) from southwestern China. This species is macroscopically similar to *R. fuscus* (Hongo) N.K. Zeng & Zhu L. Yang, *R. griseus* (Frost) Manfr. Binder & Bresinsky and *R. pseudogriseus* N.K. Zeng & Zhu L. Yang. However, *R. fuscus* is characterized by the relatively large basidiospores measuring 9–12 × 3.5–4.5 μm and narrower pileipellis hyphae (4–8 μm wide) (Zeng *et al.* 2016). *Retiboletus griseus* has a reticulum over the upper 2/3 of the stipe, a cream to grayish brown stipe staining orange-yellow when bruised, and a distribution in North/Central America (Smith and Thiers 1971; Ortiz-Santana *et al.* 2007). *Retiboletus pseudogriseus* differs from *R. ater* in its grayish white pileus covered with brown to blackish brown squamules, broad basidiospores measuring 9.5–11 × 4–4.5 μm, and narrow hyphae in pileipellis (up to 8 μm wide) (Zeng *et al.* 2016). Phylogenetically, *R. ater* forms a distinct lineage within *Retiboletus*. However, its phylogenetically related species are yet unknown (Liu *et al.* 2020).

18.2 *Retiboletus brunneolus* Yan C. Li & Zhu L. Yang, The Boletes of China: *Tylopilus* s.l. 213 (2021)

MycoBank: MB 842389

The validating description and diagnosis were previously published by Wu *et al.* [2016 (81): 77].

Retiboletus brunneolus Yan C. Li & Zhu L. Yang was originally described based on material from Fujian Province, China (Wu *et al.* 2016a). However, this publication was not valid in accordance with Article F.5.1 in the International Code of Nomenclature for algae, fungi, and plants (Shenzhen Code) (Turland *et al.* 2017), because the identifier of the name provided in the original description was incorrectly cited as MB 810364 [the identifier of the name Lanmaoa angustispora G. Wu & Zhu L. Yang published in 2016 by Wu et al. (2016b)]. To enable its formal use, the name is validated here, with correct citation of the identifier. As the name and validating diagnosis are ascribed to Yan C. Li & Zhu L. Yang (Wu *et al.* 2016a), under the Article 46.2, Yan C. Li & Zhu L. Yang is the author of the name *Retiboletus brunneolus*.

Etymology: The epithet "*brunneolus*" refers to the brownish color of the basidioma. Characteristics of this species observed from the type and additional specimens are as follows.

Basidioma very small to medium-sized. Pileus 2.5–7 cm in diam., hemispherical to subhemispherical, blackish gray to dark gray when young, grayish brown to yellowish brown, slightly darker in the center; surface dry with tomentose squamules, without discoloration when bruised; context whitish to grayish or

Fig. 18.2 a *Retiboletus brunneolus* Yan C. Li & Zhu L. Yang. Photos by Q. Zhao (KUN-HKAS 81007).

Fig. 18.2 b *Retiboletus brunneolus* Yan C. Li & Zhu L. Yang. Photos by F. Li (KUN-HKAS 106427).

grayish white, without discoloration when bruised. Hymenophore adnate when young, adnate to slightly depressed around apex of stipe when mature; surface whitish to cream when young, dingy white to grayish white when mature, without discoloration when bruised; pores round to roundish or angular, 0.3–1 mm wide; tubes 4–12 mm long, concolorous or a little paler than hymenophoral surface, without discoloration when bruised. Spore print pale brown to brownish. Stipe 4–9 × 0.6–1 cm, cylindrical to clavate; surface gray to grayish brown but grayish white to grayish yellow at apex, with indistinct or prominent reticulum over upper 1/3 part, without discoloration when touched; context solid, white to grayish always with yellowish to yellow tinge at base, without discoloration when hurt; basal mycelium white. Taste and odor mild.

Basidia 35–40 × 10–12 µm, clavate to narrowly clavate, 4-spored, hyaline in KOH. Basidiospores (9) 10–12.5 × 4.5–5 µm (Q = 2–2.5, Q_m = 2.21 ± 0.16), elongated to subcylindrical and inequilateral in side view with slight suprahilar depression, elongated to cylindrical in ventral view, smooth, yellowish to brownish yellow in KOH, yellow to yellow-brown in Melzer's reagent. Hymenophoral trama boletoid composed of 5–14 µm wide. Cheilocystidia 30–40 × 9–11 µm, subfusiform to fusoid-ventricose with a long pedicel, yellow to brownish yellow in KOH, thin-walled. Pleurocystidia much bigger than cheilocystidia, 50–80 × 9–14.5 µm, subfusiform to fusoid-ventricose, sometimes with a long pedicel, thin-walled. Pileipellis a palisadoderm composed of 5–15 µm wide interwoven filamentous hyphae, yellowish to pale brownish in KOH and yellowish brown to brownish in Melzer's reagent; terminal cells 17–36 × 6.5–15 µm, clavate to cylindrical. Pileal trama composed of 5–10 μm wide interwoven hyphae, hyaline to yellowish in KOH and yellowish to brownish in Melzer's reagent. Clamp connections absent in all tissues.

Habitat: Scattered on soil in tropical forests dominated by plants of the family Fagaceae.

Known distribution: Currently known from southeastern and southern China.

Specimens examined: CHINA, FUJIAN PROVINCE: Sanming, Sanyuan National Forest Park, alt. 260 m, 24 August 2007, Y.C. Li 993 (KUN-HKAS 52680, type). GUANGDONG PROVINCE: Fengkai County, Heishiding Nature Reserve, alt. 800 m, 3 August 2012, F. Li 764 (KUN-HKAS 106427), the same location and date, alt. 800 m, 4 September 2012, F. Li 993 (KUN-HKAS 77764); Shaoguan, Ruyuan County, Nanling National Nature Reserve, alt. 1000 m, 11 July 2013, Q. Zhao 1897 (KUN-HKAS 81007).

Commentary: *Retiboletus brunneolus* is characterized by the blackish gray to dark gray and then grayish brown to yellowish brown pileus, the white to grayish or grayish white pileal context without discoloration when bruised, the whitish to cream or dingy white to grayish white hymenophore, the prominent and coarse reticulum over upper 1/3 part, the white to grayish context in stipe but always yellowish to yellow at base, and the palisadoderm pileipellis composed of 5–15 μm wide interwoven filamentous hyphae. *Retiboletus brunneolus* can be easily confused with *R. ater* and *R. fuscus* (Hongo) N.K. Zeng & Zhu L. Yang because of the color of basidioma and their distribution in tropical areas. However, *R. ater* has a much darker pileus and small basidiospores measuring 7–11 × 3–4.5 (5) μm (Liu *et al.* 2020). *Retiboletus fuscus* differs from *R. brunneolus* in its entirely reticulate stipe, relatively narrow basidiospores measuring 8.5–12.5 × 3.5–4.5 μm and narrow hyphae in pileipellis which are 4–8 μm wide (Zeng *et al.* 2016). In our multi-locus phylogenetic analysis (Fig. 4.1), *R. brunneolus* forms an independent lineage. Its phylogenetic relationships with other species in this genus remain unresolved.

18.3 *Retiboletus fuscus* (Hongo) N.K. Zeng & Zhu L. Yang, Mycologia 108: 365 (2016)

Basionym: *Boletus griseus* var. *fuscus* Hongo, J Jap Bot 4(10): 301 (1974a).

Basidioma small to medium-sized. Pileus 3.5–8 cm in diam., subhemispherical to hemispherical when young, applanate when mature, grayish brown to grayish black, darker in the center; surface dry, tomentose, without discoloration when touched; context grayish white to white, without discoloration when bruised. Hymenophore adnate when young, slightly depressed around apex of stipe when mature; surface whitish to cream when young and then dingy white to grayish yellow when mature, without discoloration when bruised; pores subangular, 0.3–1 mm wide; tubes 5–12 mm long, concolorous or much paler than hymenophoral surface, without discoloration when bruised. Spore print pale brown to brownish. Stipe 4–11 × 0.8–3.5 cm, subcylindrical to clavate; surface gray to grayish white when young, grayish yellow when mature, entirely reticulate; reticulum grayish brown to grayish black; context solid to soft, white to grayish white in the upper part, but yellowish to grayish yellow in the base, without discoloration when hurt; basal mycelium grayish white to cream. Taste and odor mild.

Basidia 20–32 × 7–9 μm, clavate to narrowly clavate, 4-spored, hyaline in KOH. Basidiospores 8.5–12.5 × 3.5–4.5 μm (Q = 2.1–3.4, Q_m = 2.8 ± 0.17), subcylindrical to subfusiform in side view, cylindrical to fusiform in ventral view, smooth, yellowish to brownish yellow in KOH, yellow to yellow-brown in Melzer's reagent. Hymenophoral trama boletoid composed of 3.5–10 μm wide. Cheilo- and pleurocystidia 30–45 × 6–11 μm, subfusiform to fusoid-ventricose, usually yellowish brown to brownish in KOH, yellow-brown to brown in Melzer's reagent, thin-walled. Pileipellis a trichoderm composed of 4–8 μm wide interwoven filamentous hyphae, hyaline to yellowish or yellowish brown in KOH and yellow to brownish yellow in Melzer's reagent; terminal cells 25–51 × 5–8 μm, clavate to cylindrical. Pileal trama composed

Fig. 18.3 a *Retiboletus fuscus* (Hongo) N.K. Zeng & Zhu L. Yang. Photos by Y.J. Hao (KUN-HKAS 96891).

Fig. 18.3 b *Retiboletus fuscus* (Hongo) N.K. Zeng & Zhu L. Yang. Photo by Y.J. Hao (KUN-HKAS 96890).

of 3–10 μm wide interwoven hyphae, colorless to yellowish in KOH and yellowish to pale yellow in Melzer's reagent. Clamp connections absent in all tissues.

Habitat: Scattered to solitary on soil in tropical to subtropical forests dominated by plants of the families Fagaceae and Pinaceae.

Known distribution: Known from central and southwestern China, and Japan.

Specimens examined: CHINA, GUIZHOU PROVINCE: Tongren, Fanjing Mountain, alt. 2000 m, 26 October 1983, X.L. Wu 1253 (KUN-HKAS 14521). HENAN PROVINCE: Luanchuan County, Laojun Mountain, alt. 2000 m, 13 August 2015, B. Li 54 (KUN-HKAS 89814).YUNNAN PROVINCE: Ning'er County, alt. 1350 m, 1 August 2008, B. Feng 274 (KUN-HKAS 55385); Lanping County, Tongdian Town, alt. 2400 m, 15 August 2010, B. Feng 840 (KUN-HKAS 68621); Lanping County, Hexi Town, alt. 2600 m, 16 August 2010, X.T. Zhu 184 (KUN-HKAS 68360); Nanhua County, Wild Mushroom Market, altitude unknown, 25 July 2012, Y.Y. Cui 47 (KUN-HKAS 79727); Lancang County, Zhutang Town, Zhanmapo Village, alt. 1450 m, 22 August 2016, Y.J. Hao LC-53 (KUN-HKAS 96890), the same location and date, Y.J. Hao LC-54 (KUN-HKAS 96891) and Y.J. Hao LC-57 (KUN-HKAS 96894). SICHUAN PROVINCE: Xichang, Puge County, Luoji Mountain, alt. 2000 m, 28 July 2012, T. Guo 498 (76190). HUBEI PROVINCE: Macheng County, Wunaoshan National Forest Park, alt. 200 m, 26 June 2013, T. Guo 639 (KUN-HKAS 81841).

Commentary: *Retiboletus fuscus* was originally described as *Boletus griseus* var. *fuscus* by Hongo (1974a) from Japan. Zeng *et al.* (2016) transferred the variety to the genus *Retiboletus* and elevated it to species rank, namely *R. fuscus* (Hongo) N.K. Zeng & Zhu L. Yang based on the morphology and molecular data. Morphologically, *R. fuscus* generally shares the color of the pileus, hymenophore and spore deposit with *R. griseus* (Frost) Manfr. Binder & Bresinsky and *R. pseudogriseus* N.K. Zeng & Zhu L. Yang. However, *R. griseus* differs in its reticulum over the upper 2/3 of the stipe, cream to grayish brown stipe staining grayish yellow to orange-yellow when bruised, and distribution in North/Central America (Smith and Thiers 1971; Ortiz-Santana *et al.* 2007). *Retiboletus pseudogriseus* differs in its grayish white pileus covered with brown to blackish brown squamules, broad basidiospores measuring 9.5–11 × 4–4.5 μm, and narrow hyphae in pileipellis (up to 8 μm wide) (Zeng *et al.* 2016).

18.4 *Retiboletus nigrogriseus* N.K. Zeng, S. Jiang & Zhi Q. Liang, Phytotaxa 367: 48 (2018)

Basidioma small to medium-sized. Pileus 3–7 cm in diam., subhemispherical to hemispherical when young, subhemispherical to applanate when mature, black when young, then gray to grayish black when mature, slightly darker in the center; surface tomentose, dry, without discoloration when touched; context white, staining brownish to fuliginous when bruised. Hymenophore adnate to slightly decurrent; surface white to gray-white, staining brownish to fuliginous when bruised; pores subangular, 0.3–0.5 mm wide; tubes 3–6 mm long, concolorous or a little paler than hymenophoral surface, staining brownish to fuliginous when bruised. Spore print pale brown to brownish. Stipe 4–6 × 1.5–2.5 cm, cylindrical, clavate; surface gray to black, but grayish white at apex; surface entirely covered with black to dark gray reticulum, staining brownish to fuliginous when bruised; context solid, white in the upper part, but with yellowish green tinge downwards, staining brownish to fuliginous when bruised; basal mycelium white. Taste and odor mild.

Basidia 25–32 × 6–9 µm, clavate to narrowly clavate, 4-spored, hyaline in KOH. Basidiospores 8–11 × 3.5–4.5 µm (Q = 2–2.6, Q_m = 2.3 ± 0.16), elongated to subcylindrical and inequilateral in side view with slight suprahilar depression, elongated to cylindrical in ventral view, smooth, yellow-brown to ocherous in KOH, brown to dark brownish in Melzer's reagent. Hymenophoral trama boletoid composed of 3–8 µm wide. Cheilocystidia 25–38 × 5–6 µm, subfusiform to fusoid-ventricose, usually pale yellowish brown in KOH, yellow-brown to brown in Melzer's reagent, thin-walled. Pleurocystidia 30–37 × 6–10 µm, subfusiform to fusoid-ventricose, concolorous with cheilocystidia. Pileipellis a trichoderm to palisadoderm composed of 3.5–10 µm wide interwoven filamentous hyphae, brownish to yellowish brown or blackish brown in KOH and yellow-brown to dark brown in Melzer's reagent; terminal cells 18–50 × 5–9 µm, clavate to cylindrical. Pileal trama composed of 5–10 µm wide interwoven filamentous hyphae, pale yellowish to brownish yellow in KOH and yellow to brownish in Melzer's reagent. Clamp connections absent in all tissues.

Habitat: Scattered to solitary on soil in tropical forests dominated by plants of the family Fagaceae.

Known distribution: Currently known from southern and southwestern China and Thailand.

Specimens examined: CHINA, HAINAN PROVINCE: Qiongzhong County, Yinggeling National Nature Reserve, alt. 800 m, 28 July 2009, N.K. Zeng 369 (FHMU 213). YUNNAN PROVINCE: Pu'er, Jiangcheng County, Baozang Town, Liangmahe Village, alt. 1200 m, 28 June 2019, B. Feng 2226 (KUN-HKAS 107185).

Commentary: *Retiboletus nigrogriseus* was originally described by Zeng *et al.* (2018) from southern China and is characterized by the black to grayish black pileus which is always with olivaceous tinge, the white context staining brownish to fuliginous when bruised, the white to grayish white hymenophore without lilac to purplish tinge, the fine hymenophore pores which are 0.3–0.5 mm wide, the entirely reticulate stipe, the yellowish green context at base of stipe, the trichoderm to palisadoderm pileipellis composed of 3.5–10 µm wide interwoven filamentous hyphae, and the distribution in tropical forests.

Retiboletus nigrogriseus is phylogenetically related and morphologically similar to *R. nigerrimus* (R.

Heim) Manfr. Binder & Bresinsky and *R. zhangfeii* N.K. Zeng & Zhu L. Yang. However, *R. nigerrimus* originally described from Papua New Guinea (Heim 1963), has a distinctive blue tinged pileus, a lemon-yellow context of the pileus, orange context of the base of stipe, and long basidiospores measuring 11.5–14.5 × 3.6–4.6 μm (Heim 1963). *Retiboletus zhangfeii*, described from tropical to subtropical China, has a pileus with a distinctive purple tinge, pores grayish pink to grayish purple when mature, and relatively broad basidiospores measuring 9.5–11.5 × 4–5 μm (Zeng *et al*. 2016).

Fig. 18.4 *Retiboletus nigrogriseus* N.K. Zeng, S. Jiang & Zhi Q. Liang. Photos by B. Feng (KUN-HKAS 107185).

18.5 *Retiboletus pseudogriseus* N.K. Zeng & Zhu L. Yang, Mycologia 108: 370 (2016)

Basidioma small to medium-sized. Pileus 3–9 cm in diam., subhemispherical to convexo-applanate, grayish brown to blackish brown when young, brownish to yellowish brown when mature, darker in the center; surface tomentose, dry, without discoloration when touched; context white, without discoloration when bruised. Hymenophore adnate when young and then depressed around apex of stipe when mature; surface whitish to cream when young, grayish yellow to yellowish when mature, without discoloration when bruised; pores subangular, 0.5–1 mm wide; tubes 4–10 mm long, concolorous or a little paler than hymenophoral surface, without discoloration when bruised. Spore print pale brown to brownish. Stipe 6–9 × 0.8–1.5 cm, cylindrical or clavate, gray to grayish brown but whitish to cream at apex; surface entirely covered with black to blackish brown reticulum, without discoloration when touched; context solid, white to grayish white upwards, but yellowish downwards, without discoloration when hurt; basal mycelium white. Taste and odor mild.

Basidia 25–40 × 8–10 µm, clavate, 4-spored, hyaline in KOH. Basidiospores 9–12 × 4–4.5 µm (Q = 2.25–2.85, Q_m = 2.57 ± 0.13), subfusiform to subcylindrical with shallow suprahilar depression in side view, fusiform to cylindrical in ventral view, smooth, yellowish to brownish yellow in KOH, yellow to yellow-brown in Melzer's reagent. Hymenophoral trama boletoid composed of 3.5–10 µm wide. Cheilo- and pleurocystidia 32–60 × 7–11 µm, subfusiform to fusoid-ventricose, yellow to brownish yellow in KOH, thin-walled. Pileipellis a trichoderm to palisadoderm composed of 5–13 µm wide interwoven filamentous hyphae, brownish to yellowish brown in KOH and yellow-brown to dark brown in Melzer's reagent; terminal cells 30–85 × 4–13 µm, clavate to subcylindrical. Pileal trama composed of 3.5–10 µm wide interwoven hyphae, yellowish to pale brownish in KOH and yellowish brown to brownish in Melzer's reagent. Clamp connections absent in all tissues.

Fig. 18.5 a *Retiboletus pseudogriseus* N.K. Zeng & Zhu L. Yang. Photos by F. Li (KUN-HKAS 77694).

Fig. 18.5 b *Retiboletus pseudogriseus* N.K. Zeng & Zhu L. Yang. Photo by N.K. Zeng (KUN-HKAS 83950, type).

Habitat: Scattered on soil in tropical forests dominated by plants of the family Fagaceae.

Known distribution: Currently known from southeastern and southern China.

Specimens examined: FUJIAN PROVINCE: Zhangping County, Xinqiao Town, Chengkou Village, alt. 360 m, 1 September 2009, N.K. Zeng 647 (KUN-HKAS 106467), the same location, 3 September 2009, N.K. Zeng 668 (KUN-HKAS 83950, type). GUANGDONG PROVINCE: Fengkai County, Heishiding Nature Reserve, alt. 250 m, 13 August 2012, F. Li 775 (KUN-HKAS 77694).

Commentary: *Retiboletus pseudogriseus* originally described from southeastern China by Zeng *et al.* (2018) is characterized by the grayish brown to blackish brown and then grayish brown to yellowish brown pileus, the grayish to white context without discoloration when bruised, the whitish to cream and then grayish yellow to yellowish hymenophore, the entirely reticulate stipe, the trichoderm to palisadoderm pileipellis composed of 3.5–10 μm wide interwoven hyphae, and the distribution in tropical forests. *Retiboletus pseudogriseus* is similar to *R. fuscus* in the color of pileus and context. However, *R. fuscus* has a relatively dark pileus when young, relatively narrow (4–8 μm wide) hyphae in pileipellis, and a distribution in temperate to subtropical forests. Species to which *R. pseudogriseus* is phylogenetically related remain as yet unknown (Fig. 4.1).

18.6 *Retiboletus zhangfeii* N.K. Zeng & Zhu L. Yang, Mycologia 108: 370 (2016)

Basidioma medium-sized to large. Pileus 5–12 cm in diam., subhemispherical to convex or plano-convex to applanate, gray to smoky-gray, always with dark purplish green tinge, slightly darker in the center; surface dry, nearly glabrous to fibrillose; context white to grayish white when young, dingy white to pale greenish when mature or aged, staining at first purplish and then smoky-gray when injured. Hymenophore adnate when young, adnate to depressed around apex of stipe when mature; surface grayish when young and becoming grayish pink when mature, staining purplish brown at first and then smoky-gray when injured; pores roundish to angular, 0.5–1 mm wide; tubes concolorous or a little paler than hymenophoral surface, up to 5–8 mm long, staining at first purplish brown and then smoky-gray when hurt. Spore print grayish red to grayish brown. Stipe 7–12 × 0.8–1.5 cm, clavate to subcylindrical, always enlarged downwards, olivaceous green to grayish olivaceous or olivaceous yellow; surface entirely covered with olivaceous green to blackish green reticulum, staining at first reddish brown and then smoky-gray when injured; context grayish green to olivaceous green, staining at first reddish brown to purplish brown and then smoky-gray when hurt; basal mycelium grayish white. Taste and odor mild.

Basidia 18–43 × 7–10 μm, clavate, 4-spored, rarely 2-spored, hyaline in KOH. Basidiospores 9.5–11.5 × 4–5 μm (Q = 2.11–2.75, Q_m = 2.44 ± 0.19), subcylindrical with shallow suprahilar depression in side view, cylindrical to fusiform in ventral view, smooth, olivaceous brown to brownish in KOH, yellow-brown to dark brown in Melzer's reagent. Hymenophoral trama boletoid; hyphae cylindrical. Cheilo- and pleurocystidia 25–40 × 9–12 μm, abundant, fusiform to subfusiform, yellow to brownish yellow in KOH, brown to yellow-brown in Melzer's reagent, thin-walled. Caulocystidia forming the reticulum over the surface of stipe, clavate to subfusiform or ventricose. Pileipellis a trichoderm composed of 3–6 μm wide filamentous hyphae, gray to grayish yellow in KOH and yellowish brown to dark brown in Melzer's reagent; terminal cells 9–54 × 3–6 μm, clavate to subcylindrical. Clamp connections absent in all tissues.

Habitat: Solitary on soil in subtropical to tropical forests dominated by plants of the family Fagaceae.

Known distribution: Currently known from China.

Specimens examined: CHINA, ZHEJIANG PROVINCE: Lishui, Baiyunshan Forest Park, alt. 200 m, 23 August 2015, Y.K. Li 158 (RITF 2703). YUNNAN PROVINCE: Chuxiong, Agriculture Market, alt. 1900 m, 25 August 2007, Z.L. Yang 4917 (KUN-HKAS 52234); Nanhua County, Wild Mushroom Market, altitude unknown, 2 August 2009, Y.C. Li 1951 (KUN-HKAS 59699). HUNAN PROVINCE: Yizhang County, Mangshan National Forest Park, alt. 1800 m, 3 September 2007, Y.C. Li 1073 (KUN-HKAS 53418), the same location and date, Y.C. Li 1075 (KUN-HKAS 53420). JIANGXI PROVINCE: Fuzhou, Dagang Town, Shuyuan Village, alt. 40 m, 18 June 2012, G. Wu 912 (KUN-HKAS 77084).

Commentary: *Retiboletus zhangfeii* originally described from China by Zeng *et al.* (2016) is characterized by the gray to smoky-gray pileus always with dark purplish green tinge, the grayish to grayish pink hymenophore, the initially white to dingy white and then pale greenish context which is at first staining purplish and then smoky-gray when injured, the entirely reticulate stipe, the trichoderm pileipellis composed of 3–6 μm wide hyphae, and the distribution in tropical to subtropical forests. In China, *R. zhangfeii* has been previously misidentified as *R. nigerrimus* originally described from Papua New Guinea (Heim 1963). However, *R. nigerrimus* differs from *R. zhangfeii* in its blue tinged pileus, lemon-yellow context of pileus and orange context of the base of stipe, relatively large basidiospores measuring 11.5–14.5 × 3.6–4.6 μm.

In our multi-locus phylogenetic analysis (Fig. 4.1), *R. zhangfeii* is closely related to the sympatric species *R. nigrogriseus*. However, *R. nigrogriseus* differs from *R. zhangfeii* in its black to grayish black pileus which is always with olivaceous tinge, white context staining brownish to fuliginous when bruised, white to grayish white hymenophore without lilac to purplish tinge, fine hymenophore pores which are 0.3–0.5 mm wide, and yellowish green context at base of stipe.

Fig. 18.6 *Retiboletus zhangfeii* N.K. Zeng & Zhu L. Yang. Photos by Y.C. Li (KUN-HKAS 53418).

Chapter 19
Royoungia Castellano, Trappe & Malajczuk

Royoungia Castellano, Trappe & Malajczuk, Aust Syst Bot 5: 614 (1992)

Type species: *Royoungia boletoides* Castellano, Trappe & Malajczuk, Aust Syst Bot 5(5): 614 (1992).

Diagnosis: Species in this genus is stipitate-pileate or gasteroid. If stipitate-pileate, it is different from all the other genera in the Boletaceae in its white to cream pileal context and yellow to bright yellow context of stipe but chrome-yellow to golden yellow at base, without discoloration when injured, pink to reddish scabrous squamules on the surface of stipe, initially white to pinkish and then pinkish to pink hymenophore, trichoderm to hyphoepithelium pileipellis. If gasteroid, it differs from other gasteroid genera in the Boletaceae in its yellow to bright yellow or chrome-yellow peridium, columella and rhizomorph without discoloration when bruised, and brown to dark chocolate-brown gleba.

Basidioma stipitate-pileate or gasteroid. If stipitate-pileate, pileus dry, nearly glabrous, or fibrillose to tomentose, or sometimes rugose, without discoloration when bruised; context whitish to white, without discoloration when hurt. Hymenophore initially adnate and then adnate to depressed around apex of stipe when mature; surface initially white and then pinkish to pink, without discoloration when injured; pores roundish to subangular; tubes concolorous or much paler than hymenophoral surface, without discoloration when injured. Spore print pale red to grayish red. Stipe central, yellowish to yellow upwards, bright yellow to chrome-yellow downwards; basal mycelium bright yellow to chrome-yellow. Basidiospores smooth, elongated to cylindrical or fusiform. Hymenial cystidia fusiform to subfusiform or fusoid-ventricose. Pileipellis a trichoderm composed of interwoven filamentous hyphae, or a hyphoepithelium with an outer layer composed of filamentous hyphae and an inner layer composed of globose to subglobose concatenated cells. Clamp connections absent in all tissues. If gasteroid, basidioma subglobose, yellow to bright yellow, without discoloration when bruised, nearly glabrous to finely tomentose; gleba initially brownish to brown and then dark brown to dark chocolate-brown when mature; loculi irregularly shaped. Rhizomorphs yellow to bright yellow. Clamp connections absent in all tissues.

Commentary: *Royoungia* was originally described from Australia by Castellano *et al.* (1992) based on a gasteroid species, namely *Ro. boletoides* Castellano, Trappe & Malajczuk. This genus has been traditionally considered to be a gasteroid genus. However, our present and previous molecular phylogenetic studies indicate that *Royoungia* include species with both boletoid (stipitate-pileate) and sequestrate phenotypes (Wu *et al.* 2016a). Currently, four species are known from China, and all of them with stipitate-pileate phenotypes.

Key to the species of the genus *Royoungia* in China

1. Basidioma distributed in tropical forests; stipe surface without reticulum; pileus with green or red tinge, without yellow tinge; pileipellis a hyphoepithelium composed of filamentous hyphae and inflated cells, or a palisadoderm composed of broad vertically arranged hyphae ·············· 2
1. Basidioma distributed in subtropical forests; stipe surface distinctly reticulate, or sometimes without reticulum; pileus pale yellow, grayish green, yellowish olive or grayish olive; pileipellis a trichoderm composed of interwoven filamentous hyphae ·············· *Ro. reticulata*
2. Basidioma very small; pileal surface with dark red, brownish red or crimson tinge, without green or olive tinge ·············· 3
2. Basidioma small; pileal surface greenish yellow to grayish green or olive to olive-gray ·············· *Ro. grisea*
3. Pileal surface dark red or brownish red; hymenial cystidia present, relatively narrow (3.5–6 μm wide); pileipellis a palisadoderm composed of 7–23 μm wide vertically arranged filamentous hyphae ·············· *Ro. rubina*
3. Pileal crimson to dark red; hymenial cystidia rare, relatively broad (7–11 μm wide); pileipellis a hyphoepithelium, with the outer layer consisting of 2.5–6 μm wide interwoven hyphae and inner layer made up of 10–16 μm wide subglobose concatenated cells ·············· *Ro. coccineinana*

19.1 *Royoungia coccineinana* (Corner) Yan C. Li & Zhu L. Yang, Fungal Divers 81: 119 (2016)

Basionym: *Boletus coccineinanus* Corner, *Boletus* in Malaysia 152 (1972).

Synonym: *Tylopilus coccineinanus* (Corner) E. Horak, Malay Fores Rec 51: 77 (2011).

Basidioma very small. Pileus 1.2–1.5 cm in diam., subhemispherical to convex, dark red to blackish red or crimson; surface rugose to roughened, viscid when wet and tomentose when dry; context whitish to cream, without color change when injured. Hymenophore adnate to depressed around apex of stipe; surface initially white and then pinkish; pores angular to roundish, 1–1.5 mm wide; tubes up to 3 mm long, concolorous or a little paler than hymenophoral surface, without discoloration when injured. Spore print pale red to grayish red. Stipe 2–5 × 0.2–0.3 cm, clavate to subcylindrical, red to dark red, but cream to yellowish at apex and chrome-yellow or golden yellow at base; surface covered with concolorous scabrous squamules, without color change when hurt; context yellowish to yellow, but chrome-yellow or golden yellow at base, without color change when bruised; basal mycelium chrome-yellow or golden yellow. Taste and odor mild.

Basidia 26–43 × 8–11 μm, clavate, 4-spored, hyaline in KOH. Basidiospores 9–11.5 × 4.5–5 μm (Q = 2–2.33, Q_m = 2.13 ± 0.08), elongated to subcylindrical with shallow suprahilar depression in side view, elongated to cylindrical in ventral view, smooth, hyaline to yellowish in KOH, yellow to brownish yellow in Melzer's reagent. Hymenophoral trama boletoid composed of 4.5–11 μm wide filamentous hyphae, hyaline to yellowish in KOH, yellowish to yellow in Melzer's reagent. Cheilo- and pleurocystidia 31–50 × 7–11 μm, rare, clavate (Fig. 67c in Wu et al. 2016a). Caulocystidia forming the scabrous squamules over the surface of stipe, subfusiform to lanceolate or clavate (Fig. 67e in Wu *et al.* 2016a). Pileipellis a hyphoepithelium (Fig. 67d in Wu *et al.* 2016a): outer layer consisting of 2.5–6 μm wide interwoven hyphae with clavate to subcylindrical terminal cells 18–53 × 2.5–4.5 μm, pale yellowish to brownish yellow in KOH and yellow to yellow-brown in Melzer's reagent; inner layer made up of 10–16 μm wide subglobose concatenated cells arising from 4–7 μm wide radially arranged filamentous hyphae. Pileal trama composed of broad (up to 15 μm wide) interwoven hyphae, hyaline to yellowish in KOH and yellowish to brownish in Melzer's reagent. Clamp connections absent in all tissues.

Habitat: Scattered on soil in tropical forests dominated by plants of the family Fagaceae.

Known distribution: Currently known from Malaysia and southwestern China.

Specimens examined: CHINA, YUNNAN PROVINCE: Ruili, Ruili Botanic Garden, alt. 900 m, 7 October 2010, Y.C. Li 2165 (KUN-HKAS 68927); Menghai County, Bulangshan Town, alt. 1200 m, 22 October 2019, L.K. Jia 645 (KUN-HKAS 107311). GUIZHOU PROVINCE: Leishan County, Leigong Mountain, alt. 700 m, 22 June 2017, R. Zhang 400 (KUN-HKAS 101830).

Commentary: *Royoungia coccineinana* originally described as *Boletus coccineinanus* by Corner (1972) is characterized by the crimson to dark red pileus, the white to cream pileal context without discoloration when bruised, the initially white and then pinkish hymenophore, the red to dark red stipe with a bright yellow to chrome-yellow base, the rare and clavate hymenial basidia, and the hyphoepithelium pileipellis. This species is phylogenetically related and morphologically similar to *Ro. rubina* Yan C. Li & Zhu L. Yang. However, *Ro. rubina* has a dark red or brownish red pileus which is always with brownish tinge, a palisadoderm pileipellis, and abundant hymenial cystidia.

194 The Boletes of China: *Tylopilus* s.l.

Fig. 19.1 *Royoungia coccineinana* (Corner) Yan C. Li & Zhu L. Yang. Photos by L.K. Jia (KUN-HKAS 107311).

19.2 *Royoungia grisea* Yan C. Li & Zhu L. Yang, Fungal Divers 81: 120 (2016)

Basidioma small. Pileus 3–4.5 cm in diam., subhemispherical or convex to plano-convex, initially greenish yellow to grayish green and then olive to olive-gray, darker in the center; surface dry, tomentose to fibrillose, sometimes minutely cracked with age; context grayish white to whitish, without discoloration when injured. Hymenophore adnate to depressed around apex of stipe; surface initially pallid to whitish and then pinkish to pink, without discoloration when bruised; pores subangular to roundish, 0.5–1 mm wide; tubes up to 10 mm long, concolorous or much paler than hymenophoral surface, without discoloration when injured. Spore print pale red to grayish red. Stipe 3–5 × 0.8–1.2 cm, clavate, always enlarged downwards, white to cream or yellowish, but chrome-yellow or golden yellow at base; surface densely covered with red to pink-red scabrous squamules, without color change when hurt; context whitish to cream in the upper part, yellowish to yellow downwards, but chrome-yellow or golden yellow at base, without color change when bruised; basal mycelium golden yellow to chrome-yellow. Taste and odor mild.

Fig. 19.2 *Royoungia grisea* Yan C. Li & Zhu L. Yang. Photos by Y.J. Hao (KUN-HKAS 80110).

Basidia 25–37 × 10–12 μm, clavate, 4-spored, hyaline in KOH. Basidiospores 11–14 × 5.5–6.5 μm (Q = 1.82–2.35, Q_m = 2.1 ± 0.11), elongated to subcylindrical and inequilateral in side view with slight suprahilar depression, elongated to cylindrical in ventral view, smooth, hyaline to yellowish in KOH, yellow to brownish yellow in Melzer's reagent. Hymenophoral trama boletoid composed of 4–9 μm wide filamentous hyphae, hyaline to yellowish in KOH, yellowish to yellow in Melzer's reagent. Cheilo- and pleurocystidia 49–70 × 5–7.5 μm, narrowly fusiform to subfusiform or ventricose, sometimes with 1–2 constrictions in the middle parts, hyaline to yellowish in KOH, thin-walled. Caulocystidia forming the scabrous squamules over the surface of stipe, subfusiform, lanceolate, clavate or ventricose. Pileipellis a hyphoepithelium: outer layer composed of 4–8 μm wide interwoven hyphae with clavate to subcylindrical terminal cells 32–46.5 × 4.5–8.5 μm, pale yellowish to brownish yellow in KOH and yellow to brownish in Melzer's reagent; inner layer composed of 10–20 μm wide subglobose concatenated cells arising from radially arranged filamentous hyphae, hyaline to yellowish in KOH and yellowish to brownish in Melzer's reagent. Pileal trama composed of broad interwoven hyphae (up to 12 μm wide), hyaline to yellowish in KOH and yellowish to brownish in Melzer's reagent. Clamp connections absent in all tissues.

Habitat: Scattered on soil in tropical forests dominated by plants of the family Fagaceae.

Known distribution: Currently known from central, southwestern, and southern China.

Specimens examined: CHINA, GUIZHOU PROVINCE: Suiyang County, Kuankuoshui National Nature Reserve, alt. 1100 m, 14 July 1983, X.L. Wu 820 (KUN-HKAS 14470). HUBEI PROVINCE: Xingshan County, Muyu Town, alt. 1800 m, 16 July 2012, J. Qin 554 (KUN-HKAS 77955). GUANGDONG PROVINCE: Fengkai County, Heishiding Nature Reserve, alt. 250 m, 10 May 2012, F. Li 217 (KUN-HKAS 90205), the same location, 2 June 2013, Y.J. Hao 830 (KUN-HKAS 80110). HENAN PROVINCE: Nanyang, Neixiang County Xiaguan Town, Qinggangshu Village, Taohuayuan, alt. 300 m, 31 July 2010, X.F. Shi 426 (KUN-HKAS 90183, type).

Commentary: *Royoungia grisea*, originally described by Wu *et al.* (2016a) from China, is characterized by the initially greenish yellow to grayish green and then olive to olive-gray pileus, the white pileal context without discoloration when injured, the initially white and then pinkish hymenophore, the white to cream context of stipe but yellowish to yellow downwards and chrome-yellow or golden yellow at base, the red to pink-red scabrous stipe, and the hyphoepithelium pileipellis. This species is very similar to *Chiua virens* in macroscopic appearance. However, *C. virens* has a yellow to yellowish scabrous stipe without any red tinge, a trichoderm pileipellis composed of filamentous hyphae, and narrow basidiospores measuring 11.5–13.5 × 5–5.5 μm.

In our multi-locus phylogenetic analysis (Fig. 4.1), *Ro. grisea* is closely related to *Ro. reticulata* Yan C. Li & Zhu L. Yang. Indeed, they are morphologically similar to each other. However, *Ro. reticulata* differs from *Ro. grisea* in its large basidiospores measuring 13–16 × 5.5–6.5 μm, short hymenial cystidia measuring 28–42× 4.5–6.5 μm, and trichoderm pileipellis.

19.3 *Royoungia reticulata* Yan C. Li & Zhu L. Yang, Fungal Divers 81: 121 (2016)

Basidioma small. Pileus 3–5 cm in diam., subhemispherical to convex or plano-convex to applanate, grayish green to yellowish olive when young, grayish olive to pale yellow when mature, darker in the center, staining pale red to brownish red when bitten by insects; surface dry, fibrillose to tomentose, without discoloration when bruised; context whitish without discoloration when hurt. Hymenophore adnate when young, depressed around apex of stipe when mature; surface initially white and then pinkish, without discoloration when injured; pores subangular to roundish 1–1.5 mm wide; tubes 6–15 mm long, concolorous or much paler than hymenophoral surface, without discoloration when injured. Spore print pale red to grayish red (Fig. 3.4 j). Stipe 3–6 × 1–1.5 cm, clavate to subcylindrical, usually enlarged downwards, white to cream, bright yellow to chrome-yellow or golden yellow at base; surface sometimes longitudinally striated over upper 1/3, and anastomosing downwards, middle parts covered with pink to red-pink scabrous squamules, without color change when hurt; context white to cream, but yellowish to yellow downwards and chrome-yellow or golden yellow at base, without color change when hurt; basal mycelium chrome-yellow or golden yellow. Taste and odor mild.

Basidia 20–38 × 8–12 μm, clavate, 4-spored, hyaline in KOH. Basidiospores 13–16 × 5.5–6.5 μm [Q = 2–2.42 (2.67), Q_m = 2.39 ± 0.13], elongated to subcylindrical with slight suprahilar depression in side view, elongated to cylindrical in ventral view, smooth, hyaline to yellowish in KOH, yellow to brownish yellow in Melzer's reagent. Hymenophoral trama boletoid composed of 3–13 μm wide filamentous hyphae, hyaline to yellowish in KOH, yellowish to yellow in Melzer's reagent. Cheilo- and pleurocystidia rare, 28–42 × 4.5–6.5 μm, fusiform to subfusiform or ventricose, hyaline to yellowish in KOH, yellowish to yellow in Melzer's reagent, thin-walled. Caulocystidia forming the scabrous squamules over the surface of stipe, fusiform to subfusiform or clavate to pyriform, concolorous with cheilo- and pleurocystidia. Pileipellis

Fig. 19.3 a *Royoungia reticulata* Yan C. Li & Zhu L. Yang. Photo by Z.L. Yang (KUN-HKAS 52253, type).

Fig. 19.3 b *Royoungia reticulata* Yan C. Li & Zhu L. Yang. Photo by Y.C. Li (KUN-HKAS 107181).

a trichoderm composed of 3–6 μm wide interwoven hyphae; terminal cells 22–36 × 3–5.5 μm, clavate, yellowish to pale brownish in KOH and yellowish brown to brownish in Melzer's reagent. Pileal trama composed of 7–10 μm wide interwoven hyphae, colorless to yellowish in KOH and yellow to yellowish brown in Melzer's reagent. Clamp connections absent in all tissues.

Habitat: Scattered on soil in subtropical forests dominated by plants of the family Fagaceae.

Known distribution: Currently known from southwestern China.

Specimens examined: CHINA, YUNNAN PROVINCE: Kunming, Kunming Botanic Garden, alt. 1980 m, 31 August, 2007, Z.L. Yang 4936 (KUN-HKAS 52253, type), the same location, 30 July, 2007, Y.C. Li 960 (KUN-HKAS 52647), the same location, 14 July, 2007, Z.L. Yang 4773 (KUN-HKAS 90196), the same location, 5 August 2005, Z.L. Yang 4604B (KUN-HKAS 90197), the same location, 27 September 2015, Y.C. Li 2842 (KUN-HKAS 107181); Kunming, Yeya Lake, alt. 2100 m, 13 May 2012, B. Feng 1335 (KUN-HKAS 82477), the same location, 13 October 2012, G. Wu 1099 (KUN-HKAS 77271), the same location, 22 October 2011, Z.L. Yang 5604 (KUN-HKAS 71089); Nanhua County, Wild Mushroom Market, altitude unknown, 3 August 2009, Y.C. Li 1957 (KUN-HKAS 59704).

Commentary: *Royoungia reticulata*, originally described by Wu *et al.* (2016a) from China, is characterized by the grayish green to yellowish olive or pale yellow pileus, the whitish context without discoloration when injured, the initially whitish to pallid and then pinkish to pink hymenophore, the reticulate stipe with pink to red-pink scabrous squamules, the cream to yellowish context of stipe but chrome-yellow or golden yellow at base, and the trichoderm pileipellis composed of 3–6 μm wide interwoven hyphae.

Royoungia reticulata is phylogenetically sister and morphologically similar to *Ro. grisea*. However, *Ro. grisea* differs from *Ro. reticulata* in that it has small basidiospores 11–14 × 5.5–6.5 μm, long hymenial cystidia 49–70 × 5–7.5 μm, and hyphoepithelium pileipellis.

19.4 *Royoungia rubina* Yan C. Li & Zhu L. Yang, Fungal Divers 81: 122 (2016)

Basidioma very small. Pileus 1–1.2 cm in diam., subhemispherical to broadly convex, dark red or brownish red always with brownish tinge, darker in the center; surface dry or slightly viscid when wet, finely rugose; context whitish to pallid or cream, without color change when injured. Hymenophore depressed around apex of stipe; surface initially whitish to pallid and then pinkish to pink, without discoloration when injured; pores angular to roundish, 0.5–1 mm wide; tubes up to 3 mm long, concolorous or much paler than hymenophoral surface, without discoloration when injured. Spore print pale red to grayish red. Stipe 1.5–2 × 0.2 cm, clavate, red to brownish red, but yellowish to yellow at apex and chrome-yellow to bright yellow at base; surface covered with red scabrous squamules, without color change when hurt; context yellowish to yellow, but chrome-yellow to golden yellow at base, without color change when hurt; basal mycelium chrome-yellow to golden yellow. Taste and odor mild.

Basidia 22–30 × 7–10 µm, clavate, 4-spored, hyaline in KOH. Basidiospores 10–11.5 × 4–5 µm (Q = 2.2–2.63, Q_m = 2.36 ± 0.15), subcylindrical to subfusiform in side view with slight suprahilar depression, cylindrical to fusiform in ventral view, smooth, hyaline to yellowish in KOH, yellow to brownish yellow in Melzer's reagent. Hymenophoral trama boletoid composed of 2.5–7 µm wide filamentous hyphae, hyaline to yellowish in KOH, yellowish to yellow in Melzer's reagent. Cheilo- and pleurocystidia 45–62 × 3.5–6 µm, narrowly clavate to narrowly subfusiform or lanceolate, with 3–5 constrictions in middle parts, colorless to yellowish in KOH, yellowish to yellow in Melzer's reagent, thin-walled. Pileipellis a palisadoderm composed of broad (7–23 µm wide) vertically arranged filamentous hyphae, hyaline to yellowish in KOH and yellow to brownish yellow in Melzer's reagent; terminal cells 28–150 × 7–16 µm, clavate. Pileal trama composed of 3.5–12 µm wide interwoven hyphae, colorless to yellowish in KOH and yellow to yellowish brown in Melzer's reagent. Clamp connections absent in all tissues.

Habitat: Solitary on soil in tropical forests dominated by *Castanopsis kawakamii*.

Known distribution: Currently known from southeastern China.

Specimen examined: CHINA, FUJIAN PROVINCE: Sanming, Sanyuan National Forest Park, alt. 260 m, 27 August 2007, Y.C. Li 1034 (KUN-HKAS 53379, type).

Commentary: *Royoungia rubina*, originally described by Wu *et al.* (2016a) from China, is characterized by the dark red or brownish red pileus which is always with brownish tinge, the white to cream context without discoloration when hurt, the initially whitish to pallid and then pinkish to pink hymenophore, the red to brownish red stipe which is yellowish to yellow at apex and chrome-yellow at base, and the palisadoderm pileipellis.

Royoungia rubina is phylogenetically sister and morphologically similar to *Ro. coccineinana*. However, *Ro. coccineinana* differs from *Ro. rubina* in its crimson to dark red pileus which is without brownish tinge, hyphoepithelium pileipellis with an upper layer composed of filamentous hyphae and an inner layer composed of erect moniliform hyphae, and rare and broad hymenial cystidia.

Fig. 19.4 *Royoungia rubina* Yan C. Li & Zhu L. Yang. Photos by Y.C. Li (KUN-HKAS 53379, type).

Chapter 20

Sutorius Halling, Nuhn & N.A. Fechner

Sutorius Halling, Nuhn & N.A. Fechner, Mycologia 104(4): 955 (2012)

Type species: *Boletus eximius* Peck, J Mycol 3: 54 (1887).

Diagnosis: This genus differs from other genera in Boletaceae in its differently colored hymenophoral surface and tubes, red to reddish brown discoloration of the basidiomata when injured, and darkened scabrous squamules over the surface of stipe.

Pileus hemispherical to subhemispherical or applanate; surface dry, nearly glabrous to subtomentose, staining red to reddish brown when injured; context white to pallid or grayish, always staining red or reddish brown when injured. Hymenophore adnate when young, adnate to slightly depressed around apex of stipe when mature; surface brown, dark brown, reddish brown or vinaceous brown, sometimes black, grayish black to blackish red, blackish brown, staining red to reddish brown when injured; pores nearly round to subangular; tubes whitish to grayish or pinkish gray, becoming ochraceous to reddish brown when injured. Spore print reddish brown to brown. Stipe central, covered with darkened scabrous squamules, staining reddish brown, sometimes without discoloration when touched in aged basidioma; context grayish to whitish or pallid, staining reddish brown when bruised, sometimes without discoloration when injured in aged basidioma; basal mycelium white to cream. Basidiospores smooth, cylindrical to fusiform. Hymenial cystidia abundant or rare, subfusiform to subfusoid-ventricose, sometimes clavate. Pileipellis a trichoderm composed of interwoven filamentous hyphae.

Commentary: *Sutorius*, typified by *S. eximius* (Peck) Halling, Nuhn & Osmundson, was proposed by Halling *et al.* (2012a) and included two species, namely *S. australiensis* (Bougher & Thiers) Halling & N.A. Fechner and *S. eximius*. In this study, we follow the circumscription of *Sutorius* in Halling *et al.* (2012a), Gelardi (2018) and Gelardi *et al.* (2019). The type species, *S. eximius*, is one of the most widely reported species in this genus. It was originally described from North America (Peck 1887), then reported from Australia, Costa Rica, China, Guyana, Japan, Papua New Guinea, Thailand, and Zambia (Singer 1947; Heim 1963; Smith and Thiers 1971; Hongo 1973, 1974b; Bougher and Thiers 1991; Halling and Mueller 2005; Fulgenzi *et al.* 2007; Wu *et al.* 2016a). Our phylogenetic study indicates that the Chinese collections identified as *S. eximius* include at least six species including two new species.

Key to the species of *Sutorius* in China

1. Species with a distribution in temperate forests associated with trees of the family Fagaceae, or in subtropical to alpine forests associated with trees of the family Pinaceae ·············· 2
1. Species with a distribution in tropical forests associated with trees of the family Fagaceae, or in tropical to subtropical forests associated with trees of the family Fagaceae or with trees of the families Fagaceae and Pinaceae ·············· 4
2. Basidiospores not more than 13 μm long; basidiomata distributed in subalpine to alpine forests ·············· *S. microsporus*
2. Basidiospores more than 13 μm long; Basidiomata distributed in temperate forests, or in subtropical to alpine forests ·············· 3
3. Pileus hemispherical to subhemispherical; hymenial cystidia abundant, subfusiform to ventricose; occurring in subtropical to alpine forests and associated with trees of the family Pinaceae ·············· *S. alpinus*
3. Pileus subhemispherical to applanate; hymenial cystidia rare; occurring in temperate forests and associated with trees of the families Fagaceae and Pinaceae ·············· *S. eximius*
4. Hymenophore pale reddish brown, brown to yellowish brown, without black to grayish black or blackish brown tinge; basidiospores mostly ⩾ 11.5 μm long ·············· 5
4. Hymenophore black to grayish black when young, blackish red to blackish brown when mature; basidiospores ⩽ 11.5 μm long ·············· *S. obscuripellis*
5. Basidiomata distributed throughout tropical forests dominated by plants of the family Fagaceae; pileus brown to dark brown and then reddish brown; hymenophore brownish red to brown without dark or blackish tinge ·············· *S. subrufus*
5. Basidiomata distributed throughout tropical to subtropical mixed forests dominated by plants of the families Fagaceae and Pinaceae; pileus dark violet-brown, grayish brown to reddish brown; hymenophore brownish gray, dark brown, blackish brown or brownish orange ···· *S. pseudotylopilus*

20.1 *Sutorius alpinus* Yan C. Li & Zhu L. Yang, The Boletes of China: *Tylopilus* s.l. 239 (2021)

MycoBank: MB 834745

Etymology: The epithet "*alpinus*" refers to the habitat of this species.

Type: CHINA, YUNNAN PROVINCE: Kunming, Xishan Mountain, alt. 2050 m, 10 August 2007, Y.C. Li 985 (KUN-HKAS 52672, GenBank Acc. No.: KF112399 for nrLSU, KF112207 for *tef1-α*, KF112584 for *rpb1*, KF112802 for *rpb2*).

Basidioma small to large. Pileus 4–15 cm in diam., hemispherical to subhemispherical when young, subhemispherical to applanate or plano-convex when mature, reddish brown (9E7–8) to red-brown (8D5–7) when young, brown (7E6–7) to brownish (7D5–6) when mature; surface covered with finely matted squamules, dry, margin with a slight sterile extension; context grayish white (13B1) to whitish (14B1), without color change or sometimes staining reddish brown (9E7–8) when injured. Hymenophore adnate when young, adnate to slightly depressed around apex of stipe when mature; surface grayish red (12C2) to dark reddish purple (13D3) or reddish purple (14D3), sometimes violet-brown (11E5) to ruby red (12E8) or grayish red (11D3–4) when young, and then becoming reddish brown (9D4–5) to brownish (7D4–5) or pale brownish (6D5–6) when mature, staining reddish brown when injured; pores angular, 0.3–1 mm wide; tubes 6–15 mm long, purplish gray (13C2) to dark purplish gray (14C2), staining reddish brown. Spore print reddish brown (7C5) to brownish (7D6) (Fig. 3.4 k). Stipe 5–12 × 1.2–3 cm, subcylindrical or attenuate downwards, pale grayish (10C1), grayish (10D1) or gray (11C1), but deeper in color downwards, covered with blackish squamules, staining reddish brown when bruised; context white to grayish, staining reddish brown when injured; basal mycelium white. Taste and odor mild.

Basidia 28–50 × 11–15 μm, clavate, 4-spored, hyaline in KOH. Basidiospores (12) 14–15.5 (18) × (4) 4.5–5.5 (6.5) μm [Q = (2.23) 2.73–3.57 (3.75), Q_m = 3.27 ± 0.35], subcylindrical to subfusiform in side view with slight suprahilar depression, cylindrical to fusiform in ventral view (Fig. 3.3 g), smooth, yellowish to brownish yellow in KOH, yellow to yellow-brown in Melzer's reagent. Hymenophoral trama boletoid composed of 3–12 μm wide filamentous hyphae, hyaline to yellowish in KOH, yellowish to yellow in Melzer's reagent. Cheilocystidia 26–42 × 5.5–7.5 μm, fusiform to subfusiform or ventricose, usually pale yellow to brownish in KOH, thin-walled. Pleurocystidia similar to cheilocystidia in form but much bigger, 40–56 × 10–15 μm. Caulocystidia forming the scabrous squamules over the surface of stipe, ventricose, fusoid-ventricose, clavate or subfusiform with a sharp apex. Pileipellis a trichoderm composed of 3–6 μm wide filamentous hyphae with clavate to subcylindrical terminal cells 31–77 × 3–7.5 μm, yellowish brown to brownish in KOH and yellow-brown to dark brown in Melzer's reagent. Pileal trama composed of 7–15 μm wide interwoven hyphae, yellowish to pale brownish in KOH and yellowish brown to brownish in Melzer's reagent. Clamp connections absent in all tissues.

Habitat: Scattered on soil in subtropical to alpine forests dominated by plants of the family Pinaceae.

Known distribution: Known from southwestern China.

Additional specimens examined: CHINA, YUNNAN PROVINCE: Jianchuan County, Laojun Mountain, alt. 2700 m, 2 August 2005, Y.C. Li 293 (KUN-HKAS 48526); Heqing County, alt. 3000 m,

Fig. 20.1 a *Sutorius alpinus* Yan C. Li & Zhu L. Yang. Photos by B. Feng (KUN-HKAS 68595).

28 July 2006, Y.C. Li 661 (KUN-HKAS 50415); Yulong County, Yulong Reservoir, alt. 3100 m, 29 July 2006, Y.C. Li 666 (KUN-HKAS 50420); Yulong County, Shigu Town, alt. 3000 m, 17 July 2008, Y.C. Li 1242 (KUN-HKAS 56096); Kunming, Xishan Mountain, alt. 2050 m, 10 August 2007, Y.C. Li 985 (KUN-HKAS 52672); Yongping County, on the way from Yongping to Baoshan, at the landmark of 3295 km, alt. 2080 m, 3 July 2009 Y.C. Li 1909 (KUN-HKAS 59657); Lanping County, Tongdian Town, alt. 2400 m, 13 August 2010, B. Feng 814 (KUN-HKAS 68595).

Fig. 20.1 b *Sutorius alpinus* Yan C. Li & Zhu L. Yang. Photo by Y.C. Li (KUN-HKAS 52672, type).

Commentary: *Sutorius alpinus* is characterized by the reddish brown or red-brown then brown to brownish pileus, the grayish red to dark reddish purple or reddish purple hymenophoral surface, the purplish gray to dark purplish gray tubes, the blackish scabrous stipe, the distribution in subtropical to alpine forests in southwestern China, and the association with trees of the family Pinaceae. For line drawings of *S. alpinus* see that of *S. eximius* in Wu *et al.* (2016a). *Sutorius alpinus* is macroscopically similar to the other species in the *S. eximius* group, viz. *S. microsporus* Yan C. Li & Zhu L. Yang, *S. eximius*, *S. obscuripellis* Vadthanarat, Raspé & Lumyong, *S. pseudotylopilus* Vadthanarat, Raspé & Lumyongg and *S. subrufus* N.K. Zeng, H. Chai & S. Jiang. However, *S. microsporus* is mainly characterized by the short basidiospores (11–13 × 4–5 μm), and the rare and small hymenial cystidia (see our description of *S. microsporus* below). *Sutorius eximius* is characterized by the rare and small hymenial cystidia, the distribution in temperate forests in central and northeastern China and the association with trees of the family Fagaceae (Halling *et al.* 2012a). *Sutorius obscuripellis* is characterized by the black to grayish black then blackish red to blackish brown pileus, the short basidiospores (9–11.5 × 3.5–4.5 μm), the distribution in tropical, and the association with trees of the family Fagaceae. *Sutorius pseudotylopilus* is characterized by the small basidiospores (11–13 × 4–4.5 μm), the rare and small hymenial cystidia, the distribution in tropical to subtropical forests, and the association with trees of the families Fagaceae and Pinaceae. *Sutorius subrufus* is characterized by the slender basidiospores measuring 8–14 × 3–4.5 μm, the distribution in tropical forests and associated with trees of the family Fagaceae.

20.2 *Sutorius eximius* (Peck) Halling, Nuhn & Osmundson, Mycologia 104 (4): 955 (2012)

Basionym: *Boletus eximius* Peck, J Mycol 3(5): 54 (1887).

Synonyms: *Suillus eximius* (Peck) Kuntze, Revisio generum plantarum 3: 535 (1898); *Tylopilus eximius* (Peck) Singer, Am Midl Nat 37: 109 (1947); *Leccinum eximium* (Peck) Singer, Persoonia 7(2): 319 (1973).

Fig. 20.2 a *Sutorius eximius* (Peck) Halling, Nuhn & Osmundson. Photo by J. Li (KUN-HKAS 91261).

Basidioma small to large. Pileus 4–10 cm in diam., hemispherical to applanate, brown, dark brown to reddish brown; surface matted to tomentose, dry, margin with a slight sterile extension; context white to grayish white, always with reddish brown to brownish tinge when injured. Hymenophore adnate when young, adnate to depressed around apex of stipe; surface lilac-white, brownish to violet-brown when young, brown to chestnut-brown with age, staining reddish brown when injured; pores angular, 0.3–1 mm wide; tubes 4–12 mm long, concolorous or a little paler than hymenophoral surface, staining reddish brown when injured. Spore print reddish brown to brownish. Stipe 3–8 × 0.8–2.5 cm, cylindrical to clavate, grayish white to blackish white, densely covered with dark brown to blackish brown dotted squamules; context white to grayish, staining reddish brown when injured; basal mycelium white. Taste and odor mild.

Basidia 24–30 × 9–11 μm, clavate, 4-spored, hyaline in KOH. Basidiospores (9) 10–16 (17) × 4–5.5 (6) μm [Q = (2.22) 2.25–3.4 (3.5), Q_m = 2.84 ± 0.24], subcylindrical to subfusiform in side view with slight suprahilar depression, cylindrical to fusiform in ventral view, smooth, somewhat thick-walled (up to 1 μm thick), yellowish to brownish yellow in KOH, yellow to yellow-brown in Melzer's reagent. Hymenophoral trama boletoid composed of 2–8 μm wide filamentous hyphae, hyaline to yellowish in KOH, yellowish to yellow in Melzer's reagent. Cheilocystidia 16–21 × 7–9 μm, rare, subfusiform to clavate, hyaline to yellowish in KOH. Pleurocystidia 25–38 × 6.5–10 μm, rare, subfusiform to fusoid-ventricose. Caulocystidia forming the scabrous squamules over the surface of stipe, clavate, subfusiform, or subfusoid-ventricose. Pileipellis a trichoderm composed of 4–7 μm wide filamentous hyphae with clavate to subcylindrical terminal cells 21–40 × 5–7 μm, pale yellowish brown to pale brownish in KOH and yellowish brown to brownish in Melzer's reagent. Pileal trama composed of 2–7 μm wide interwoven hyphae, hyaline to yellowish in KOH and yellowish to brownish in Melzer's reagent. Clamp connections absent in all tissues.

Habitat: Scattered on soil in temperate forests dominated by plants of the families Fabaceae, Fagaceae, Dipterocarpaceae, and Pinaceae.

Known distribution: Known from North America and central and northeastern China.

Specimens examined: CHINA, HENAN PROVINCE: Nanyang, Neixiang County, alt. 300 m, 31 July 2010, X.F. Shi 417 (KUN-HKAS 107184). HUBEI PROVINCE: Xingshan County, Muyu Town, alt. 1700 m, 16 July 2012, J. Qin 544 (KUN-HKAS 77945). LIAONING PROVINCE: Benxi County, Xiamatang Town, Majia Village, alt. 320 m, 22 August 2015, J. Li 233 (KUN-HKAS 91261).

Commentary: *Sutorius eximius* has been well documented by Halling *et al.* (2012a) and references therein. Wu *et al.* (2016a) reported the presence of *S. eximius* in southwestern China. However, our multi-locus phylogenetic analysis (Fig. 4.1) indicates that collections from southwestern China represent two independent species, namely *S. alpinus* and *S. microsporus*. Indeed, *S. eximius* occurs in China, but it is restricted to central and northeastern China. For comparisons of *S. alpinus*, *S. microspores*, and *S. eximius* see the commentary on *S. alpinus*.

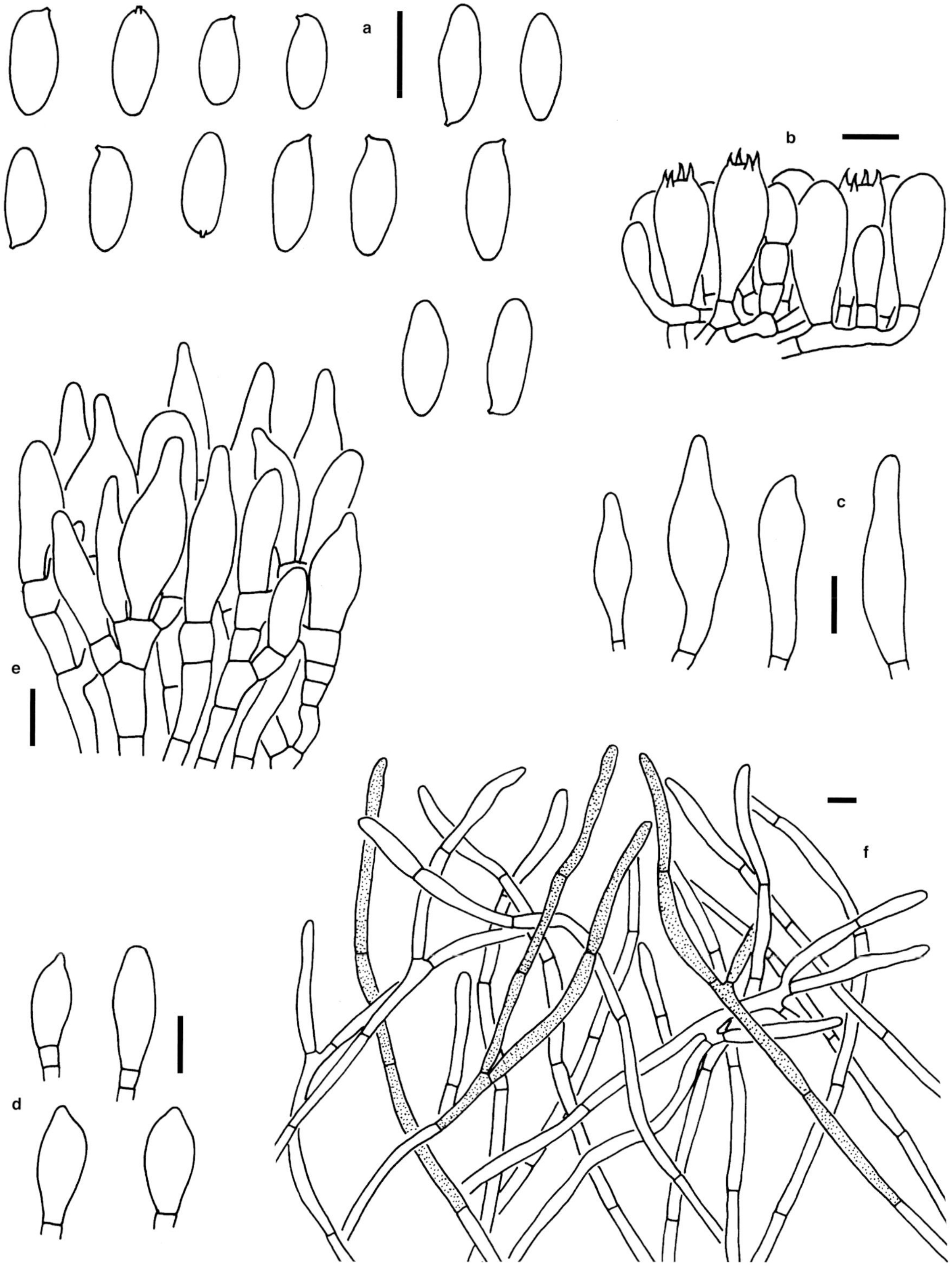

Fig. 20.2 b *Sutorius eximius* **(Peck) Halling, Nuhn & Osmundson (KUN-HKAS 91261).**
a. Basidiospores; b. Basidia; c. Pleurocystidia; d. Cheilocystidia; e. Squamules on the surface of stipe; f. Pileipellis. Scale bars = 10 μm.

20.3 *Sutorius microsporus* Yan C. Li & Zhu L. Yang, The Boletes of China: *Tylopilus* s.l. 245 (2021)

MycoBank: MB 834747

Etymology: The epithet "*microsporus*" refers to the size of the basidiospores.

Type: CHINA, YUNNAN PROVINCE: Yulong County, Taian Town, alt. 2600 m, 20 August 2010, B. Feng 939 (KUN-HKAS 68720, GenBank Acc. No.: MT154773 for nrLSU).

Basidioma small to large. Pileus 4–12 cm in diam., hemispherical to subhemispherical; dark brown (8F7–8), black-brown (9F7–8), violet-brown (10F7–8) when young, brownish (7D4–5), reddish brown (8D4–5) to brownish orange (6C3–4) when mature; surface covered with tomentose to matted squamules, dry, margin with a distinct sterile extension; context dingy white to grayish, becoming brownish orange (5C3–4) to grayish orange (5B3–4) when injured. Hymenophore adnate when young, adnate to depressed around apex of stipe when mature; surface pale reddish brown (9E4–5) to violet-brown (10F6–7), staining reddish brown when bruised; pores subangular, 0.3–1 mm wide; tubes 6–12 mm long, brownish orange (5C4–5) to pale brownish gray (6C2), staining reddish brown when bruised. Spore print reddish brown (7C5) to brownish (7D6). Stipe 4–8 × 1.8–2.5 cm, cylindrical to clavate, pale brown (7E4–5) to grayish brown (8E3–4), covered with dark brown (7F4–5) to blackish brown (8F5–6) dotted squamules; staining reddish brown when bruised; context white to grayish, staining reddish brown when bruised; basal mycelium white to cream. Taste and odor mild.

Basidia 22–35 × 10–13 μm, clavate, 4-spored, hyaline in KOH. Basidiospores [60/3/2] 11–13 × 4–4.5 (5) μm [Q = 2.44–3 (3.25), Q_m = 2.79± 0.17], subcylindrical to subfusiform in side view with slight suprahilar depression, cylindrical to fusiform in ventral view, smooth, yellowish to brownish yellow in KOH, yellow to yellow-brown in Melzer's reagent. Hymenophoral trama boletoid composed of 3–8 μm wide filamentous hyphae, hyaline to yellowish in KOH, yellowish to yellow in Melzer's reagent.

Fig. 20.3 a *Sutorius microsporus* Yan C. Li & Zhu L. Yang. Photos by B. Feng (KUN-HKAS 68720, type).

Cheilo- and pleurocystidia 21–31 × 4.5–8 µm, rare, subfusiform, fusoid-ventricose, or lanceolate. Caulocystidia forming the squamules on the surface of stipe, morphologically similar to hymenial cystidia. Pileipellis a trichoderm composed of 3.5–5 µm wide filamentous hyphae, yellowish brown to brownish in KOH and yellow-brown to dark brown in Melzer's reagent; terminal cells 18–39 × 3–5 µm, clavate to subcylindrical. Pileal trama composed of 7–10 µm wide interwoven hyphae, pale yellowish to brownish yellow in KOH and yellow to brownish in Melzer's reagent. Clamp connections absent in all tissues.

Habitat: Scattered on soil in subalpine to alpine forests dominated by plants of the family Pinaceae.

Known distribution: Known from southwestern China.

Additional specimens examined: CHINA, YUNNAN PROVINCE: Shangri-La County, Geza Town, Geza Village, alt. 3200 m, 26 July 2006, Z.W. Ge 1092 (KUN-HKAS 50673); Shangri-La County, Haba Snow Mountain, alt. 3750 m, 12 August 2008, Y.C. Li 1451 (KUN-HKAS 56291). SICHUAN PROVINCE: Batang County, alt. 2600 m, 1 August 2006, Z.W. Ge 1134 (KUN-HKAS 50714). XIZANG AUTONOMOUS REGION: Pulan County, Baga Town, alt. 3000 m, 28 July 1975, Z. Mu 401 (KUN-HKAS 28950).

Commentary: *Sutorius microsporus* has a dark brown, black-brown or violet-brown, then brownish, reddish brown to brownish orange pileus, a pale reddish brown to violet-brown hymenophoral surface, brownish orange to pale brownish gray tubes, a distribution in subalpine to alpine forests, and an association with trees of the family Pinaceae.

Fig. 20.3 b *Sutorius microsporus* Yan C. Li & Zhu L. Yang. Photo by Y.C. Li (KUN-HKAS 56291).

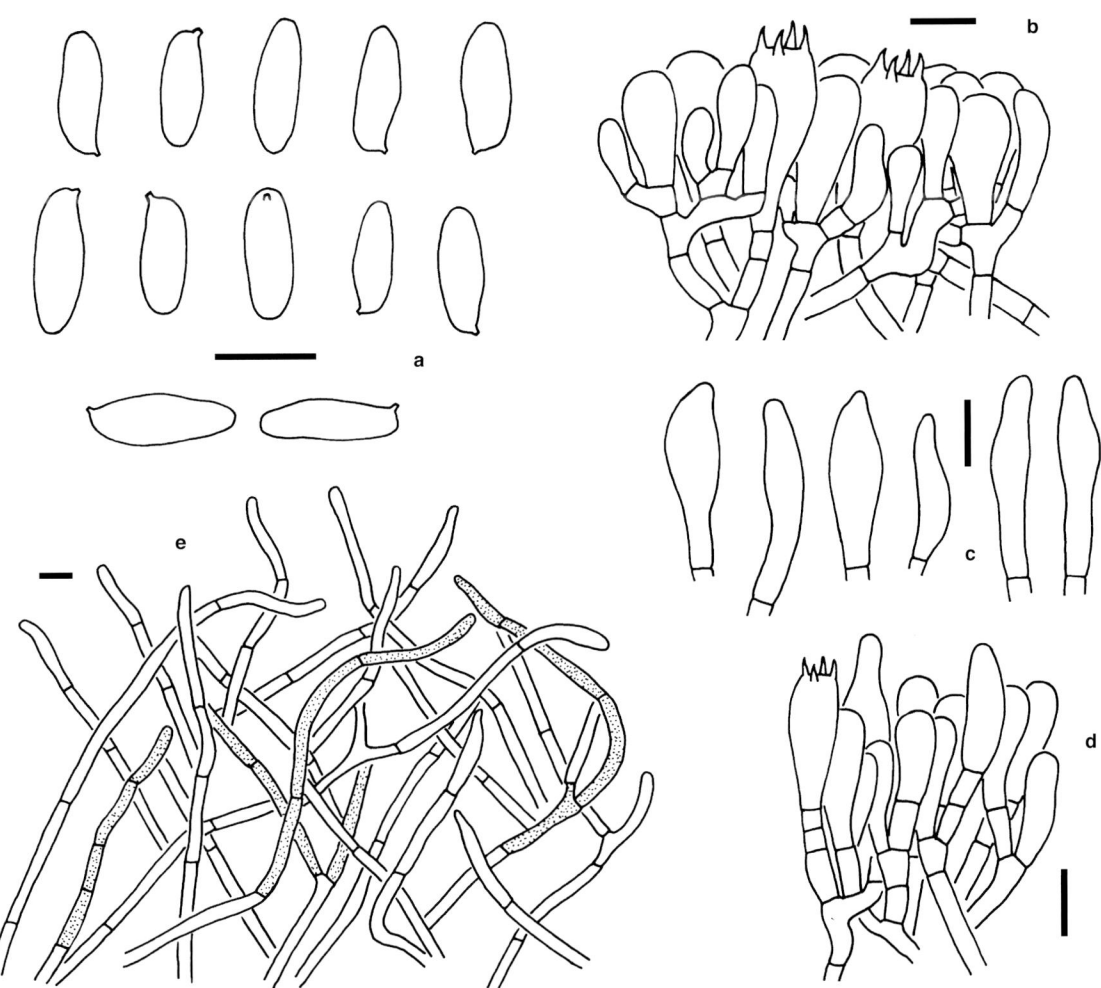

Fig. 20.3 c *Sutorius microsporus* Yan C. Li & Zhu L. Yang (KUN-HKAS 68720, type).
a. Basidiospores; b. Basidia; c. Cheilo- and pleurocystidia; d. Squamules on the surface of stipe; e. Pileipellis. Scale bars = 10 μm.

Sutorius microsporus is phylogenetically sister and morphologically similar to *S. alpinus* and *S. australiensis* (Bougher & Thiers) Halling & N.A. Fechner. However, *S. alpinus* differs from *S. microsporus* in its large basidiospores measuring (12) 14–15.5 (18) × (4) 4.5–5.5 (6.5) μm, and abundant and large hymenial cystidia. *Sutorius australiensis*, described from Australia, has relatively large basidiospores measuring 11.9–16.8 × 3.5–4.9 μm, a distribution in tropical forests, and an association with broad-leaved trees of the families Myrtaceae and Casuarinaceae. For detailed descriptions, comparisons with similar species and images of *S. australiensis* see Halling *et al.* (2012a).

20.4 *Sutorius obscuripellis* Vadthanarat, Raspé & Lumyong, Front Microbiol 12 (no. 643505): 10 (2021)

Basidioma small to medium-sized. Pileus 4–6 cm in diam., hemispherical or convex when young, subhemispherical to plano-convex or applanate when mature; surface dark gray (12F1), brownish gray (7F2–4), or dark brown (8F3–4) when young, and brownish gray (11D2–3), reddish gray (12D2–3), purplish gray (13D2–3), or reddish white (10A2) when mature, covered with finely matted squamules, dry, margin with a slight sterile extension; context grayish (9C1) to pale gray (8C1), staining indistinct reddish brown when injured. Hymenophore adnate when young, adnate to depressed around apex of stipe when mature; surface dark gray (11F1) to blackish (12F1) when young, grayish brown (10E3) to brownish gray (11E2) when mature, staining reddish brown when injured; pores angular, 0.3–1 mm wide; tubes 4–9 mm long, gray (11C1) to dark gray (10E1), staining indistinct reddish brown when injured. Spore print reddish brown (7C5) to brownish (7D6). Stipe 4–8 × 1.2–2.5 cm, cylindrical, clavate, white (11A1) or whitish (12A1), covered with brownish gray (10D2), dark brownish gray (11D2–3), or reddish gray (12D2–3) dotted squamules; surface staining reddish brown when bruised; context grayish brown (11E2–3) to dark grayish brown (10E2) at the apical part, but brownish gray (8D2–3) to brownish (7D3–4) downwards, staining indistinct reddish brown when injured; basal mycelium white. Taste and odor mild.

Basidia 23–32 × 7–10 μm, clavate, 4-spored, hyaline in KOH. Basidiospores [60/3/3] 9–11.5 × 3.5–4.5 μm (Q = 2.5–3.14, Q_m = 2.74 ± 0.11), subcylindrical to subfusiform in side view with slight suprahilar depression, cylindrical to fusiform in ventral view, smooth, yellowish to brownish yellow in KOH, yellow to yellow-brown in Melzer's reagent. Hymenophoral trama boletoid composed of 3–10 μm wide filamentous hyphae, hyaline to yellowish in KOH, yellowish to yellow in Melzer's reagent. Cheilocystidia abundant, 24–35 × 6–10 μm, fusiform to subfusiform or ventricose, usually yellowish to pale brownish

Fig. 20.4 a *Sutorius obscuripellis* Vadthanarat, Raspé & Lumyong. Photos by S.P. Jian (KUN-HKAS 101158).

in KOH, thin-walled. Pleurocystidia rare, similar to cheilocystidia but much bigger 42–53 × 7–8 μm. Caulocystidia forming the scabrous squamules over the surface of stipe, ventricose, fusoid-ventricose, clavate or subfusiform with a sharp apex. Pileipellis a trichoderm composed of 3–4 μm wide filamentous hyphae with subcylindrical terminal cells 29–52 × 3–4 μm, yellowish to pale brownish in KOH and yellow to brownish in Melzer's reagent. Pileal trama composed of 3.5–8 μm wide interwoven hyphae, hyaline to yellowish in KOH and yellowish to brownish in Melzer's reagent. Clamp connections absent in all tissues.

Habitat: Scattered on soil in tropical forests dominated by plants of the family Fagaceae.

Known distribution: Known from Thailand and southern and southwestern China.

Additional specimens examined: CHINA, GUANGDONG PROVINCE: Fengkai County, Heishiding Nature Reserve, alt. 250 m, 5 June 2012, F. Li 432 (KUN-HKAS 106335). YUNNAN PROVINCE: Malipo County, Mengdong Town, alt. 1200 m, 28 June 2017, S.P. Jian 532624MF-205-93 (KUN-HKAS 101158); Malipo County, Tiechang Town, Shuitangzi Village, alt. 1200 m, 24 June 2017, G. Wu 2070 (KUN-HKAS 107150, GenBank Acc. No.: MT154772 for nrLSU, MW165273 and OK557209 for *tef1-α*). GUIZHOU PROVINCE: Leishan County, Wudong Miaozhai, alt. 1300 m, 25 June 2017, R.

Fig. 20.4 b *Sutorius obscuripellis* Vadthanarat, Raspé & Lumyong. Photos by G. Wu (KUN-HKAS 107150).

214 The Boletes of China: *Tylopilus* s.l.

Zhang 439 (KUN-HKAS 101867).

Commentary: *Sutorius obscuripellis* was originally described from Thailand by Vadthanarat et al. (2021) This species is characterized by the dark gray, brownish gray, dark brown then brownish gray, reddish gray to purplish gray pileus, the dark gray to blackish then grayish brown to brownish gray hymenophoral surface, the gray to dark gray tubes, the distribution in tropical forests, and the association with trees of the family Fagaceae. This species can be easily confused with *S. pseudotylopilus* and *S. subrufus*. However, *S. pseudotylopilus* and *S. subrufus* both have a pale reddish brown, brown, or yellowish brown pileus without any black to grayish black or blackish brown tinge, and relatively large basidiospores (mostly ⩾ 11.5 μm long). In our multi-locus phylogenetic analysis (Fig. 4.1), *S. obscuripellis* forms a distinct lineage. Its phylogenetic relationships with other species in this genus remain unresolved.

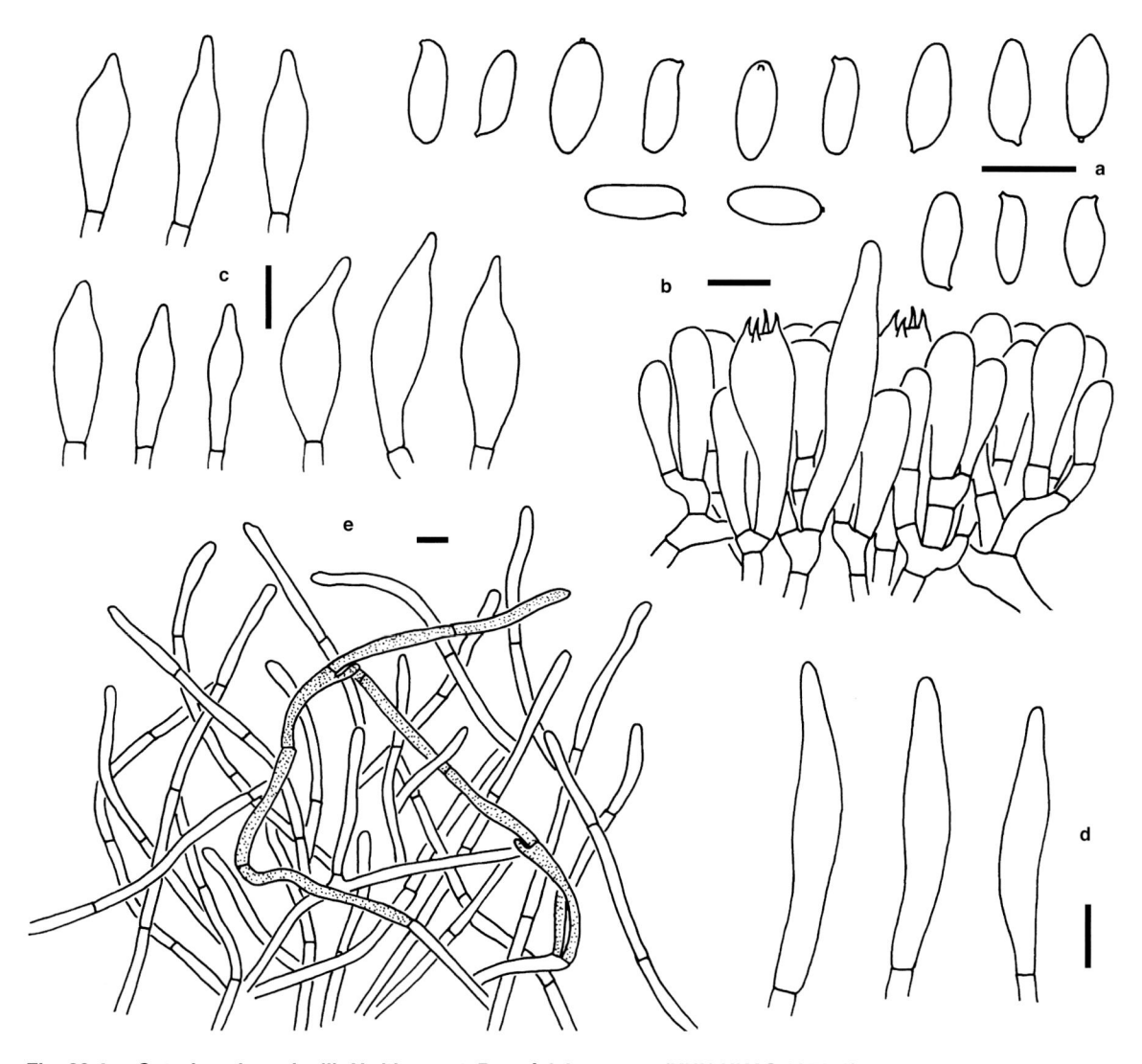

Fig. 20.4 c *Sutorius obscuripellis* **Vadthanarat, Raspé & Lumyong (KUN-HKAS 107150).**
a. Basidiospores; b. Basidia and pleurocystidium; c. Cheilocystidia; d. Pleurocystidia; e. Pileipellis. Scale bars = 10 μm.

20.5 *Sutorius pseudotylopilus* Vadthanarat, Raspé & Lumyong, Front Microbiol 12 (no. 643505): 16 (2021)

Basidioma small to medium-sized. Pileus 4–9 cm in diam., hemispherical, plano-convex or applanate; surface dark violet-brown (10F4–5) to grayish brown (9F3–4) in the center, pale reddish brown (9D4–5) to reddish brown (8D5–6) towards margin; surface covered with tomentose to matted squamules, dry; margin with a slight sterile extension; context whitish (14A1) to pale grayish (14B1), becoming indistinct reddish brown when injured. Hymenophore adnate when young, adnate to depressed around apex of stipe when mature; surface pale brownish gray (7E2) or brownish gray (8E2) to dark brown (7F6) or blackish brown (8F6–7) when young, pale brown (6D6) to brownish orange (6C6) with age, staining reddish brown when bruised; pores subangular to roundish, 0.3–1 mm wide; tubes 5–20 mm long, rose-white (7A2–3) to reddish white (8A2) or orange-gray (6B2) staining reddish brown when bruised. Spore print reddish brown (7C5) to brownish (7D6). Stipe 5–12 × 1.2–3 cm, subcylindrical to clavate, dark gray (9C1), grayish (8C1) or gray (8D1), covered with black (9F1) to dark gray (8E1) scabrous squamules; surface staining reddish brown when bruised; context white to grayish always with reddish brown tinge, staining reddish brown when bruised; basal mycelium white. Taste and odor mild.

Basidia 25–33 × 8–11 μm, clavate, 4-spored, hyaline in KOH. Basidiospores [60/3/3] (10.5) 11–13 × 4–4.5 μm [Q = (2.44) 2.56–3.13 (3.25), Q_m = 2.81± 0.19], subcylindrical to subfusiform in side view with slight suprahilar depression, cylindrical to fusiform in ventral view, smooth, yellowish to brownish yellow in KOH, yellow to yellow-brown in Melzer's reagent. Hymenophoral trama boletoid composed of 3.5–12 μm wide filamentous hyphae, hyaline to yellowish in KOH, yellowish to yellow in Melzer's reagent. Cheilo- and pleurocystidia 23–44 × 4.5–8 μm, rare, subfusiform, ventricose, or subfusoid-ventricose. Caulocystidia forming the squamules on the surface of stipe, morphologically similar to hymenial cystidia. Pileipellis a trichoderm composed of 2.5–6 μm wide filamentous hyphae, pale yellowish brown to pale brownish in KOH and yellowish brown to brownish in Melzer's reagent; terminal cells 25–52 × 3–6 μm, subcylindrical. Pileal trama composed of 7–15 μm wide interwoven hyphae, yellowish to pale brownish in KOH and yellowish brown to brownish in Melzer's reagent. Clamp connections absent in all tissues.

Habitat: Scattered on soil in tropical to subtropical forests dominated by plants of the families Fagaceae and Pinaceae.

Known distribution: Known from Thailand and southeastern and southern China.

Additional specimens examined: CHINA, JIANGXI PROVINCE: Fuzhou County, Zhanping Town, Qianfang Village, alt. 260 m, 19 June 2012, G. Wu 983 (KUN-HKAS 77110). GUANGDONG PROVINCE: Fengkai County, Heishiding Nature Reserve, alt. 250 m, 9 May 2012, F. Li 1331 (KUN-HKAS 79188). YUNNAN PROVINCE: Fugong County, Zilijia Town, Ekeluo Village, alt. 1800 m, 4 August 2011, G. Wu 499 (KUN-HKAS 74813, GenBank Acc. No.: MT154774 for nrLSU, OK557210 for *atp6*).

Commentary: *Sutorius pseudotylopilus* was originally described from Thailand by Vadthanarat *et al.* (2021). This species has a dark violet-brown to grayish brown to pale reddish brown or reddish brown

Fig. 20.5 a *Sutorius pseudotylopilus* Vadthanarat, Raspé & Lumyong. Photos by G. Wu (KUN-HKAS 74813).

pileus, a brownish gray to dark brown, or blackish brown to brownish orange hymenophoral surface, rose-white to reddish white then orange-gray tubes, a distribution in tropical to subtropical forests and an association with trees of the families Fagaceae and Pinaceae.

Sutorius pseudotylopilus has a sympatric distribution as *S. subrufus*. These species are not only similar in macroscopic appearance, but also related to each other in our multi-locus phylogenetic analysis (Fig. 4.1). However, *S. pseudotylopilus* differs from *S. subrufus* in its distinct ruby discoloration in the hymenophoral surface on injury, rare and small hymenial cystidia, and association with trees in mixed forests dominated by Fagaceae and Pinaceae.

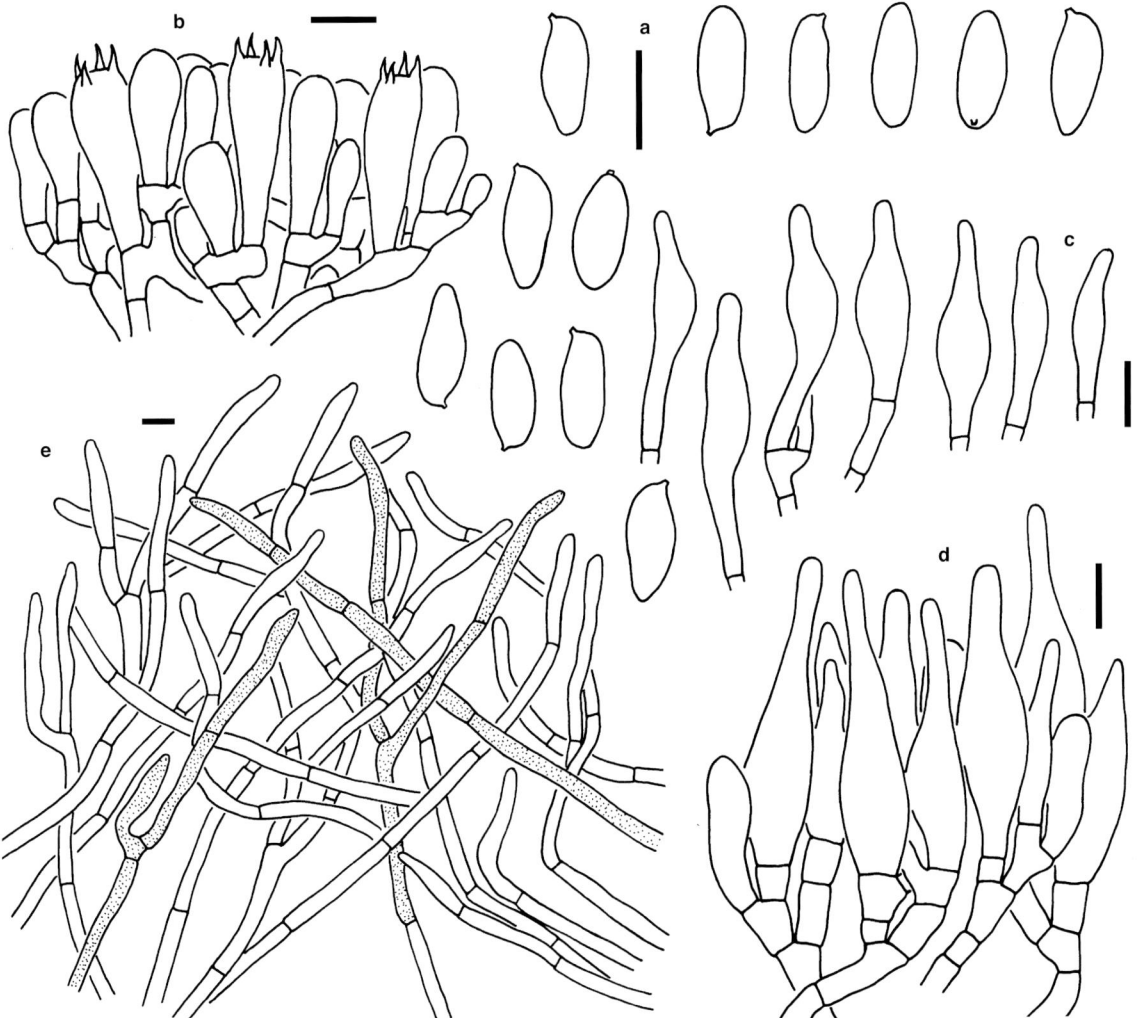

Fig. 20.5 b *Sutorius pseudotylopilus* Vadthanarat, Raspé & Lumyong (KUN-HKAS 74813).
a. Basidiospores; b. Basidia; c. Cheilo- and pleurocystidia; d. Squamules on the surface of stipe; e. Pileipellis. Scale bars = 10 μm.

20.6 *Sutorius subrufus* N.K. Zeng, H. Chai & S. Jiang, MycoKeys 46: 80 (2019)

Basidioma small to medium-sized. Pileus 4–8 cm in diam., subhemispherical to applanate, brown to dark brown and then reddish brown; surface dry, covered with finely matted squamules, margin with a slight sterile extension; context white to pallid, staining reddish to grayish red when injured. Hymenophore adnate when young, adnate to depressed around apex of stipe when mature; surface pale brownish red to brown, usually without discoloration or staining indistinct rufescent tinge when injured; pores angular, 0.3–1 mm wide; tubes 6–10 mm long, concolorous or a little paler than hymenophoral surface, staining indistinct rufescent tinge when injured. Spore print reddish brown to brownish. Stipe 5–12 × 1.5–2.5 cm, cylindrical or clavate, gray-white, but brownish yellow at base, covered with pale reddish brown to blackish brown dotted squamules, becoming reddish when injured; context white to grayish white, becoming reddish to brownish red when injured; basal mycelium white. Taste and odor mild.

Basidia 25–30 × 7–11 μm, clavate, 4-spored, hyaline in KOH. Basidiospores 8–14 × 3–4.5 μm [Q = (2.2) 2.25–3.75, Q_m = 2.57 ± 0.2], subcylindrical to subfusiform in side view with slight suprahilar depression, cylindrical to fusiform in ventral view, smooth, thick-walled (up to 0.5 μm thick), olive-brown to yellowish brown in KOH, brown to ochreous brown in Melzer's reagent. Hymenophoral trama boletoid composed of 3–10 μm wide filamentous hyphae, hyaline to yellowish in KOH, yellowish to yellow in Melzer's reagent. Cheilocystidia 26–48 × 6–10 μm, fusiform to subfusiform or ventricose, usually yellowish to brownish in KOH, thin-walled. Pleurocystidia similar to cheilocystidia but much bigger 40–

Fig. 20.6 a *Sutorius subrufus* N.K. Zeng, H. Chai & S. Jiang. Photos by N.K. Zeng (KUN-HKAS 107155).

Fig. 20.6 b *Sutorius subrufus* N.K. Zeng, H. Chai & S. Jiang. Photos by L.H. Han (KUN-HKAS 84694).

55 × 7–10 µm. Caulocystidia forming the scabrous squamules over the surface of stipe, ventricose, fusoid-ventricose, clavate or subfusiform. Pileipellis a trichoderm composed of 3–6 µm wide filamentous hyphae, yellowish brown to brownish in KOH and yellow-brown to dark brown in Melzer's reagent; terminal cells 31–40 × 3.5–6 µm, subcylindrical. Pileal trama composed of 4–12 µm wide interwoven hyphae, pale yellowish to brownish yellow in KOH and yellow to brownish in Melzer's reagent. Clamp connections absent in all tissues.

Habitat: Scattered on soil in tropical forests dominated by plants of the family Fagaceae.

Known distribution: Known from southern and southwestern China.

Specimens examined: CHINA, HAINAN PROVINCE: Qiongzhong County, Limu Mountain, alt. 850 m, 3 August 2010, N.K. Zeng 821 (KUN-HKAS 59841); Baisha County, Yinggeling National Nature Reserve, alt. 750 m, 26 July 2009, N.K. Zeng 330 (KUN-HKAS 107155); Ledong County, Jianfengling, alt. 800 m, 17 August 1999, M.S. Yuan 4306 (LUN-HKAS 34877). YUNNAN PROVINCE: Mengla County, Xishuangbanna National Natural Reserve, alt. 1000 m, 8 July 2014, L.H. Han 398 (KUN-HKAS 84694).

Commentary: *Sutorius subrufus* originally described from China by Chai *et al.* (2019) is characterized by the brown to dark brown and then reddish brown pileus, the white to pallid context staining reddish to grayish red when injured, the pale brownish red to brown hymenophoral surface which is usually without discoloration or staining indistinct rufescent tinge when injured.

Our multi-locus phylogenetic analysis (Fig. 4.1) indicates that *S. subrufus* is closely related to *S. pseudotylopilus*. However, *S. pseudotylopilus* differs from *S. subrufus* in its dark violet-brown to grayish brown and then reddish brown pileus, brownish gray to dark brown or blackish brown and then pale brown to brownish orange hymenophoral surface, and rare and small hymenial cystidia. For detailed descriptions, comparisons with similar species, line drawings and images of *S. subrufus* see Chai *et al.* (2019).

Chapter 21
Tylocinum Yan C. Li & Zhu L. Yang

Tylocinum Yan C. Li & Zhu L. Yang, Fungal Divers 81: 147 (2016)

Type species: *Tylocinum griseolum* Yan C. Li & Zhu L. Yang, Fungal Divers 81: 147 (2016), Figs. 90 a–c, 91.

Diagnosis: This genus can be distinguished from all the other genera in the Boletaceae by its dark scabrous stipe surface, white to pallid context without discoloration or staining indistinct red tinge when injured, white to pallid hymenophore staining reddish brown when injured, relatively big hymenophoral pores up to 2 mm wide, palisadoderm pileipellis, and smooth basidiospores.

Pileus hemispherical to convex when young, plano-convex to applanate when mature; surface rimose when dry, always cracked into small squamules on a whitish to pallid background; context solid when young, soft when mature, white to pallid, without discoloration or staining indistinct reddish tinge when injured. Hymenophore adnate to depressed around apex of stipe or sometimes slightly decurrent; surface white to pallid, staining reddish or reddish brown tinge when injured; pores angular to roundish, up to 2 mm wide; tubes concolorous with hymenophoral surface, becoming reddish brown when injured. Spore print brown to brownish (Fig. 3.4 l). Stipe central, concolorous with pileal surface or much deeper; surface covered with concolorous verrucose or granular squamules; context white to pallid, staining indistinct reddish brown when injured; basal mycelium white to pallid. Basidiospores smooth. Pileipellis a palisadoderm composed of broad vertically arranged hyphae (up to 22 μm wide). Hymenial cystidia present, fusiform to subfusiform often with a sharp apex and a long pedicel. Clamp connections absent in all tissues.

Commentary: *Tylocinum* was originally described by Wu *et al*. (2016a) and harbored a single species. In the protologue, discoloration of the basidioma was not observed when injured. However, careful examination of additional collection (KUN-HKAS 84548) based on molecular and morphological data showed that the basidiomata are indeed becoming reddish brown when injured. For detailed descriptions and comparisons with its similar genera see Wu *et al*. (2016a). Currently, there is only a single species in this genus.

21.1 *Tylocinum griseolum* Yan C. Li & Zhu L. Yang, Fungal Divers 81: 147 (2016)

Basidioma very small to small. Pileus 1.8–4 cm in diam., hemispherical to subhemispherical when young, plano-convex to applanate when mature, gray to dark gray, slightly darker in the center; surface rimose when dry, cracked into small granular squamules on a whitish to pallid background, margin always extended; context soft when mature, white to pallid, without color change or staining red when bruised. Hymenophore adnate when young, depressed around apex of stipe or sometimes slightly decurrent; surface white when young, pallid to dingy white when mature, staining reddish brown to red when injured; pores angular to roundish, up to 2 mm wide; tubes up to 5 mm long, concolorous or a little paler than hymenophoral surface, staining reddish brown to red when injured. Spore print brownish. Stipe 3.5–7 × 0.4–1.2 cm, cylindrical, concolorous with pileal surface; surface covered with concolorous granular squamules, staining red to reddish brown when touched; context white to pallid, becoming indistinct red when bruised; basal mycelium pallid. Taste and odor mild.

Basidia 28–45 × 9–14 μm, clavate, 4-spored, hyaline in KOH. Basidiospores 11–14.5 (16) × 4.5–5.5 μm (Q = 2.6–3.22, Q_m = 2.81 ± 0.13), subcylindrical to subfusiform in side view with slight suprahilar depression, cylindrical to fusiform in ventral view, smooth, yellowish to brownish yellow in KOH, yellow to yellow-brown in Melzer's reagent. Hymenophoral trama boletoid composed of 3–11 μm wide filamentous hyphae, hyaline to yellowish in KOH, yellowish to yellow in Melzer's reagent. Cheilo- and pleurocystidia 45–60 × 10–14 μm, fusiform to subfusiform or clavate often with sharp apex and long pedicel, yellowish to brownish yellow in KOH, thin-walled. Pileipellis a palisadoderm composed of 8–22 μm wide vertically arranged hyphae, yellowish to pale brownish in KOH and yellowish brown to brownish in Melzer's reagent; terminal cells 40–72 × 10–21 μm, clavate to subcylindrical. Pileal trama composed of 4–7 μm wide interwoven hyphae, hyaline to yellowish in KOH and yellowish to brownish in Melzer's reagent. Clamp connections absent in all tissues.

Fig. 21.1 a *Tylocinum griseolum* Yan C. Li & Zhu L. Yang. Photos by Y.C. Li (KUN-HKAS 50209).

Fig. 21.1 b *Tylocinum griseolum* Yan C. Li & Zhu L. Yang. Photos by L.H. Han (KUN-HKAS 84548).

Habitat: Scattered on soil in tropical forests dominated by plants of the family Fagaceae.

Known distribution: Currently known from southwestern China.

Specimens examined: CHINA, YUNNAN PROVINCE: Jinghong County, Dadugang Town, alt. 1350 m, 7 July 2006, Y.C. Li 455 (KUN-HKAS 50209), the same location, 14 July 2006, Y.C. Li 527 (KUN-HKAS 50281, type), 22 July 2007, Y.C. Li 925 (KUN-HKAS 52612); Longling County, Longshan Town, Yiwanshui Village, alt. 1600 m, 13 June 2014, L.H. Han 252 (KUN-HKAS 84548).

Commentary: *Tylocinum griseolum*, originally described by Wu et al. (2016a) from China, has a gray to dark gray tomentose pileus, a white to pallid pileal context, a palisadoderm pileipellis, big hymenophoral pores (up to 2 mm wide), a verrucose stipe, a distribution in tropical forests, and an association with fagaceous plants. In the protologue, this species was described without discoloration when injured. However, detailed examination of additional collection (KUN-HKAS 84548) based on fresh basidiomata and molecular data showed that the hymenophore is indeed becoming reddish brown when injured and the context of pileus and stipe also staining indistinct reddish brown when injured. For detailed descriptions, line drawings and images of *Ty. griseolum* see Wu et al. (2016a).

Chapter 22
Tylopilus P. Karst.

Tylopilus P. Karst., Revue mycol 3(9): 16 (1881)

Type species: *Boletus felleus* Bull., Herbier de la France, 8. Tab. 379 (1788).

Diagnosis: The genus can be distinguished from other genera in the Boletaceae by its bitter taste in most species; white to pallid or yellowish context without discoloration or staining red to reddish brown, or blue when hurt; white to cream then pinkish to pink or yellowish hymenophoral surface; cutis, trichoderm, hymeniderm, or palisadoderm pileipellis; and smooth basidiospores.

Pileus hemispherical or convex when young, subhemispherical to plano-convex or applanate when mature; surface dry to gelatinous, tomentose to fibrillose or nearly glabrous; context white to pallid or cream to yellowish, without color change or staining reddish brown or bluish when injured, mostly with bitter taste. Hymenophore adnate, slightly decurrent, or depressed around apex of stipe; surface whitish, cream or yellowish when young, and becoming whitish, pinkish or pale purplish pink when mature; pores subangular to roundish; tubes concolorous or much paler than hymenophoral surface, without color change or staining reddish brown or rarely bluish when injured. Stipe central, nearly glabrous, farinose, tomentose, fibrillose or distinctly reticulate; basal mycelium white. Basidiospores subglobose to broadly ellipsoid or ellipsoid, elongated, cylindrical, or fusiform. Hymenial cystidia abundant, subfusiform. Pileipellis a cutis, trichoderm, hymeniderm, or palisadoderm. Clamp connections absent in all tissues.

Commentary: *Tylopilus* shares the color of the hymenophore and context and the structure of the pileipellis with some species in *Porphyrellus*. But *Porphyrellus* differs from *Tylopilus* in its umber basidiomata, orange-red to brownish red or brown spore print, mild taste of context, and bluish or initially bluish and then reddish brown discoloration when injured. Currently, 32 species, including eleven new species and one new combination, are known from China.

Key to the species of the genus *Tylopilus* in China

1. Surface of stipe distinctly reticulate ··· 2
1. Surface of stipe without reticulum, or with indistinct reticulum restricted to the apical part ········ 7
2. Species with a distribution in temperate forests; context white, without color change when injured ·· 3
2. Species without a distribution in temperate forests; context white, becoming brownish red when injured ·· 4
3. Pileus gray to grayish brown or grayish white ··· *T. felleus*
3. Pileus brownish violet or reddish brown to light brown with magenta tinge ······ *T. violaceobrunneus*
4. Species with a distribution in subalpine to alpine, or in tropical to subtropical forests; pileus dark olive-green to olive-brown when young, brown to dark brown when mature ·················· 5
4. Species with a distribution in tropical, or in tropical to subtropical forests; pileus without dark olive-green to olive-brown tinge when young, but with brown to dark brown tinge when mature ·················· 6
5. Species with a distribution in subalpine to alpine forests; basidiospores 13–15.5 × 3.5–5 μm ··· *T. alpinus*
5. Species with a distribution in tropical to subtropical forests; basidiospores relatively small 8.5–12.5 × 3.5–4.5 μm ·· *T. brunneirubens*
6. Species with a distribution in tropical forests; pileus brown to brownish orange when young, and then brownish to brownish yellow when mature; basidiospores 10.5–13.5 × 3.5–4 μm; hymenial cystidia 50–84 × 13–22 μm ··· *T. rufobrunneus*
6. Species with a distribution in subtropical forests; pileus grayish ruby to grayish red or brownish red; basidiospores relatively small 9–11.5 × 3.5–4.5 μm; hymenial cystidia relatively small 27–57 × 9–16 μm ··· *T. pseudoalpinus*
7. Pileus with pink, purple or violaceous tinge ··· 8
7. Pileus without pink, purple or violaceous tinge ··· 17
8. Context of pileus and stipe becoming rufescent when injured ··· 9
8. Context of pileus and stipe without discoloration when injured ·· 12
9. Pileus inviscid; pileipellis a palisadoderm composed of 3.5–10 μm wide vertically arranged hyphae ·· 10
9. Pileus viscid or inviscid; pileipellis a trichoderm or an ixotrichoderm composed of 3–7 μm wide interwoven hyphae ··· 11
10. Pileus grayish red to violet-brown or reddish brown, always with black-violaceous tinge; basidiospores 10–14 × 4–5 μm ··· *T. atroviolaceobrunneus*
10. Pileus dark ruby to violet-brown or gray-red to gray-ruby; basidiospores relatively small 8–10.5 × 3.5–4.5 μm ··· *T. violaceorubrus*
11. Pileus vinaceous to deep violaceous-brown or purple-chestnut; basidiospores 8.5–12 × 3–4 μm; pileipellis an ixotrichoderm to a trichoderm composed of gelatinous 3.5–5 μm wide interwoven filamentous hyphae ··· *T. plumbeoviolaceoides*

11. Pileus grayish purple to dull purple; basidiospores 7.5–9.5 × 3.5–4.5 µm; pileipellis a trichoderm composed of 3–7 µm wide interwoven filamentous hyphae ········· *T. purpureorubens*
12. Pileus without brownish to soil brown color when mature; hymenophoral surface pinkish to whitish without purple tinge in younger basidiomata; pileipellis a palisadoderm, or a trichoderm ················13
12. Pileus purplish brown when young, brownish to soil brown when mature; hymenophore surface purplish pink in younger basidiomata; pileipellis a cutis to intricate trichoderm composed of narrow (5–10 µm wide) filamentous hyphae···*T. neofelleus*
13. Pileus with purplish to drab purple, or dark purple to blackish purple tinge; pileipellis a trichoderm or a palisadoderm composed of 5–13 µm wide vertically arranged hyphae ············ 14
13. Pileus pink to grayish pink, without purple tinge; pileipellis a trichoderm composed of 4.5–7 µm wide filamentous hyphae ···*T. griseipurpureus*
14. Species with a distribution in tropical forests; pileipellis a palisadoderm composed of 5–13 µm wide vertically arranged hyphae ·· 15
14. Species with a distribution in temperate forests; pileipellis a trichoderm composed of 3.5–7 µm wide filamentous hyphae···*T. fuligineoviolaceus*
15. Pileus much deeper in color, dark purple to blackish purple; basidiospores mostly more than 8 µm long ·· 16
15. Pileus pallid purplish, grayish purplish, drab purple or sometimes purplish pink; basidiospores relatively short 6.5–8 × 3.5–4.5 µm ·· *T. obscureviolaceus*
16. Basidiospores 10.5–13.5 × 3.5–5 µm; pleurocystidia and cheilocystidia morphologically similar 38–60 × 5–10 µm ···*T. albopurpureus*
16. Basidiospores 8–10.5 (12) × 3.5–4 µm; pleurocystidia relatively wide 45–60 × 12–14 µm, cheilocystidia narrow 30–51 × 6–9 µm··*T. atripurpureus*
17. Pileus and stipe reddish brown to vinaceous, or brown to dark brown, or castaneous············· 18
17. Pileus and stipe without reddish brown to vinaceous, or brown to dark brown, or castaneous tinges ···22
18. Context of pileus and stipe becoming rufescent when injured, but never bluish when injured 19
18. Context of pileus and stipe without discoloration when injured, or staining blue when injured·········20
19. Pileus reddish brown to salmon; basidiospores 7.5–10 × 3.5–5 µm······················· *T. argillaceus*
19. Pileus castaneous; basidiospores relatively large 9.5–14 × 3–4 µm······················ *T. castanoides*
20. Basidioma staining greenish blue when bruised; stipe always has a greenish blue ring at apex; pileipellis a palisadoderm··· *T. virescens*
20. Basidioma without discoloration when bruised; stipe without greenish blue ring at apex; pileipellis a palisadoderm, or a trichoderm ··· 21
21. Pileus reddish brown to vinaceous; basidiospores 10–12 × 3.5–4.5 µm; pileipellis an intricate trichoderm composed of 4–6 µm wide interwoven filamentous hyphae ···········*T. vinaceipallidus*
21. Pileus dark reddish brown; basidiospores relatively large 12.5–14.5 ×4.5–5.5 µm; pileipellis a palisadoderm composed of 5–10 µm wide somewhat vertically arranged hyphae ······*T. phaeoruber*

22. Pileus with greenish to green or olive-green to olivaceous tinge ·· 23
22. Pileus without greenish to green or olive-green to olivaceous tinge ·· 28
23. Context of pileus and stipe without discoloration when injured······················· *T. subotsuensis*
23. Context of pileus and stipe becoming rufescent when injured ··· 24
24. Species with distributions in temperate, or tropical, or subtropical forests; pileipellis a palisadoderm, or a cutis to trichoderm; basidiospores cylindrical to fusiform, or elongated to subcylindrical ··· 25
24. Species with distributions in tropical forests; pileipellis an epithelium; basidiospores subglobose to broadly ellipsoid or ellipsoid··· *T. otsuensis*
25. Species with distributions in tropical, or subtropical forests; basidiospores relatively short, mostly not more than 11 μm long; cheilo- and pleurocystidia relatively slender and not more than 13 μm wide; pileipellis a palisadoderm, or a cutis to trichoderm ·· 26
25. Species with a distribution in temperate forests; basidiospores 11–14 × 3.5–4.5 μm; cheilo- and pleurocystidia up to 17 μm wide; pileipellis a palisadoderm························· *T. himalayanus*
26. Pileipellis a palisadoderm; cheilo- and pleurocystidia relatively slender and not more than 10 μm wide; basidiospores relatively slender and not more than 4 μm wide ···················· 27
26. Pileipellis a cutis to trichoderm; cheilo- and pleurocystidia 29–58 × 6–13 μm; basidiospores 9–13 × 3.5–5 μm ··· *T. jiangxiensis*
27. Species with a distribution in subtropical forests; pileus yellowish olivaceous with brownish to pale brown tinge··· *T. olivaceobrunneus*
27. Species with a distribution in tropical forests; pileus greenish to grayish green or pale green then olive-gray to olive-brown·· *T. griseiviridus*
28. Pileus without gray to dark gray tinge; basidiospores broadly ellipsoid to ovoid ···················· 29
28. Pileus gray to dark gray; basidiospores subfusiform to oblong························· *T. griseolus*
29. Pileus with orange-red to yellow-red tinge ·· 30
29. Pileus with yellow then orange to orange-brown tinge, or with fawn-ochraceous to tawny-ochraceous tinge, without red tinge ·· 31
30. Pileus orange-red to brownish red; stipe concolorous with pileal surface; pileipellis a cutis composed of 4–6 μm wide interwoven hyphae··· *T. aurantiacus*
30. Pileus yellowish red; stipe yellow; pileipellis a palisadoderm composed of 5–9 μm wide vertically arranged hyphae ·· *T. rubrotinctus*
31. Pileus yellowish when young and then fawn-ochraceous to tawny-ochraceous when mature, becoming fuscous-fawn to dark brown or dingy tan when injured or aged; stipe yellowish to yellowish ochraceous or tan, but always white to pallid at apex, staining fuscous on injury··················· *T. fuscatus*
31. Pileus yellow to orange-yellow when young, and then orange to orange-brown when mature, without discoloration when touched; stipe concolorous with pileal surface, without discoloration when touched ··· *T. pseudoballoui*

22.1 *Tylopilus albopurpureus* Yan C. Li & Zhu L. Yang, The Boletes of China: *Tylopilus* s.l. 269 (2021)

MycoBank: MB 834750

Etymology: The epithet "*albopurpureus*" refers to the color of the basidiomata.

Type: CHINA, YUNNAN PROVINCE: Mengla County, Xishuangbanna National Natural Reserve, alt. 1000 m, 6 July 2014, G. Wu 1261 (KUN-HKAS 88998, GenBank Acc. No.: MT154740 for nrLSU, MT110353 for *tef1-α*, MT110391 for *rpb1*, MT110426 for *rpb2*).

Basidioma small to medium-sized. Pileus 3–7 cm in diam., subhemispherical, plano-convex or applanate, purplish red (13F3–4), dark purplish red (14F3–5), pale purplish red (12F2–3) or blackish purple (13F7–8) when young, dark purple (14E4–5), dark purplish gray (13D3), grayish ruby (12D3), purplish gray (14D2–3), or pale purplish gray (14D3–4) when mature, slightly darker in the center; surface covered with fibrillose to tomentose squamules, dry, without discoloration when touched; context solid, whitish (10C1) to pallid (11B1), without discoloration when bruised. Hymenophore adnate when young, adnate to rarely depressed around apex of stipe when mature; surface whitish (8A1) to pinkish (8A2), without discoloration when bruised; pores subangular to roundish, 0.3–1 mm wide; tubes 4–6 mm long, concolorous or a little paler than hymenophoral surface, without discoloration when bruised. Spore print pale pinkish (12B2) to pinkish (13B2) or reddish (11B2). Stipe 5–9 × 1.1–2.2 cm, subcylindrical or clavate, always enlarged downwards, grayish magenta (14E3–4) or concolorous with pileal surface, but grayish (9C1), pale gray (10C1), brownish gray (6D2–3) or brownish orange (5C3–4) with age, without discoloration when touched; context solid, without discoloration when injured; basal mycelium white. Taste bitter, odor mild.

Basidia 20–32 × 7–9 μm, clavate to narrowly clavate, 4-spored, hyaline in KOH. Basidiospores 10.5–13.5 × 3.5–5 μm (Q = 2.5–3.38, Q_m = 2.89 ± 0.27), subcylindrical to subfusiform in side view with slight suprahilar depression, cylindrical to fusiform in ventral view, smooth, hyaline to yellowish in

Fig. 22.1 a *Tylopilus albopurpureus* Yan C. Li & Zhu L. Yang. Photos by G. Wu (KUN-HKAS 88998, type).

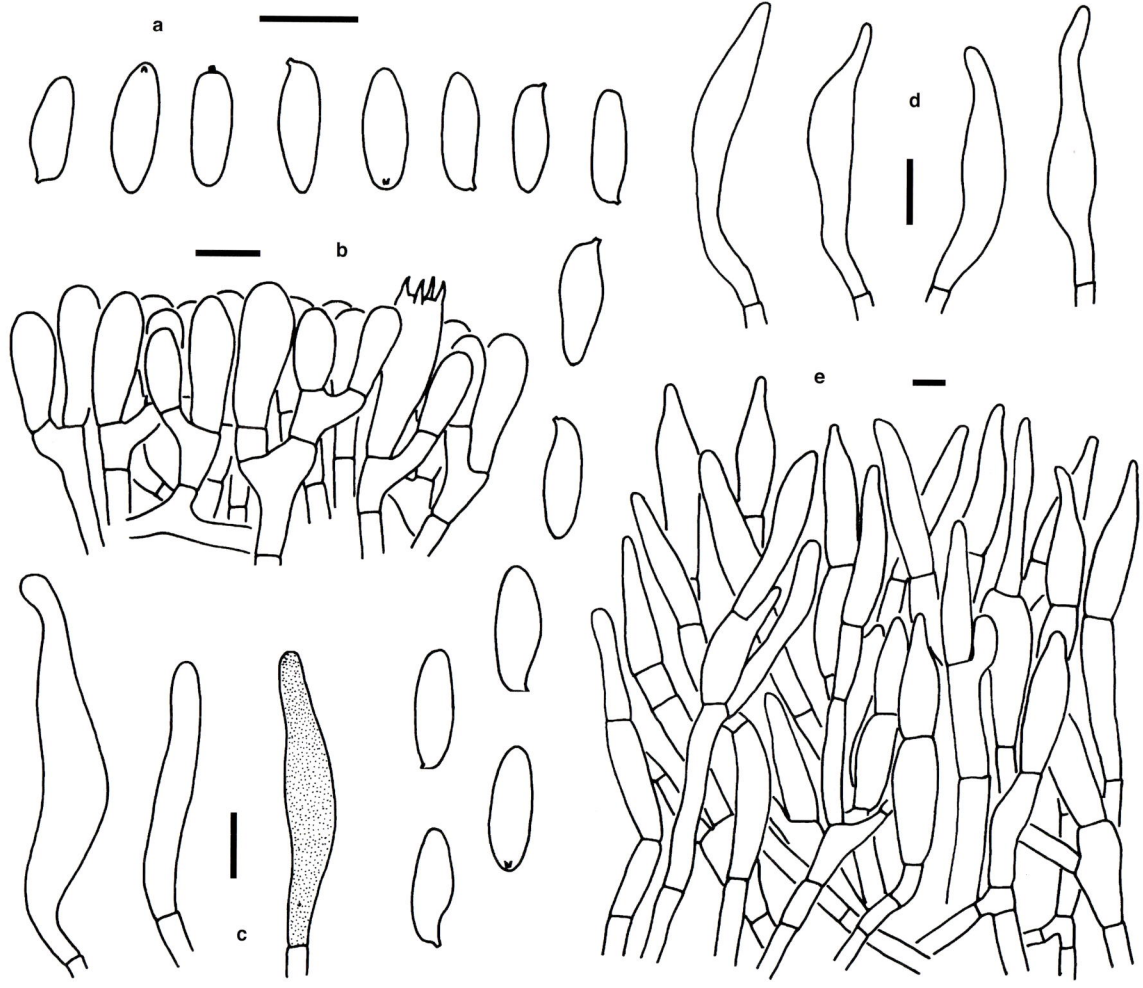

Fig. 22.1 b *Tylopilus albopurpureus* Yan C. Li & Zhu L. Yang (KUN-HKAS 88998, type).
a. Basidiospores; b. Basidia; c. Pleurocystidia; d. Cheilocystidia; e. Pileipellis. Scale bars = 10 μm.

KOH, yellow to brownish yellow in Melzer's reagent. Hymenophoral trama boletoid composed of 3–8 μm wide filamentous hyphae, hyaline to yellowish in KOH, yellowish to yellow in Melzer's reagent. Cheilocystidia 38–60 × 5–10 μm, subfusiform to fusoid-ventricose, yellow to brownish in KOH, thin-walled. Pleurocystidia similar to cheilocystidia. Pileipellis a palisadoderm composed of 4–13 μm wide vertically arranged hyphae, yellowish to pale brownish in KOH and yellowish brown to brownish in Melzer's reagent; terminal cells 34–85 × 5–13 μm, pyriform to clavate or cystidioid. Pileal trama composed of 4–10 μm wide interwoven hyphae, hyaline to yellowish in KOH and yellowish to brownish in Melzer's reagent. Clamp connections absent in all tissues.

Habitat: Solitary on soil in tropical forests dominated by plants of the family Fagaceae.

Known distribution: Currently known from southern and southwestern China.

Specimens examined: CHINA, HAINAN PROVINCE: Qiongzhong County, Limu Mountain, alt. 630 m, 5 August 2010, N.K. Zeng 853 (KUN-HKAS 107162). GUANGDONG PROVINCE: Fengkai

County, Heishiding Nature Reserve, alt. 250 m, 3 July 2012, F. Li 581 (KUN-HKAS 106353), the same location, 5 July 2012, F. Li 624 (KUN-HKAS 106418). YUNNAN PROVINCE: Mengla County, Xishuangbanna National Nature Reserve, alt. 1000 m, 8 July 2014, L.H. Han 397 (KUN-HKAS 84693).

Commentary: *Tylopilus albopurpureus* is characterized by the dark purplish red to blackish purple then dark purple to dark purplish gray or grayish ruby pileus, the adnate and whitish to pinkish hymenophore, the white to pallid context without discoloration when bruised, the palisadoderm pileipellis, and the distribution in tropical forests in southwestern China. Macroscopically, this species is somewhat similar to the purple or violaceous species in *Tylopilus*, namely *T. atripurpureus* (Corner) E. Horak, *T. atroviolaceobrunneus* Yan C. Li & Zhu L. Yang, *T. violaceorubrus* Yan C. Li & Zhu L. Yang, *T. fuligineoviolaceus* Har. Takah., *T. griseipurpureus* (Corner) E. Horak, *T. neofelleus* Hongo, *T. obscureviolaceus* Har. Takah., *T. plumbeoviolaceoides* T.H. Li, B. Song & Y.H. Shen, *T. purpureorubens* Yan C. Li & Zhu L. Yang, and *T. violaceobrunneus* Yan C. Li & Zhu L. Yang. However, *T. atripurpureus* differs from *T. albopurpureus* in its dark pileus, relatively small basidiospores measuring (8) 9–10.5 (12) × 3.5–4 µm, and broad pleurocystidia measuring 45–60 × 12–14 µm (Corner 1972; Horak 2011). *Tylopilus atroviolaceobrunneus* differs in its somewhat dark colored pileus, and reddish to reddish brown discoloration of the basidiomata when injured (Wu *et al*. 2016a). *Tylopilus violaceorubrus* differs in its dark ruby to violet-brown or gray-ruby pileus, small basidiospores measuring 8–10.5 × 3.5–4.5 µm, and reddish to reddish brown discoloration of the basidiomata when injured. *Tylopilus fuligineoviolaceus* differs in its relatively small basidiospores measuring 8.5–10.5 × 3–4 µm, and distribution in temperate forests in northeastern China and Japan (Takahashi 2007). *Tylopilus griseipurpureus* differs in its grayish tinged pileus, and relatively small basidiospores measuring 8–11 × 3–4.5 µm (Corner 1972; Horak 2011; Wu *et al*. 2016a). *Tylopilus neofelleus* differs in its brownish to soil brown pileus when aged, small basidiospores measuring 8–9 (10) × 3–4 µm, large pleurocystidia measuring 70–85 × 10–15 µm and intricate trichoderm pielipellis (Hongo 1967; Fu *et al*. 2006; Wu *et al*. 2016a). *Tylopilus obscureviolaceus* differs in its small basidiospores measuring 6.5–8 × 3.5–4.5 µm, palisadoderm pileipellis composed of 5–11 µm wide vertically arranged hyphae, and broad pleurocystidia measuring 34–76 × 9–16 µm (Takahashi 2004; see our description below). *Tylopilus plumbeoviolaceoides* differs in its pinkish to purplish or reddish brown discoloration in the basidiomata when injured, relatively small basidiospores measuring 8.5–10.5 × 3–4 µm, and ixotrichoderm to trichoderm pileipellis (Li *et al*. 2002). *Tylopilus purpureorubens* differs in its reddish brown or rufescent discoloration of the basidiomata on injury, small basidiospores measuring 7.5–9.5 × 3.5–4.5 µm, trichoderm pileipellis composed of 3–7 µm wide filamentous hyphae, and broad hymenial cystidia measuring 37–71 × 9–16 µm (see our description below). *Tylopilus violaceobrunneus* differs in its distinctly reticulate stipe, brownish violet or reddish brown to light brown pileus always with magenta tinge, and trichoderm pileipellis (Wu *et al*. 2016a).

Our multi-locus phylogenetic analysis (Fig. 4.1) indicates that *T. albopurpureus* is related to *T. himalayanus* D. Chakr., K. Das & Vizzini and *T. rubrobrunneus* Mazzer & A.H. Sm. However, *T. himalayanus* has a green to yellowish green or olivaceous pileus, and a white context that stains reddish brown when bruised (Chakraborty *et al*. 2018). *Tylopilus rubrobrunneus* has a dark vinaceous brown to chocolate-colored pileus, a white context staining olivaceous when injured, and a trichoderm pileipellis (Mazzer and Smith 1967).

22.2 *Tylopilus alpinus* Yan C. Li & Zhu L. Yang, Fungal Divers 81: 150 (2016)

Basidioma small to very large. Pileus 4–18 cm in diam., hemispherical when young, subhemispherical to plano-convex or applanate, dark olive to olive-green when young, olive-brown to brown when aged; surface dry, nearly glabrous or with fine tomentose to fibrillose squamules; context whitish to grayish white, staining pale red to grayish red when injured. Hymenophore adnate to slightly decurrent when young, adnate to slightly depressed around apex of stipe when mature; surface initially grayish white to whitish and then pinkish to grayish pink, staining a brownish red to grayish red or orange-brown tinge when bruised; pores subangular to roundish, 0.3–0.5 mm wide; tubes 6–15 mm long, concolorous or a little paler than hymenophoral surface, staining brownish red to grayish red or orange-brown tinge when hurt. Spore print pinkish to reddish. Stipe 6–18 × 2–2.5 cm, clavate to subcylindrical, always enlarged downwards, whitish to grayish staining pale orange to pale brownish red when bruised; surface with distinct concolorous reticulum at upper part; context solid, whitish to pallid or grayish, staining grayish red to grayish orange when injured; basal mycelium whitish to pallid. Taste bitter.

Basidia 25–32 × 9–11 μm, clavate to narrowly clavate, 4-spored, hyaline in KOH. Basidiospores 13–14.5 (15.5) × (3.5) 4–5 μm (Q = 2.7–3.5, Q_m = 3.01 ± 0.22), subcylindrical to subfusiform in side view with slight suprahilar depression, cylindrical to fusiform in ventral view, smooth, hyaline to yellowish in KOH, yellow to brownish yellow in Melzer's reagent. Hymenophoral trama boletoid; hyphae cylindrical, 4–11 μm wide, hyaline to yellowish in KOH, yellowish to yellow in Melzer's reagent. Cheilocystidia 43–56 × 8–12.5 μm, fusiform to subfusiform or fusoid-ventricose, brown to yellowish brown in KOH, with brown to dark brown encrustations, somewhat thick-walled (up to 1 μm thick). Pleurocystidia much bigger than cheilocystidia, 62–87 × 13–16 μm, fusiform to subfusiform or fusoid-ventricose, yellowish brown to brown in KOH, brown to dark brown or dark olive-brown in Melzer's reagent, somewhat thick-walled (up to 1 μm thick). Caulocystidia forming the reticulum over the surface of stipe, subfusoid-ventricose, clavate, or lanceolate. Pileipellis a trichoderm composed of 3.5–7 μm wide interwoven filamentous hyphae, yellowish to pale brownish in KOH and yellowish brown to brownish in Melzer's reagent; terminal cells 26–111 × 4.5–6.5 μm, clavate or cylindrical. Pileal trama composed of 4–12 μm wide interwoven hyphae, hyaline to yellowish in

Fig. 22.2 a *Tylopilus alpinus* Yan C. Li & Zhu L. Yang. Photo by Q. Zhao (KUN-HKAS 87955).

Fig. 22.2 b *Tylopilus alpinus* Yan C. Li & Zhu L. Yang. Photo by Y.C. Li (KUN-HKAS 56369).

KOH and yellowish to brownish in Melzer's reagent. Clamp connections absent in all tissues.

Habitat: Scattered to solitary on soil in subalpine to alpine forests dominated by plants of the family Pinaceae.

Known distribution: Currently known from southwestern China.

Specimens examined: CHINA, XIZANG AUTONOMOUS REGION: Linzhi, Bayi District, Lulang Town, alt. 3400 m, 23 July 2019, L.K. Jia 170 (KUN-HKAS 112947). YUNNAN PROVINCE: Deqin County, Benzilan Town, Baima Snow Mountain, alt. 3500 m, 18 August 2008, B. Feng 327 (KUN-HKAS 55438), the same location, 19 August 2008, Y.C. Li 1528-1 (KUN-HKAS 56369), the same location, 6 September 2008, X.F. Tian 385 (KUN-HKAS 90204); Shangri-La County, Daxue Mountain, alt. 3100 m, 21 August 2014, Q. Zhao 2189 (KUN-HKAS 87955) and Q. Zhao 2198 (KUN-HKAS 87964).

Commentary: *Tylopilus alpinus* originally described from China by Wu *et al*. (2016a) is characterized by the dark olive to olive-green then olive-brown to brown pileus, the initially whitish to pallid or grayish context staining pale red to grayish red when injured, the initially whitish to grayish white and then pinkish to grayish pink hymenophore staining a brownish red to grayish red or orange-brown tinge when bruised, the whitish to grayish or pale orange stipe which is reticulate on upper half, and the distribution in subalpine to alpine forests. This species shares the color of the basidiomata, the discoloration of the context when hurt, and the ornamentation of the stipe with *Tylopilus brunneirubens* (Corner) Watling & E. Turnbull. However, *T. brunneirubens* has a small basidioma, small basidiospores (8.5–12.5 × 3.5–4.5 μm), and distribution in tropical to subtropical forests.

22.3 *Tylopilus argillaceus* Hongo, J Jap Bot 60 (12): 372 (1985)

Basidioma small to medium-sized. Pileus 4–8.5 cm in diam., subhemispherical to convex or applanate, dark castaneous brown to reddish salmon, darker in the center; surface viscid, nearly glabrous or with fibrillose to fine tomentose squamules; context solid, grayish white to white, staining red or reddish brown when injured. Hymenophore adnate when young, adnate to depressed around apex of stipe when mature; surface grayish white to whitish when young and pinkish to pink when mature, becoming reddish when injured; pores angular to roundish, 0.5–1 mm wide; tubes up to 20 mm long, concolorous or a little paler than hymenophoral surface, staining red or reddish brown when injured. Spore print pinkish to reddish. Stipe 8–12 × 1.5–1.8 cm, clavate, enlarged downwards; surface concolorous or much darker than pileal surface, nearly glabrous to fibrillose; context whitish to cream, staining red to brownish red when bruised; basal mycelium whitish to pallid. Taste bitter.

Basidia 20–30 × 8–12 μm, clavate, 4-spored, hyaline in KOH. Basidiospores 7.5–10 × 3.5–4.5 (5) μm (Q = 1.88–2.57, Q_m = 2.13 ± 0.18), elongated to subcylindrical and inequilateral in side view with slight suprahilar depression, elongated to cylindrical in ventral view, smooth, hyaline to yellowish in KOH, yellow to brownish yellow in Melzer's reagent. Hymenophoral trama boletoid composed of 4–9 μm wide filamentous hyphae, hyaline to yellowish in KOH, yellowish to yellow in Melzer's reagent. Cheilocystidia 34–55 × 8.5–11 μm, subfusiform to subfusoid-ventricose, often spindly upwards, hyaline to yellowish in KOH, thin-walled. Pleurocystidia 55–72 × 7.5–14 μm, much bigger than cheilocystidia, subfusiform to subfusoid-ventricose, often spindly upwards, hyaline to yellowish in KOH, thin-walled. Pileipellis a trichoderm composed of 3–6.5 μm wide filamentous hyphae, pale yellowish to brownish yellow in KOH and yellow to brownish in Melzer's reagent; terminal cells 11–50 × 4–6.5 μm, clavate to subfusiform. Pileal trama composed of 2.5–6 μm wide interwoven hyphae, colorless to yellowish in KOH and yellowish to pale yellow in Melzer's reagent. Clamp connections absent in all tissues.

Habitat: Scattered to solitary on soil in tropical forests dominated by plants of the family Fagaceae.

Known distribution: Currently known from Japan and southern China.

Specimens examined: CHINA, GUANGDONG PROVINCE: Fengkai County, Heishiding Nature Reserve, alt. 250 m, 5 June 2012, F. Li 431 (KUN-HKAS 106334), the same location, 26 June 2012, F. Li 558 (KUN-HKAS 90186), the same location, 3 July 2012, F. Li 565 (KUN-HKAS 90187) and F. Li 576 (KUN-HKAS 90188), the same location, 4 July 2012, F. Li 580 (KUN-HKAS 90189) and F. Li 594 (KUN-HKAS 90190), the same location, 3 August 2012, F. Li 773 (KUN-HKAS 90191) and F. Li 778 (KUN-HKAS 90192), the same location, 14 August 2012, F. Li 795 (KUN-HKAS 90193) and F. Li 814 (KUN-HKAS 90201).

Commentary: *Tylopilus argillaceus* originally described from Japan by Hongo (1985) and then reported from China by Wu *et al.* (2016a) is characterized by the castaneous to reddish brown or salmon basidioma, the whitish context becoming red to reddish brown when injured, the whitish to pinkish or pink hymenophore staining red or reddish brown when injured, and the trichoderm pileipellis. *Tylopilus argillaceus* is macroscopically similar to *T. vinaceipallidus* (Corner) Watling & E. Turnbull. But *T.*

vinaceipallidus is characterized by its big basidiospores measuring (8) 10–12 (13.5) ×3.5–4.5 μm, and white context without discoloration when bruised.

In our multi-locus data (Fig. 4.1), *T. argillaceus* clusters, with high support, with *T. purpureorubens*. However, *T. purpureorubens* differs from *T. argillaceus* in its purplish gray to purple-gray basidioma, relatively short basidiospores measuring 7–9.5 × 3.5–4.5 μm, and sometimes septate hymenial cystidia.

Fig. 22.3 *Tylopilus argillaceus* Hongo. Photos by F. Li (KUN-HKAS 90188).

22.4 *Tylopilus atripurpureus* (Corner) E. Horak, Malay Fores Rec 51: 131 (2011)

Basionym: *Boletus atripurpureus* Corner, *Boletus* in Malaysia 166 (1972).

Basidioma small to medium-sized. Pileus 4–8 cm in diam., hemispherical to subhemispherical or applanate, blackish purple to dark purple, darker in the center; surface dry, nearly glabrous or with fibrillose squamules, without discoloration when hurt; context solid, whitish to pallid, without discoloration when bruised. Hymenophore adnate when young, adnate to depressed around apex of stipe; surface initially white to pallid and then pinkish to purplish, without discoloration when bruised;

Fig. 22.4 *Tylopilus atripurpureus* (Corner) E. Horak. Photos by K. Zhao (KUN-HKAS 89112).

pores subangular to angular or roundish, 0.5–1.5 mm wide; tubes 5–10 mm long, concolorous or a little paler than hymenophoral surface, without discoloration when bruised. Spore print pinkish to reddish. Stipe 4–11 × 1–1.8 cm, subcylindrical or clavate, always attenuate upwards, concolorous or much paler than pileal surface, almost glabrous or with fibrillose squamules, without discoloration when hurt; context solid, concolorous with that of pileus, without discoloration when hurt; basal mycelium whitish to pallid. Taste bitter.

Basidia 20–28 × 7–10 μm, clavate to narrowly clavate, 4-spored, hyaline in KOH. Basidiospores 8–10.5 (12) × 3.5–4 μm [Q = 2.1–2.94 (3.04), Q_m = 2.63 ± 0.12], subcylindrical to subfusiform in side view with slight suprahilar depression, cylindrical to fusiform in ventral view, smooth, yellowish to brownish yellow in KOH, yellow to yellow-brown in Melzer's reagent. Hymenophoral trama boletoid composed of 4–12 μm wide filamentous hyphae, hyaline to yellowish in KOH, yellowish to yellow in Melzer's reagent. Cheilocystidia 30–51 × 6–9 μm, fusiform to subfusiform or fusoid-ventricose, yellowish brown to brown yellow in KOH, brown to dark brown in Melzer's reagent, thin-walled. Pleurocystidia 45–60 × 12–14 μm, subfusiform to fusiform or fusoid-ventricose, concolorous with cheilocystidia, thin-walled. Pileipellis a palisadoderm composed of 5–12 μm wide vertically arranged hyphae, yellowish to pale brownish in KOH and yellow to brownish in Melzer's reagent; terminal cell 24–55 × 6–12 μm, pyriform to clavate or cystidioid. Pileal trama composed of 3–5 μm wide interwoven hyphae, colorless to yellowish in KOH and yellow to yellowish brown in Melzer's reagent. Clamp connections absent in all tissues.

Habitat: Scattered to solitary on soil in tropical forests dominated by plants of the family Fagaceae.

Known distribution: Known from Southeast Asia and southern China.

Specimens examined: CHINA, YUNNAN PROVINCE: Jinghong County, Dadugang Town, alt. 1360 m, 7 July 2007, Y.C. Li 454 (KUN-HKAS 50208), the same location, 14 July 2006, Y.C. Li 525 (KUN-HKAS 50279), and Y.C. Li 543 (KUN-HKAS 50297), the same location, 28 June 2014, K. Zhao 449 (KUN-HKAS 89112).

Commentary: *Tylopilus atripurpureus* was originally described as *Boletus atripurpureus* by Corner (1972) from Malaysia. This species is characterized by the blackish purple to dark purple pileus, the pinkish to purplish hymenophore, the white to pallid context without discoloration when injured, the palisadoderm pileipellis, and the distribution in tropical forests. Phylogenetically, *T. atripurpureus* is related to *T. plumbeoviolaceus* (Snell & E.A. Dick) Snell & E.A. Dick (Fig. 4.1). However, *T. plumbeoviolaceus* has a hymeniform pileipellis which is composed of a layer of cystidia-like cells measuring 14.3–45.5 × 2.5–7.8 μm, relatively small and narrow cheilocystidia measuring 16.9–22 × 5.2–7.8 μm and pleurocystidia measuring 26.3–39 × 5.2–7.8 μm (Snell and Dick, 1941; Smith and Thiers, 1971).

22.5 *Tylopilus atroviolaceobrunneus* Yan C. Li & Zhu L. Yang, Fungal Divers 81: 153 (2016)

Basidioma medium-sized to large. Pileus 5–10 cm in diam., hemispherical to subhemispherical when young, subhemispherical to nearly applanate when mature, grayish red to violet-brown or reddish brown, blackish red or darker in the center; surface dry, finely tomentose. Context solid, white, staining a rufescent or reddish tinge when injured. Hymenophore adnate when young, adnate to depressed around apex of stipe when mature; surface white when young, dull pinkish when mature, slowly becoming reddish brown when injured; pores subangular to roundish, 0.5–1 mm wide; tubes 6–8 mm long, concolorous or a little paler than hymenophoral surface, staining a rufescent or reddish brown tinge when injured. Spore print grayish red to reddish. Stipe 6–12 × 1.2–2.1 cm, clavate to subcylindrical, always enlarged downwards, concolorous with pileal surface, but much paler towards base, pallid to white at apex when young, becoming reddish to reddish brown when bruised, entirely subtomentose; context solid when young, and then spongy when mature, white, staining a reddish tinge when bruised; basal mycelium whitish to pallid. Taste bitter, odor mild.

Basidia 20–28 × 7–9 μm, clavate, 4-spored, hyaline in KOH. Basidiospores 10–14 × 4–5 μm (Q = 2.22–3.5, Q_m = 2.66 ± 0.28), subcylindrical to subfusiform in side view with slight suprahilar depression, cylindrical to fusiform in ventral view, smooth, yellowish to brownish yellow in KOH, yellow to yellow-brown in Melzer's reagent. Hymenophoral trama boletoid composed of 3–10 μm wide filamentous hyphae, hyaline to yellowish in KOH, yellowish to yellow in Melzer's reagent. Cheilocystidia 17–40 × 4–5.5 μm, lanceolate or fusiform to subfusoid-ventricose, yellowish to yellow-brown in KOH, dark yellow-brown to blackish brown in Melzer's reagent, thin-walled. Pleurocystidia 40–60 × 7.5–14 μm, much bigger than cheilocystidia, concolorous with cheilocystidia, thin-walled. Pileipellis a palisadoderm composed of 3.5–10 μm wide vertically arranged hyphae, yellowish brown to brownish in KOH and yellow-brown to dark brown in Melzer's reagent; terminal cells 17–38 × 5–11 μm, pyriform to clavate or cystidioid. Pileal trama

Fig. 22.5 a *Tylopilus atroviolaceobrunneus* Yan C. Li & Zhu L. Yang. Photos by J.W. Liu (KUN-HKAS 107143).

Fig. 22.5 b *Tylopilus atroviolaceobrunneus* Yan C. Li & Zhu L. Yang. Photo by S.P. Jian (KUN-HKAS 101143).

composed of 5.5–8 μm wide interwoven hyphae, pale yellowish to brownish yellow in KOH and yellow to brownish in Melzer's reagent. Clamp connections absent in all tissues.

Habitat: Gregarious to scattered, in tropical forests dominated by plants of the family Fagaceae.

Known distribution: Currently known from southwestern China.

Additional specimens examined: CHINA, YUNNAN PROVINCE: Jingdong County, Caiyanghe Nature Reserve, alt. 1200 m, 28 June 2014, Z.W. Ge 3514 (KUN-HKAS 84351, type), the same location and date, K. Zhao 443 (KUN-HKAS 89106); Lancang County, alt. 1000 m, 28 June 2017, J.W. Liu 530828MF0035 (KUN-HKAS 107143); Malipo County, Liuhe Town, Zhichang Village, alt. 760 m, 25 June 2017, S.P. Jian 532624MF-205-jsp-78 (KUN-HKAS 101143).

Commentary: *Tylopilus atroviolaceobrunneus* originally described from China by Wu *et al.* (2016a) is characterized by the grayish red to violet-brown or reddish brown basidioma, the white to cream context staining rufescent to reddish when injured, the white to pinkish or pink hymenophore staining reddish brown when injured, the palisadoderm pileipellis, and the distribution in tropical forests.

Our multi-locus molecular phylogenetic analysis (Fig. 4.1) indicates *T. atroviolaceobrunneus* is closely related to *T. violaceorubrus* Yan C. Li & Zhu L. Yang. Indeed, they are morphologically similar to each other. However, *T. violaceorubrus* differs from this taxon in its relatively pale basidioma, broad cheilocystidia measuring 34–62 × 6–10 μm, and small basidiospores measuring 8–10.5 × 3.5–4.5 μm (see our description below).

22.6 *Tylopilus aurantiacus* Yan C. Li & Zhu L. Yang, The Boletes of China: *Tylopilus* s.l. 281 (2021)

MycoBank: MB 834749

Etymology: The epithet "*aurantiacus*" refers to the color of the basidiomata.

Type: CHINA, YUNNAN PROVINCE: Nanhua County, Wild Mushroom Market, altitude unknown, 2 August 2009, Y.C. Li 1952 (KUN-HKAS 59700, GenBank Acc. No.: KF112458 for nrLSU, KF112223 for *tef1-α*, KF112619 for *rpb1*, KF112740 for *rpb2*).

Basidioma medium-sized to large. Pileus 5–10 cm in diam., hemispherical to subhemispherical when young, subhemispherical to plano-convex or applanate when mature, orange-red (7B7–8), brown (7E7–8), brownish red (8E7–8) or (9B4–5), or grayish red (8B4–5) when young, pale orange (5A3–4) to grayish yellow (4B6–7) when mature; surface nearly glabrous and dry, without discoloration when hurt; context solid, whitish to pallid, without discoloration when bruised. Hymenophore adnate to slightly decurrent; surface initially whitish (9A1) to pale grayish (8B1) and then pinkish (13A2) to brownish pink (10D5–6) when mature, without discoloration when hurt; pores subangular to roundish, up to 2 mm wide; tubes 5–7 mm long, concolorous or much paler than hymenophoral surface, without discoloration when bruised. Spore print reddish (12B2) to pinkish (13B2). Stipe 4–7 × 1.2–1.8 cm, subcylindrical to clavate, dark brownish orange (5C6) to brownish yellow (5C7) or pale yellow (4A3–4) when young, and brownish red (8D7–8) to dark brownish red (9D6–7) with age, without discoloration when touched; context almost white to cream, without discoloration when hurt; basal mycelium white. Taste and odor mild.

Fig. 22.6 a *Tylopilus aurantiacus* Yan C. Li & Zhu L. Yang. Photos by Y.C. Li (KUN-HKAS 59547).

Fig. 22.6 b *Tylopilus aurantiacus* Yan C. Li & Zhu L. Yang. Photos by Y.C. Li (KUN-HKAS 59700, type).

Basidia 20–28 × 5–10 μm, clavate to narrowly clavate, 4-spored, sometimes 2-spored, hyaline in KOH. Basidiospores [160/8/8] 5–7 × 3.5–4.5 μm [Q = (1.11) 1.25–1.63, Q_m = 1.45 ± 0.12], subglobose to subellipsoid and inequilateral in side view, subglobose to broadly ellipsoid or ellipsoid in ventral view, smooth, hyaline to yellowish in KOH, yellow to brownish yellow in Melzer's reagent. Hymenophoral trama boletoid composed of 3–8 μm wide filamentous hyphae, hyaline to yellowish in KOH, yellowish to yellow in Melzer's reagent. Cheilocystidia abundant, 35–50 × 6–9 μm, fusiform to fusoid-ventricose, yellowish to yellow in KOH, thin-walled. Pleurocystidia abundant, 46–60 × 7–13 μm, fusiform to fusoid-ventricose, yellowish to yellow in KOH, thin-walled. Pileipellis a cutis composed of 4–6 μm wide gelatinous interwoven filamentous hyphae, yellowish to pale brownish in KOH and yellow to brownish in Melzer's reagent; terminal cells 12–38 × 3.5–6 μm, subcylindrical. Pileal trama composed of 4–16 μm wide interwoven hyphae, hyaline to yellowish in KOH and yellowish to brownish in Melzer's reagent. Clamp connections absent in all tissues.

Habitat: Scattered on soil in tropical to subtropical forests dominated by plants of the family Fagaceae.

Known distribution: Currently known from southern and southwestern China.

Additional specimens examined: CHINA, SICHUAN PROVINCE: Mianyang, Southwest University of Science and Technology, altitude unknown, 29 July 2008, B. Feng 248 (KUN-HKAS 55358). GUANGDONG PROVINCE: Fengkai County, Heishiding Nature Reserve, alt. 250 m, 9 May 2012, F. Li 195 (KUN-HKAS 78121), the same location, 1 June 2013, K. Zhao 245 (KUN-HKAS 80670). GUANGXI PROVINCE: Ziyuan County, alt. 1000 m, 13 August 2013, C.Y. Deng (HGAS-MF01-13538). YUNNAN PROVINCE: Changning County, Huitou Village, alt. 2020 m, 25 July 2009, Y.C. Li 1800 (KUN-HKAS59547); Qiubei County, Qingping Village, alt. 1400 m, 9 August 2014, P.M. Wang 102 (KUN-HKAS 95183); Lancang County, Huimin Town, Ban'gai Village, alt. 1000 m, 24 August 2016, J.W. Liu LC104 (KUN-HKAS 97608); Kunming, Kunming Botanic Garden, alt. 1990 m, 3 September 2020, Z.L. Yang 6407 (KUN-HKAS 107679).

Commentary: *Tylopilus aurantiacus* is characterized by the orange-red to brownish red then brown to pale orange pileus, the white context without discoloration when injured, the dark brownish orange to brownish yellow or pale yellow then brownish red to dark brownish red stipe, the ellipsoid to ovoid basidiospores and the intricate trichoderm pileipellis. This species is phylogenetically sister and morphologically similar to *T. rubrotinctus* Yan C. Li & Zhu L. Yang (Fig. 4.1). However, *T. rubrotinctus* differs from this taxon in its grayish red to pale red or somewhat reddish orange to pale orange or orangish pileus, pale yellow or yellowish stipe, and palisadoderm pileipellis composed of 5–9 μm wide vertically arranged hyphae (see our description below). *Tylopilus aurantiacus* is also macroscopically similar to *Tylopilus pseudoballoui* D. Chakr., K. Das & Vizzini. However, *T. pseudoballoui* has an orange-brown to orange pileus, a yellowish to yellow then brownish yellow stipe and relatively small pleurocystidia measuring 30–45 × 8–12 μm (Chakraborty *et al.* 2018).

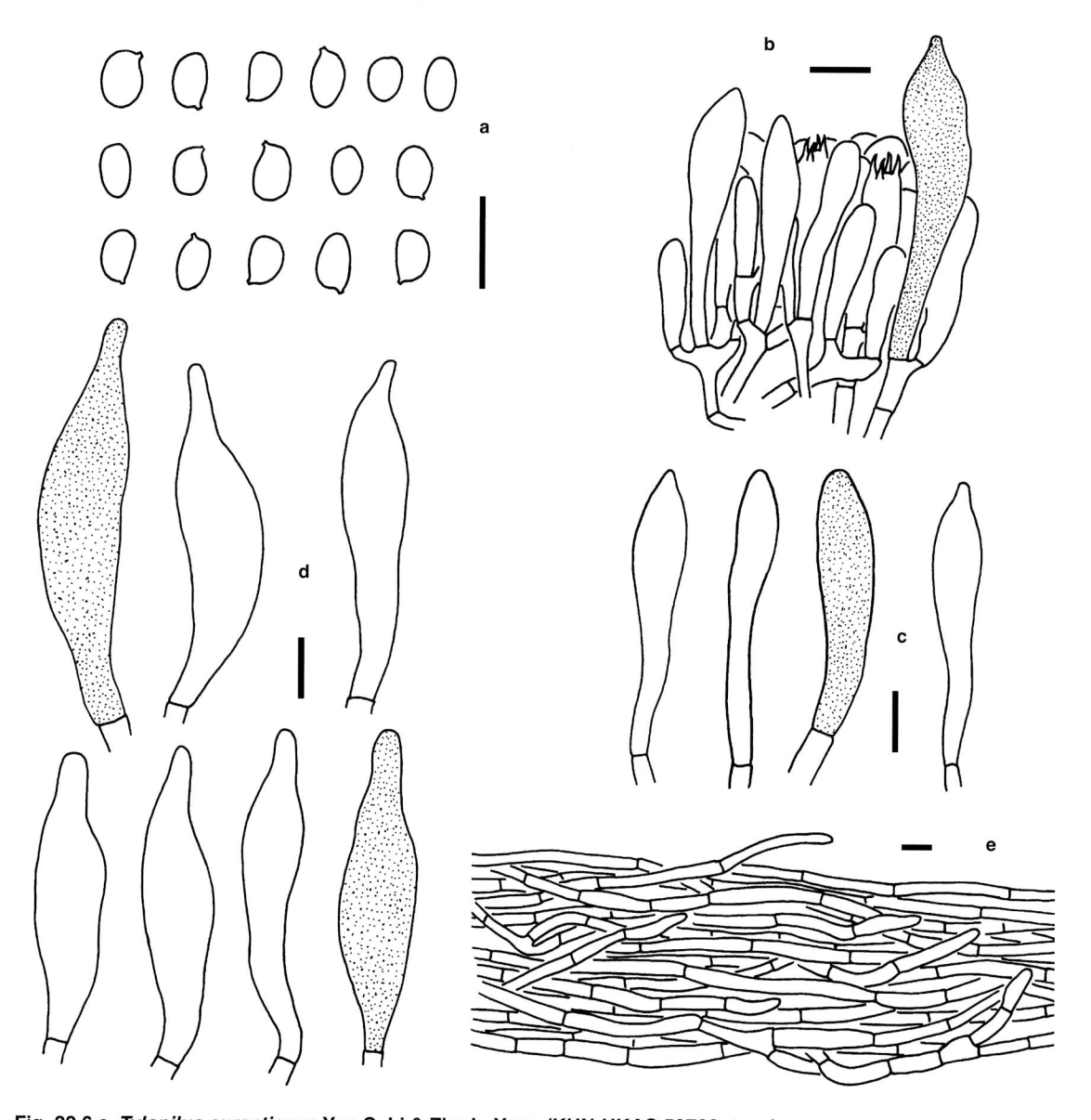

Fig. 22.6 c *Tylopilus aurantiacus* Yan C. Li & Zhu L. Yang (KUN-HKAS 59700, type).
a. Basidiospores; b. Basidia and cheilocystidia; c. Cheilocystidia; d. Pleurocystidia; e. Pileipellis. Scale bars = 10 μm.

22.7 *Tylopilus brunneirubens* (Corner) Watling & E. Turnbull, Edinb J Bot 51: 332 (1994)

Basionym: *Boletus brunneirubens*, Corner, *Boletus* in Malaysia 186 (1972).

Basidioma medium-sized. Pileus 5–9 cm in diam., hemispherical to subhemispherical when young, subhemispherical to plano-convex, olive-brown to brown when young, brownish to yellowish brown or pale brown to pale brownish when mature or aged; surface dry, nearly glabrous or with concolorous fibrillose or fine tomentose squamules, minutely cracked with age, staining rust-brown to reddish brown when hurt; context solid, white, staining a rust-brown to brownish red tinge when bruised. Hymenophore adnate when young, slightly decurrent or depressed around apex of stipe when mature; surface whitish, staining grayish red to pallid red when bruised; pores subangular to roundish, 0.5–1 mm wide; tubes 15–20 mm long, concolorous or paler than hymenophoral surface, staining grayish red to pallid red when bruised. Spore print grayish red to reddish. Stipe 5–8 × 0.8–1.5 cm, clavate to subcylindrical, always enlarged downwards, whitish to grayish white and then brown to dark brown when injured or aged, upper 1/3–2/3 with distinct reticulum, staining rust-brown or brownish red when touched; context whitish to pallid, staining a rust-brown or brownish red tinge when hurt; basal mycelium whitish to pallid. Taste and odor mild.

Basidia 18–28 × 7–10 μm, clavate to narrowly clavate, 4-spored, colorless to yellowish or pale brownish in KOH. Basidiospores 8.5–11 (12.5) × 3.5–4.5 μm [Q = 2.13–3.14 (3.57), Q_m = 2.68 ± 0.22], subcylindrical to subfusiform in side view with slight suprahilar depression, cylindrical to fusiform in ventral view, smooth, yellowish to brownish yellow in KOH, yellow to yellow-brown in Melzer's reagent. Hymenophoral trama boletoid composed of 5–11 μm wide filamentous hyphae, hyaline to yellowish in KOH, yellowish to yellow in Melzer's reagent. Cheilo- and pleurocystidia abundant, 44–69 × 11–17 μm, subfusiform to subfusoid-ventricose, often with sharp apex, thin-walled, brown to dark brown in KOH. Pileipellis a trichoderm composed of 3–6 μm filamentous hyphae, yellowish to pale brownish in KOH

Fig. 22.7 a *Tylopilus brunneirubens* (Corner) Watling & E. Turnbull. Photos by L.H. Han (KUN-HKAS 84797).

Fig. 22.7 b *Tylopilus brunneirubens* (Corner) Watling & E. Turnbull. Photos by G. Wu (KUN-HKAS 88992).

and yellowish brown to brownish in Melzer's reagent; terminal cells 22–107 × 5–7.5 μm, cylindrical. Pileal trama composed of 5–9 μm wide interwoven hyphae, colorless to yellowish in KOH and yellow to yellowish brown in Melzer's reagent. Clamp connections absent in all tissues.

Habitat: Solitary on soil in tropical to subtropical forests dominated by plants of the family Fagaceae.

Known distribution: Currently known from Southeast Asia and eastern, southeastern, southern, and southwestern China.

Specimens examined: CHINA, HUBEI PROVINCE: Macheng County, Wunaoshan National Forest Park, alt. 200 m, 26 June 2013, T. Guo 643 (KUN-HKAS 81845). FUJIAN PROVINCE: Sanming, Sanyuan National Forest Park, alt. 260 m, 27 August 2007, Y.C. Li 1043 (KUN-HKAS 53388). GUANGDONG PROVINCE: Fengkai County, Heishiding Nature Reserve, alt. 250 m, 19 July 2012, F. Li 692 (KUN-HKAS 106424), the same location, 5 September 2012, F. Li 998 (KUN-HKAS 77767), the same location, 25 September 2012, F. Li 1079 (KUN-HKAS 77790), the same location, 15 October 2012, F. Li 1082 (KUN-HKAS 77792). SICHUAN PROVINCE: Miyi County, Puwei Town, alt. 2000 m, 24 July 1986, M.S. Yuan 1159 (KUN-HKAS 18434). YUNNAN PROVINCE: Jinghong County, Dadugang Town, alt. 1350 m, 22 July 2007, Y.C. Li 922 and Y.C. Li 923 (KUN-HKAS 52609 and KUN-HKAS 52610, respectively); Yongping County, near Milestone 3260 KM of the National Road 320#, alt. 2100 m, 30 July 2009, Y.C. Li 1916 and Y.C. Li 1925 (KUN-HKAS 59664 and KUN-HKAS 59673, respectively); Kunming, Kunming Botanic Garden, alt. 1900 m, 22 July 2013, G. Wu 1143 (KUN-HKAS 80517); Mengla County, Xishuangbanna National Natural Reserve, alt. 1050 m, 6 July 2014, G. Wu 1255 (KUN-HKAS 88992); Jinghong County, Dadugang Town, alt. 1350 m, 10 July 2014, X.B. Liu 426 (KUN-HKAS 87073); Malipo County, Xinhe Village, alt. 1300 m, 3 August 2014, L.H. Han 501 and L.H. Han 506 (KUN-HKAS 84797 and KUN-HKAS 84802, respectively), the same location, 31 August 2016, G. Wu 1694 (KUN-HKAS 99813); Qiubei County, Qingping Village, alt. 1550 m, 9 August 2014, J. Li 140 (KUN-HKAS 85966); Longling County, Bizhai Town, Zhongzhai Village, alt. 1400 m, 2 August 2014, X.B. Liu 486 (KUN-HKAS 87132); Baoshan, Xiyi Town, Wojiao Village, 12 August 2015, P.M. Wang 352 (KUN-HKAS 93312).

Commentary: *Tylopilus brunneirubens*, originally described by Corner (1972) from Singapore, is characterized by the olive-brown to brown then dark brown to chestnut-brown pileus, the white context staining a rust-brown or reddish brown tinge when touched, the distinct half-reticulate stipe, the trichoderm pileipellis, and the distribution in tropical to subtropical forests.

Our multi-locus molecular phylogenetic analysis (Fig. 4.1) indicates *T. brunneirubens* is closely related to *T. alpinus* and *T. pseudoalpinus* Yan C. Li & Zhu L. Yang. Indeed, they are morphologically similar to each other. However, *T. alpinus* differs from *T. brunneirubens* in its relatively large basidioma, large basidiospores measuring 13–14.5 (15.5) × (3.5) 4–5 μm, and distribution in subalpine to alpine forests. *Tylopilus pseudoalpinus* can be easily distinguished from *T. brunneirubens* by its grayish ruby to grayish red or brownish red pileus, and dingy white to cream stipe which is entirely reticulate. In addition, in our multi-locus phylogenetic analysis (Fig. 4.1), collections labeled as *T. brunneirubens* could be split into two sublineages within a monophyletic lineage. Since few morphological differences were observed between these two phylogenetic species, we treat them as a single species in this study.

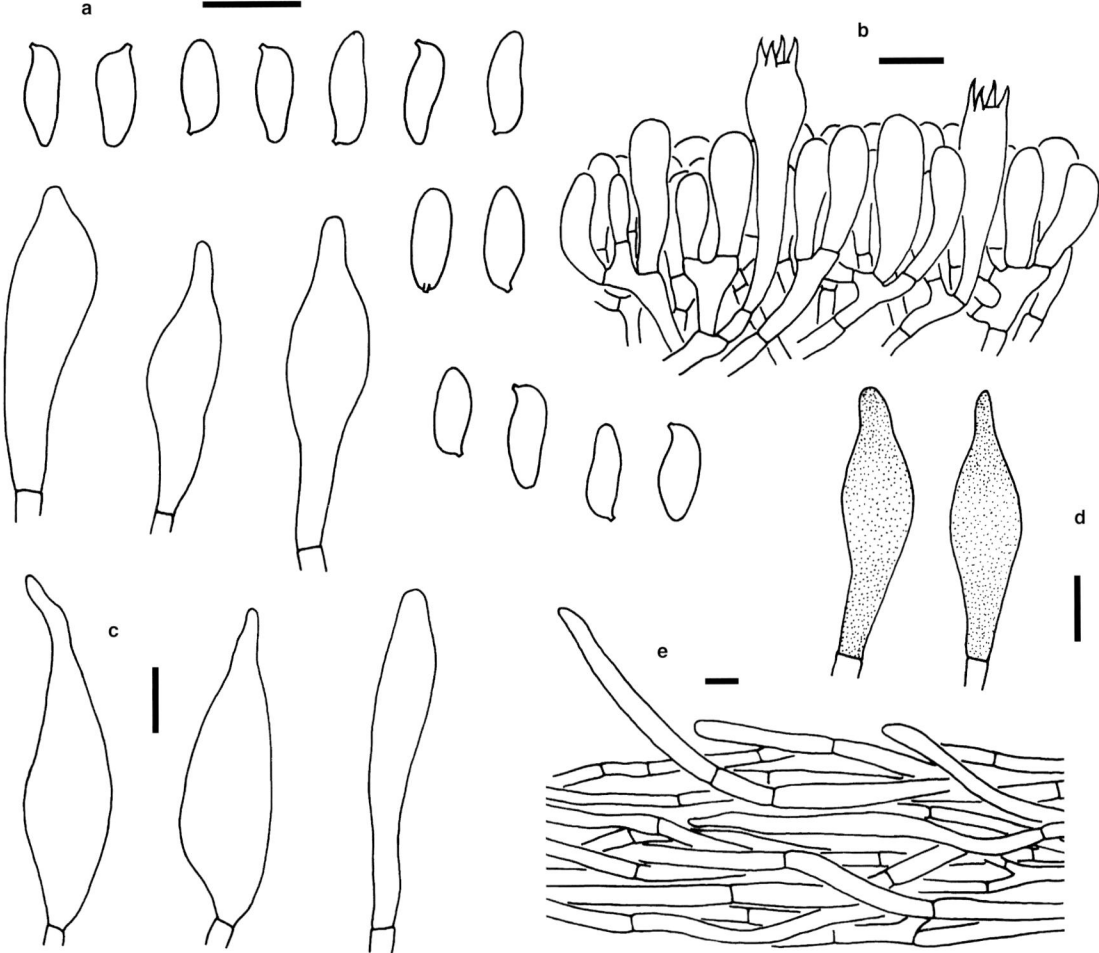

Fig. 22.7 c *Tylopilus brunneirubens* (Corner) Watling & E. Turnbull (KUN-HKAS 88992).
a. Basidiospores; b. Basidia; c. Pleurocystidia; d. Cheilocystidia; e. Pileipellis. Scale bars = 10 μm.

22.8 *Tylopilus castanoides* Har. Takah., Mycoscience 43(5): 402 (2002)

Basidioma small. Pileus 4–5 cm in diam., hemispherical to subhemispherical; dark reddish brown, chestnut-brown to brown, darker in the center; surface dry, with fine tomentose to matted squamules; context white, staining reddish brown when bruised. Hymenophore adnate when young, depressed around apex of stipe; surface white to pinkish, staining reddish brown when bruised; pores subangular to roundish, up to 2 mm wide; tubes 6–10 mm long, concolorous or a little paler than hymenophoral surface, staining pale reddish brown when bruised. Spore print grayish red to reddish. Stipe 6–10 × 1.2–2.1 cm, subcylindrical to clavate, concolorous with pileal surface or much paler, becoming brown to brownish when touched; context solid, white, becoming reddish brown when hurt; basal mycelium white to cream. Taste bitter.

Basidia 20–33 × 6–8 μm, clavate to narrowly clavate, 4-spored, hyaline in KOH. Basidiospores 9.5–14 × 3–4 μm [Q = (2.5) 2.63–3.71 (3.83), Q_m = 3.22 ± 0.33], subcylindrical to subfusiform in side view with slight suprahilar depression, cylindrical to fusiform in ventral view, smooth, yellowish to brownish yellow in KOH, yellow to yellow-brown in Melzer's reagent. Hymenophoral trama boletoid, hyphae cylindrical, 4–8 μm wide, hyaline to yellowish in KOH, yellowish to yellow in Melzer's reagent. Cheilo- and pleurocystidia 33–51 × 6–9 μm, subfusiform to fusoid-ventricose with a spindly apical part, brown to yellowish brown in KOH, with brown to dark brown encrustations, somewhat thick-walled (up to 1 μm thick). Pileipellis a trichoderm composed of 5–8 μm wide interwoven filamentous hyphae with cylindrical terminal cells 29–42 × 5–8 μm, pale yellowish to brownish yellow in KOH and yellow to brownish in

Fig. 22.8 a *Tylopilus castanoides* Har. Takah. Photos by X.B. Liu (KUN-HKAS 92325).

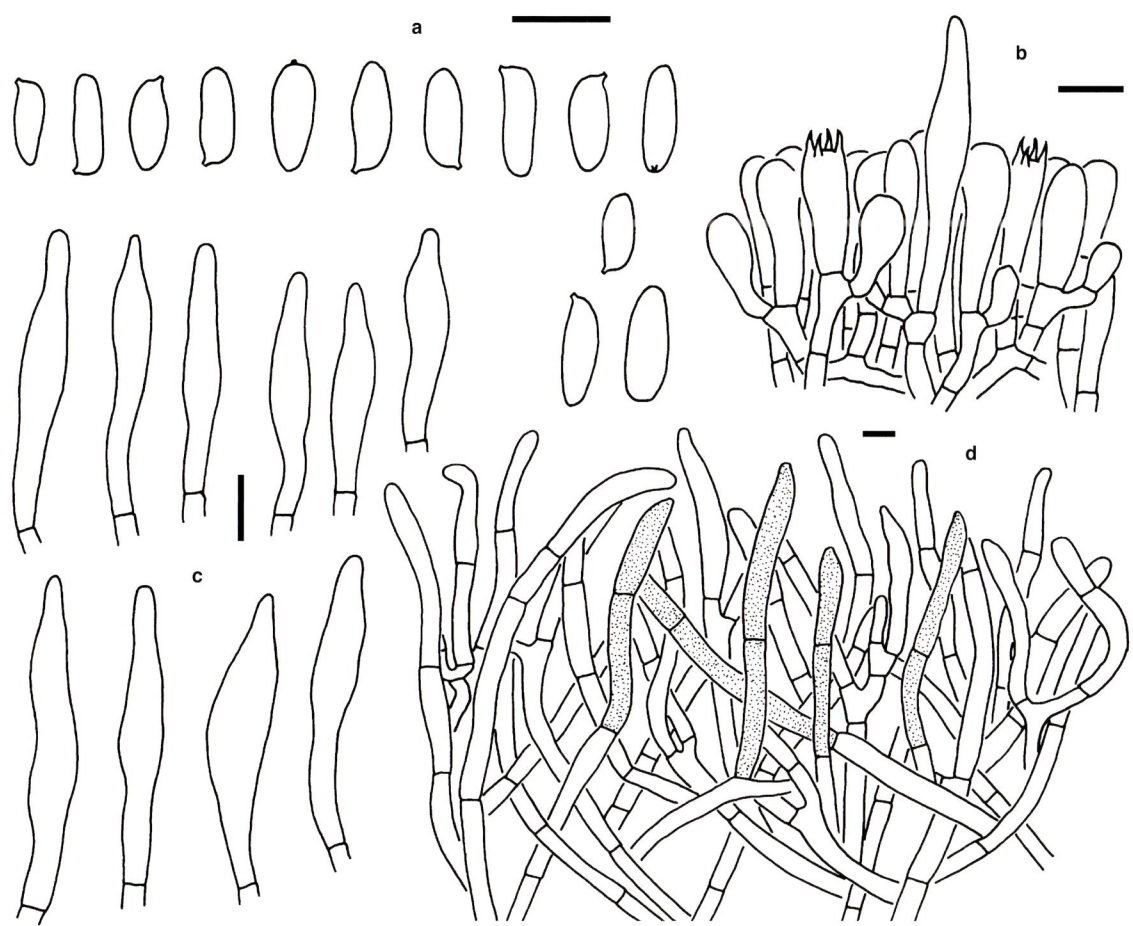

Fig. 22.8 b *Tylopilus castanoides* Har. Takah. (KUN-HKAS 92325).
a. Basidiospores; b. Basidia and pleurocystidium; c. Cheilo- and pleurocystidia; d. Pileipellis. Scale bars = 10 μm.

Melzer's reagent. Pileal trama composed of 4–11 μm wide interwoven hyphae, yellowish to pale brownish in KOH and yellowish brown to brownish in Melzer's reagent. Clamp connections absent in all tissues.

Habitat: Scattered on soil in subtropical forests dominated by plants of the family Fagaceae.

Known distribution: Currently known from central China and Japan.

Specimens examined: HUBEI PROVINCE: Shennongjia Forestry District, Hongping Town, alt. 1500 m, 12 August 2015, X.B. Liu 780 (KUN-HKAS 92325).

Commentary: *Tylopilus castanoides*, originally described by Takahashi (2002) from Japan, is characterized by the dark reddish brown, chestnut-brown or brown pileus and stipe, the white context staining reddish brown when bruised, the white to pinkish hymenophore staining reddish brown or rufescent when injured, and the trichoderm pileipellis composed of 5–8 μm wide interwoven filamentous hyphae.

Tylopilus castanoides is phylogenetically related and morphologically similar to *T. olivaceobrunneus* Yan C. Li & Zhu L. Yang. However, *T. olivaceobrunneus* can be distinguished by its yellowish olivaceous pileus becoming brownish to pale brown when aged or injured, white context without color change or staining indistinct red when injured, and brownish gray to dark brownish gray or dark brown stipe.

22.9 *Tylopilus felleus* (Bull.) P. Karst., Revue mycol 3(9): 16 (1881)

Basionym: *Boletus felleus* Bull., Herbier de la France, 8. Tab 379 (1788).

Basidioma medium-sized to large. Pileus 5–10 cm in diam., hemispherical when young, subhemispherical to applanate when mature, gray or grayish white to grayish brown, darker in the center; surface nearly glabrous or with fibrillose to fine tomentose squamules, viscid when wet; context with to grayish white, without discoloration when bruised. Hymenophore adnate when young, adnate to slightly depressed around apex of stipe when mature; surface initially white to pallid and then pinkish to pink, without discoloration when touched; pores angular to roundish, 0.5–1 mm wide; tubes 5–10 mm long, concolorous or much paler than hymenophoral surface, without discoloration when bruised. Spore print pinkish to reddish. Stipe 2.5–10 × 0.5–4 cm, clavate to subcylindrical, always enlarged downwards; surface brownish to brown, upper 1/3–1/2 part with distinct reticulum, without discoloration when injured; context solid, white to grayish white, without discoloration when injured; basal mycelium whitish. Taste bitter, odor unknown.

Fig. 22.9 *Tylopilus felleus* (Bull.) P. Karst. Photo by X.F. Shi (KUN-HKAS 90203).

Basidia 20–38 × 8–11 μm, clavate to narrowly clavate, 4-spored, hyaline in KOH. Basidiospores 14–16 (17) × 4.5–5.5 μm [Q = (2.29) 2.5–3.5 (3.88), Q_m = 3.05 ± 0.35], subcylindrical to subfusiform in side view with slight suprahilar depression, cylindrical to fusiform in ventral view, smooth, hyaline to yellowish in KOH, yellow to brownish yellow in Melzer's reagent. Hymenophoral trama boletoid composed of 3–15 μm wide filamentous hyphae, hyaline to yellowish in KOH, yellowish to yellow in Melzer's reagent. Cheilo- and pleurocystidia 50–70 × 10–18 μm, fusoid-ventricose to clavate with subacute apex, hyaline to yellowish in KOH, thin-walled. Caulocystidia forming the reticulum over the upper 1/3–1/2 of stipe, ventricose, clavate or lanceolate. Pileipellis an intricate trichoderm composed of 4.5–7 μm wide interwoven filamentous hyphae, yellowish to pale brownish in KOH and yellow to brownish in Melzer's reagent; terminal cells 20–40 × 5–8 μm, subclavate to cylindrical. Pileal trama composed of 7–15 μm wide interwoven hyphae, colorless to yellowish in KOH and yellowish to pale yellow in Melzer's reagent. Clamp connections absent in all tissues.

Habitat: Solitary on soil in temperate forests dominated by plants of the family Pinaceae or in mixed forests dominated by plants of the families Fagaceae and Pinaceae.

Known distribution: Currently known from Europe, East Asia, and North America.

Specimens examined: CHINA, JILIN PROVINCE: Antu County, Changbai Mountain, alt. 1000 m, 2 August 2008, Y.C. Li 1167 (KUN-HKAS 55832), the same location, 8 August 2010, X.F. Shi 483 (KUN-HKAS 90203). CHINA, NEI MONGGOL AUTONOMOUS REGION: Huhhot, Daqing Mountain, alt. 1500 m, 30 August 2000, Q.Z. Yao 9 (KUN-HKAS 37415). CHINA, XINJIANG UYGUR AUTONOMOUS REGION: Buerjin County, Kanas Lake, alt. 2000 m, 29 August 2004, M. Zang 14301 (KUN-HKAS 47431). GERMANY, MARBURG: University of Marburg, Rexer 6151 (KUN-HKAS 54927, MB), Rexer 8098 (KUN-HKAS 54928, MB) and Rexer 8989 (KUN-HKAS 54926).

Commentary: *Tylopilus felleus*, originally described as *Boletus felleus* Bull. by Bulliard (1788) from Europe, is characterized by the grayish to gray or grayish brown pileus, the with to grayish context without discoloration when bruised, the initially white to pallid and then pinkish to pink hymenophore, the half-reticulate stipe, the bitter taste of context, and the intricate trichoderm. *Tylopilus felleus* is widely distributed in the Holarctic. In East Asia, this species is easily confused with *T. neofelleus* Hongo as they have similar appearance and habitats. However, *T. neofelleus* has purple to purplish pink hymenophoral surface when young, relatively small basidiospores measuring 8–9 (10) × 3–4 μm, and a glabrous stipe without reticulum or with indistinct reticulum restricted to the apex of stipe.

In our multi-locus phylogenetic analysis (Fig. 4.1), *T. felleus* clusters with *T. violaceobrunneus* but without statistical support. Moreover, *T. violaceobrunneus* differs from *T. felleus* in its brownish violet or reddish brown to light brown pileus always with magenta tinge, and palisadoderm pileipellis composed of 3.5–9 μm wide nearly vertically arranged hyphae (Wu *et al*. 2016a).

22.10 *Tylopilus fuligineoviolaceus* Har. Takah., Mycoscience 48(2): 97 (2007)

Basidioma small to medium-sized. Pileus 4–8 cm in diam., hemispherical to applanate or plano-convex, violet to dark violet when young, usually dark purplish brown at maturity, much darker in the center; surface covered with fibrillose to tomentose squamules, dry, without discoloration when injured; context solid, whitish to pallid, without discoloration when bruised. Hymenophore adnate; surface initially white to pallid and then pinkish to purplish pink, without discoloration when bruised; pores subangular to roundish, 0.3–1 mm wide; tubes 4–7 mm long, concolorous or much paler than hymenophoral surface, without discoloration when bruised. Spore print reddish to pinkish. Stipe 5–8 × 1.1–2.5 cm, cylindrical or clavate, always enlarged downwards, violet to dark violet, often paler towards base, whitish at the very apex; without color change when hurt; context solid, white, without color change when injured; basal mycelium white. Taste bitter, odor mild.

Basidia 25–40 × 7–12 μm, clavate to narrowly clavate, 4-spored, hyaline in KOH. Basidiospores 8.5–10.5 × 3–4 μm (Q = 2.63–3.33, Q_m = 2.73 ± 0.26), subcylindrical to subfusiform in side view with slight suprahilar depression, cylindrical to fusiform in ventral view, smooth, hyaline to yellowish in KOH, yellow to brownish yellow in Melzer's reagent. Hymenophoral trama boletoid composed of 2–8 μm wide filamentous hyphae, hyaline to yellowish in KOH, yellowish to yellow in Melzer's reagent. Cheilocystidia 36–46 × 6–8 μm, subfusiform to fusoid-ventricose, yellowish to brownish yellow in KOH, brown to yellow-brown in Melzer's reagent, thin-walled. Pleurocystidia similar to cheilocystidia, thin-walled. Pileipellis a trichoderm composed of 3.5–7 μm wide vertically arranged hyphae, pale yellowish to brownish

Fig. 22.10 a *Tylopilus fuligineoviolaceus* Har. Takah. Photos by Q. Cai (KUN-HKAS 92013).

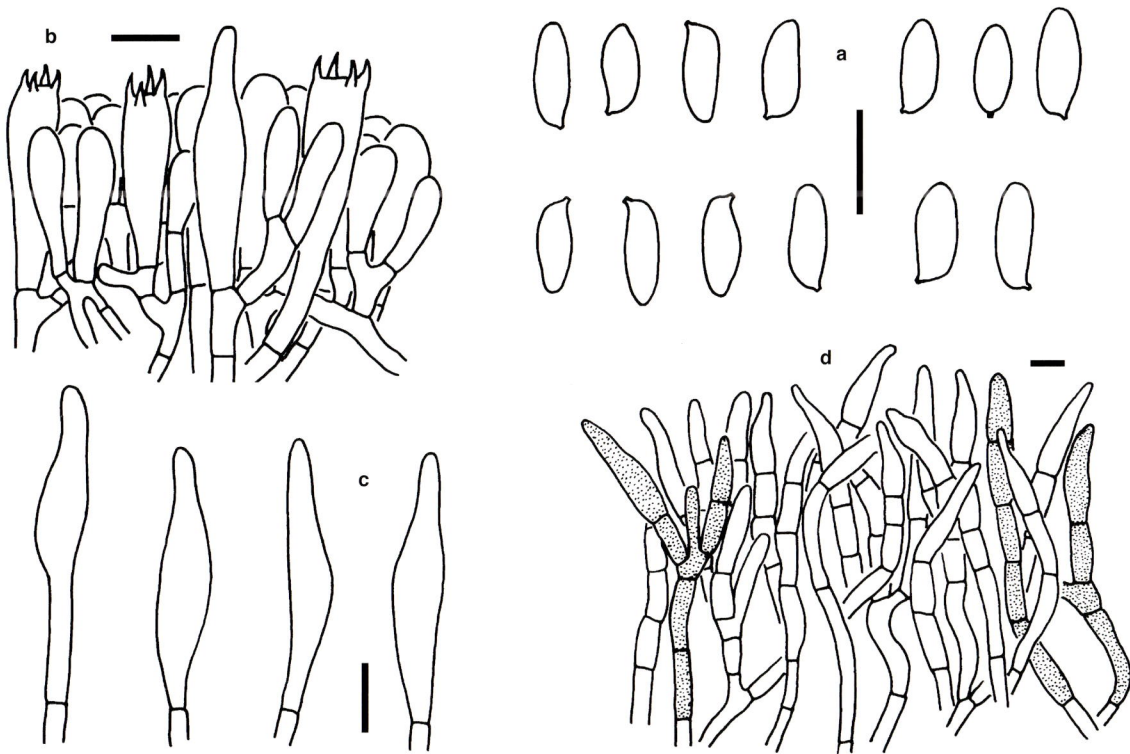

Fig. 22.10 b *Tylopilus fuligineoviolaceus* Har. Takah. (KUN-HKAS 92013).
a. Basidiospores; b. Basidia and cheilocystidium; c. Cheilo- and pleurocystidia; d. Pileipellis. Scale bars = 10 μm.

yellow in KOH and yellow to brownish in Melzer's reagent; terminal cells 18–38 × 4–9 μm, pyriform to clavate or cystidioid. Pileal trama composed of 3–6 μm wide interwoven hyphae, hyaline to yellowish in KOH and yellowish to brownish in Melzer's reagent. Clamp connections absent in all tissues.

Habitat: Solitary on soil in temperate forests dominated by plants of the family Fagaceae.

Known distribution: Currently known from northeastern China and Japan.

Specimens examined: CHINA, LIAONING PROVINCE: Benxi County, Xiamatang Town, alt. 270 m, 21 August 2015, Q. Cai 1393 (KUN-HKAS 92013).

Commentary: *Tylopilus fuligineoviolaceus* is characterized by the violet to dark violet then dark purplish brown pileus, white to pallid then pink to pinkish purple hymenophore, the violet to dark violet stipe that is much paler at the apical part, the trichoderm pileipellis composed of 3.5–7 μm wide vertically arranged hyphae, and the distribution in temperate forests. This species is phylogenetically related and macroscopically similar to the purple or violaceous species in *Tylopilus* from China, namely *T. albopurpureus*, *T. atripurpureus*, *T. atroviolaceobrunneus*, *T. griseipurpureus*, *T. neofelleus*, *T. obscureviolaceus*, *T. plumbeoviolaceoides*, *T. purpureorubens*, *T. violaceobrunneus*. For the differences among these species see the commentary on *T. albopurpureus*. *Tylopilus fuligineoviolaceus* originally described from Japan is documented and illustrated here for the first time in China (Takahashi 2007).

22.11 *Tylopilus fuscatus* (Corner) Yan C. Li & Zhu L. Yang, The Boletes of China: *Tylopilus* s.l. 296 (2021)

MycoBank: MB837921

Basionym: *Boletus balloui* var. *fuscatus* Corner [as "*ballouii*"], *Boletus* in Malaysia 194 (1972).

Synonyms: *Gyroporus fuscatus* (Corner) E. Horak, Malay Fores Rec 51: 45 (2011).; *Rubinoboletus balloui* var. *fuscatus* (Corner) Heinem. & Rammeloo, Bull Jard Bot natn Belg 53(1/2): 296 (1983).

Basidioma small to medium-sized. Pileus 4–8 cm in diam., hemispherical to convexo-applanate, yellowish when young and then fawn-ochraceous to tawny-ochraceous when mature, becoming fuscous-fawn to dark brown or dingy tan when injured or aged; margin slightly extended, viscid when wet, glabrous; context whitish to cream, staining vinaceous to dull purple when injured. Hymenophore adnate; surface white to pinkish, staining indistinct vinaceous to dull purple when touched; pores angular, 0.5–1 mm wide; tubes up to 8 mm long, concolorous or much paler than hymenophoral surface, staining indistinct vinaceous to dull purple when bruised. Stipe subcylindrical to clavate, tapering downwards, 4–8 × 0.7–1.5 cm, yellowish to yellowish ochraceous or tan, but always white to pallid at apex, staining fuscous on injury; surface always covered with brown to dark brown pruinose squamules; basal mycelium white, without discoloration when bruised; context white, staining indistinct vinaceous to dull purple when touched. Taste and odor mild.

Basidia 20–38 × 7–9 μm, clavate to narrowly clavate, 4-spored, hyaline in KOH. Basidiospores [80/4/2] 6–8 × 3.5–4.5 μm [Q = 1.5–2, Q_m = 1.72 ± 0.14], subellipsoid to elongated and inequilateral in side view with slight suprahilar depression, ellipsoid to elongated in ventral view, smooth, hyaline to yellowish in KOH, yellow to brownish yellow in Melzer's reagent. Hymenophoral trama boletoid composed of 3–8 μm wide filamentous hyphae, hyaline to yellowish in KOH, yellowish to yellow in Melzer's reagent. Cheilo- and pleurocystidia 30–53 × 7–10 μm, subfusiform to fusiform, always with a sharp apex, brown to yellowish brown in KOH, dark brown to yellow-brown in Melzer's reagent, thin-walled. Pileipellis an ixocutis composed of 3–5.5 μm wide interwoven hyphae with subcylindrical terminal cells 29–51 × 4–5.5 μm, yellowish to pale brownish in KOH and yellowish brown to brownish in Melzer's reagent. Pileal trama composed of 2–9 μm wide interwoven hyphae, hyaline to yellowish in KOH, yellowish brown to brownish in Melzer's reagent. Clamp connections absent in all tissues.

Habitat: Scattered to gregarious on soil in tropical forests dominated by plants of the family Fagaceae.

Known distribution: Currently known from Singapore and southern China.

Specimens examined: CHINA, HAINAN PROVINCE: Qiongzhong County, Limu Mountain, alt. 750 m, 2 July 2021, M.X. Li 165 (KUN-HKAS 121978), the same location, 6 May 2009, N.K. Zeng 118 (KUN-HKAS 59838).

Commentary: *Tylopilus fuscatus* was originally described as *Boletus balloui* var. *fuscatus* by Corner from Singapore (Corner 1972). It is easily recognized by its initially yellowish and then fawn-ochraceous to tawny-ochraceous pileus, whitish to cream context, vinaceous to dull purple discoloration when injured, ovoid to subovoid basidiospores, ixocutis pileipellis, and distribution in tropical forests (Corner 1972). Horak (2011) transferred *Boletus balloui* var. *fuscatus* to the genus *Gyroporus* and elevated the variety to species rank, namely *Gyroporus fuscatus* (Corner) E. Horak based on the morphology of basidiospores. However, in our phylogenetic analysis (Fig. 4.1), *Boletus ballouii* var. *fuscatus* forms an independent lineage, and nests within *Tylopilus*. Moreover, the white to pinkish hymenophore, the solid context of stipe, the discoloration when injured differ from those in the genus *Gyroporus*. Thus, a new combination is proposed.

Phylogenetically, *T. fuscatus* is related to *T. griseolus* Yan C. Li & Zhu L. Yang. However, *T. griseolus* differs from *T. fuscatus* in its pale grayish, grayish or gray then olive-gray to dark olive-gray pileus, white context staining indistinct reddish brown when bruised, adnate to slightly decurrent hymenophore, white to cream then yellowish to grayish yellow hymenophoral surface staining brown to reddish brown when bruised, white to pale yellowish stipe, and trichoderm to palisadoderm pileipellis composed of 4–9 μm wide filamentous hyphae (see our description below).

Fig. 22.11 a *Tylopilus fuscatus* (Corner) Yan C. Li & Zhu L. Yang. Photos by N.K. Zeng (KUN-HKAS 59838).

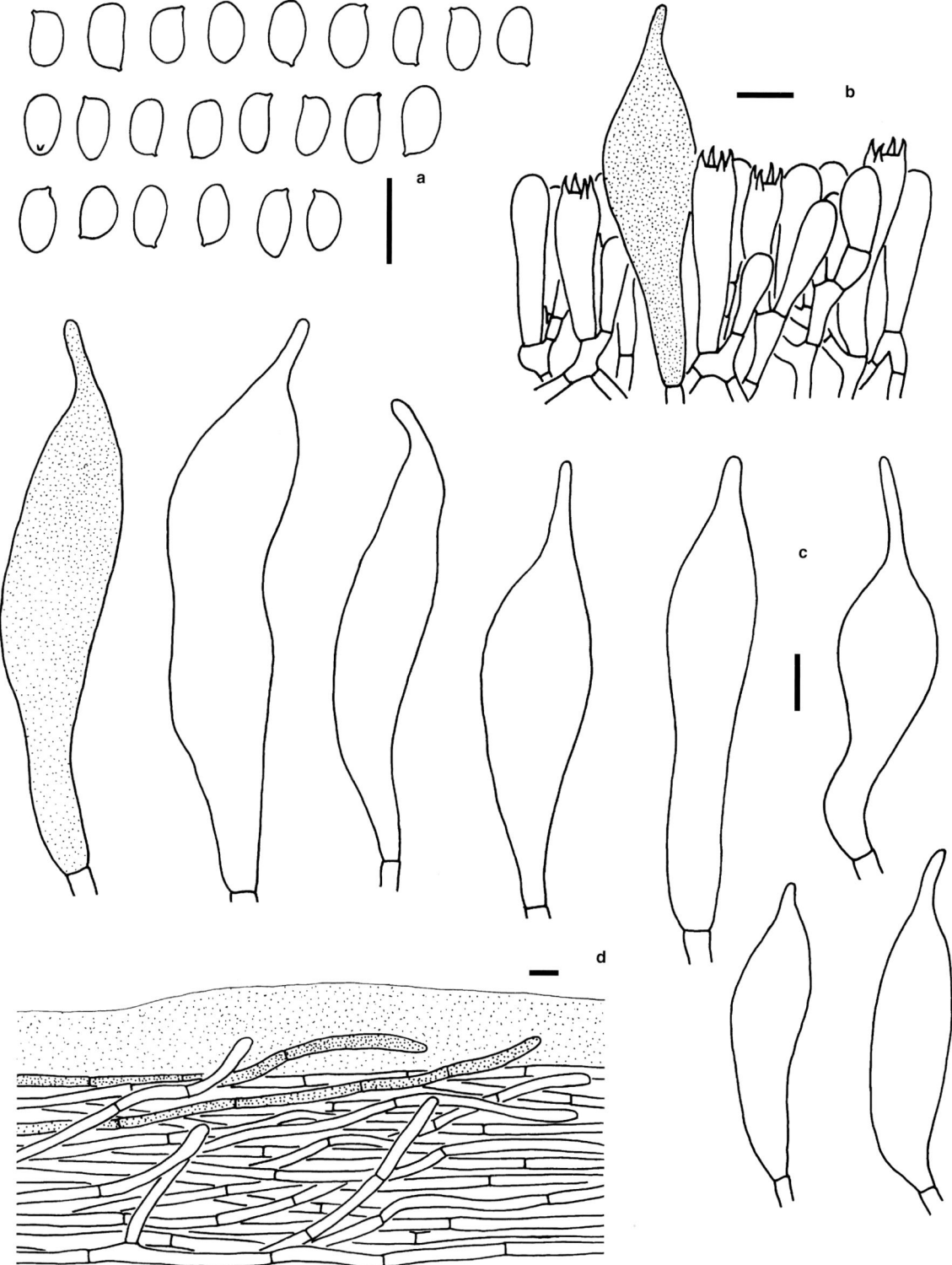

Fig. 22.11 b *Tylopilus fuscatus* (Corner) Yan C. Li & Zhu L. Yang (KUN-HKAS 59838).
a. Basidiospores; b. Basidia and cheilocystidium; c. Cheilo- and pleurocystidia; d. Pileipellis. Scale bars = 10 μm.

22.12 *Tylopilus griseipurpureus* (Corner) E. Horak, Malay Fores Rec 51: 132 (2011)

Basionym: *Boletus griseipurpureus* Corner, *Boletus* in Malaysia 168 (1972).

Basidioma small to large. Pileus 4–12 cm in diam., hemispherical when young, subhemispherical to plano-convex when mature, margin slightly decurrent, purple-mauve or grayish brown with purple tinge when young, purple-fuscous with age; surface dry, minutely tomentose, without color change when touched; context whitish to grayish white, without discoloration when bruised. Hymenophore adnate when young, adnate to slightly depressed around apex of stipe; surface whitish to cream when young, pinkish to dingy pink when mature, without discoloration when bruised; pores subangular to roundish, 0.3–0.5 mm wide; tubes 3–5 mm long, concolorous or much paler than hymenophoral surface, without discoloration when bruised. Stipe 4–7 × 0.8–1.5 cm, clavate to cylindrical, whitish, but always with a grayish purple to pinkish tinge downwards; surface nearly glabrous, without discoloration when injured; context solid, whitish, without discoloration when hurt; basal mycelium whitish. Taste and odor unknown.

Basidia 18–25 × 7–11 μm, clavate to narrowly clavate, 4-spored, hyaline in KOH. Basidiospores 8–11 × 3–4.5 μm (Q = 2.11–3.33, Q_m = 2.75 ± 0.26), subcylindrical to subfusiform in side view with slight suprahilar depression, cylindrical to fusiform in ventral view, smooth, hyaline to yellowish in KOH, yellow to brownish yellow in Melzer's reagent. Hymenophoral trama boletoid composed of 5–9 μm wide filamentous hyphae, hyaline to yellowish in KOH, yellowish to yellow in Melzer's reagent. Cheilocystidia 24–33 × 6–7 μm, clavate, subfusiform to fusiform, hyaline to yellowish in KOH, thin-walled. Pleurocystidia 54–62 × 6.5–9.5 μm, subfusiform to fusoid-ventricose, sometimes with a long pedicel, thin-walled. Pileipellis a trichoderm composed of 4.5–7 μm wide interwoven filamentous hyphae, yellowish to pale brownish in KOH and yellowish brown to brownish in Melzer's reagent; terminal cells 14.5–44 × 4.5–8 μm, clavate to fusiform or cystidioid. Pileal trama composed of 3–6 μm wide interwoven hyphae, hyaline to yellowish in KOH and yellowish to brownish in Melzer's reagent. Clamp connections absent in all tissues.

Habitat: Solitary on soil in tropical forests dominated by plants of the family Fagaceae.

Known distribution: Currently known from Southeast Asia and southern China.

Specimens examined: CHINA, HAINAN PROVINCE: Ledong County, Jianfengling, alt. 870 m, 6 August 2009, N.K. Zeng 477 (KUN-HKAS 90199), the same location, 8 August 2009, N.K. Zeng 493 (KUN-HKAS 90200).

Commentary: *Tylopilus griseipurpureus* was originally described as *Boletus griseipurpureus* from Singapore by Corner (1972) and then transferred to the genus *Tylopilus* by Horak (2011) based on the color of the hymenophore. This species is characterized by the purple-mauve to purple-fuscous pileus, the white context without discoloration when bruised, the whitish to cream and then pinkish to dingy pink hymenophore without discoloration when bruised, the glabrous and white stipe but always with grayish purple to pinkish tinge downwards, the trichoderm pileipellis, and the distribution in tropical forests. *Tylopilus griseipurpureus* differs from other species in *Tylopilus* in its purple-mauve or purple-fuscous pileus, white context without discoloration when injured, and initially whitish to pallid and then pinkish hymenophore.

Our multi-locus phylogenetic analysis (Fig. 4.1) indicates that *T. griseipurpureus* clusters with *T.*

formosus which was originally described from New Zealand by Stevenson (1962). However, *T. formosus* differs from *T. griseipurpureus* in its blackish brown basidioma, brownish orange hymenophore, and relatively slender basidiospores measuring 10–11 × 3–4 μm (Stevenson 1962).

Fig. 22.12 *Tylopilus griseipurpureus* (Corner) E. Horak. Photos by N.K. Zeng (KUN-HKAS 90199).

22.13 *Tylopilus griseiviridus* Yan C. Li & Zhu L. Yang, The Boletes of China: *Tylopilus* s.l. 301 (2021)

MycoBank: MB 837926

Etymology: The epithet "*griseiviridus*" refers to the color of the basidiomata.

Type: CHINA, YUNNAN PROVINCE: Lancang County, Nuofu Town, Dongmengsong Village, alt. 1450 m, 28 September 2016, G. Wu 1879 (KUN-HKAS 99999, GenBank Acc. No.: MW114856 for nrLSU).

Basidioma very small to small. Pileus 2–3.5 cm in diam., hemispherical to subhemispherical, greenish (30E3–4) to grayish green (30D4–5) or pale green (29D3–4) when young, olive-gray (3D3) to olive-brown (4D4) when mature; surface dry,

Fig. 22.13 a *Tylopilus griseiviridus* Yan C. Li & Zhu L. Yang. Photos by G. Wu (KUN-HKAS 99999, type).

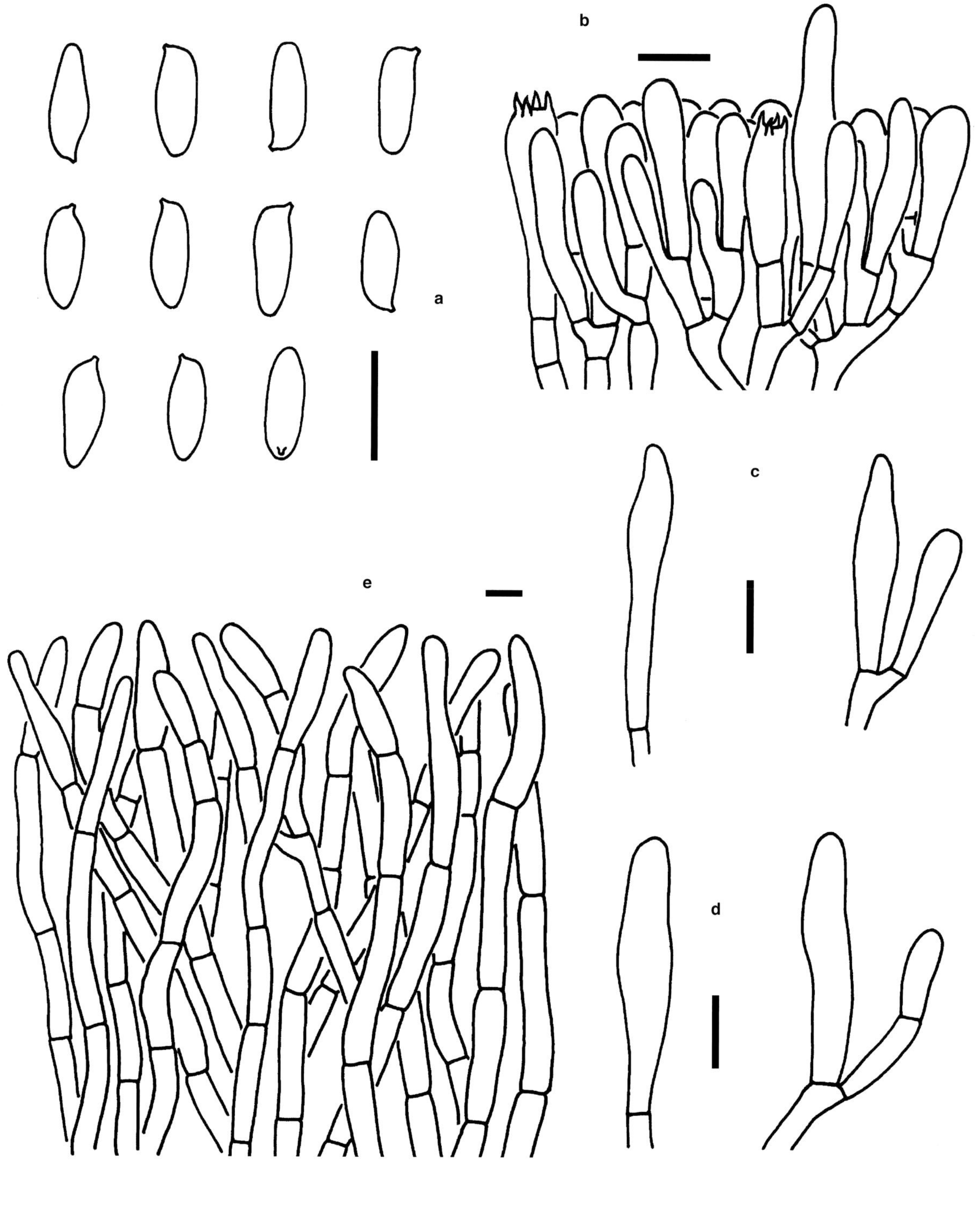

256 The Boletes of China: *Tylopilus* s.l.

Fig. 22.13 b *Tylopilus griseiviridus* Yan C. Li & Zhu L. Yang (KUN-HKAS 99999, type).
a. Basidiospores; b. Basidia and pleurocystidium; c. Cheilocystidia; d. Pleurocystidia; e. Pileipellis. Scale bars = 10 μm.

nearly smooth; context solid, white to pallid, staining indistinct reddish tinge when bruised. Hymenophore adnate; surface white (4A1) to pallid (4B1), without color change when bruised; pores subangular to roundish, 0.5–1 mm wide; tubes up to 5 mm long, concolorous or much paler than hymenophoral surface, without color change when bruised. Stipe 3–5 × 0.8–1.2 cm, subcylindrical, concolorous with pileal surface when young, but grayish red (12E2) to grayish brown (10F2–3) when aged or touched; surface nearly glabrous; context solid, white to pallid, staining indistinct reddish tinge when hurt; basal mycelium whitish to pallid. Taste bitter.

Basidia 20–30 ×5–7 μm, clavate to narrowly clavate, 4-spored, hyaline in KOH. Basidiospores [60/2/1] 7–11 × (2.5) 3–4 μm (Q = 2–3.33, Q_m = 2.77 ± 0.23), elongated to subcylindrical and inequilateral in side view with slight suprahilar depression, elongated to cylindrical in ventral view, smooth, hyaline to yellowish in KOH, yellow to brownish yellow in Melzer's reagent. Hymenophoral trama boletoid composed of 2.5–8 μm wide filamentous hyphae, hyaline to yellowish in KOH, yellowish to yellow in Melzer's reagent. Cheilo- and pleurocystidia 30–39 × 5–7 μm, abundant, fusoid-ventricose to subfusiform, hyaline to yellowish in KOH, thin-walled. Pileipellis a palisadoderm composed of 5–9.5μm wide vertically arranged hyphae, yellowish to pale brownish in KOH and yellowish brown to brownish in Melzer's reagent; terminal cells 22–63 × 6–9 μm, clavate to subcylindrical. Pileal trama composed of 3–15 μm wide interwoven hyphae, hyaline to yellowish in KOH and yellowish to brownish in Melzer's reagent. Clamp connections absent in all tissues.

Habitat: Scattered to solitary on soil in tropical forests dominated by plants of the family Fagaceae.

Known distribution: Known from southwestern China.

Commentary: *Tylopilus griseiviridus* is characterized by the initially greenish to grayish green or pale green and then olive-gray to olive-brown pileus, the white to pallid context staining indistinct reddish tinge when bruised, the white to pallid hymenophore, the greenish to grayish green or olive-gray to olive-brown stipe becoming grayish red to grayish brown when aged or touched, the palisadoderm pileipellis composed of 5–9.5 μm wide vertically arranged hyphae. Morphologically, *T. griseiviridus* is similar to *T. otsuensis* Hongo. However, *T. otsuensis* differs from *T. griseiviridus* in its dark green to olivaceous pileus, ovoid basidiospores, and epithelium pileipellis (Hongo 1966; see our description below).

Tylopilus griseiviridus is phylogenetically related to *T. olivaceobrunneus* Yan C. Li & Zhu L. Yang. However, *T. olivaceobrunneus* can be easily distinguished from *T. griseiviridus* by its relatively large basidioma, yellowish olivaceous pileus with brownish to pale brown tinge, white then pale grayish red hymenophore, and brownish gray to brown-gray or dark brown stipe (see our description below).

22.14 *Tylopilus griseolus* Yan C. Li & Zhu L. Yang, The Boletes of China: *Tylopilus* s.l. 304 (2021)

MycoBank: MB 834755

Etymology: The epithet "*griseolus*" refers to the color of the pileus.

Type: CHINA, HUNAN PROVINCE: Yizhang County, Mangshan National Forest Park, alt. 1800 m, 16 September 2016, G. Wu 1847 (KUN-HKAS 99967, GenBank Acc. No.: MT154736 for nrLSU, MW165266 for *tef1-α*, MW165284 for *rpb2*).

Basidioma small to medium-sized. Pileus 3–8 cm in diam., convex to applanate or plano-convex, pastel gray (1C1), pale grayish (2B1), grayish (2C1) or gray (5D1) when young, olive-gray (3E2) to dark olive-gray (1E3) when mature, margin much paler in color; surface dry, finely tomentose; context solid, white (5A1), staining indistinct reddish brown (9E7–8) when bruised. Hymenophore adnate when young, adnate to slightly decurrent when mature; surface white to cream (1A2) or yellowish (1A2) when young, yellowish (1A2) to pale grayish yellow (2B2–3) when mature, staining brown (6E7–8) to reddish brown (8D5–6) when bruised; pores subangular to roundish, 0.3–1 mm wide; tubes up to 8 mm long, concolorous or much paler than hymenophoral surface, staining reddish brown when bruised. Spore print brown (6D7–8) to reddish brown (8D5–6). Stipe 4–8 × 1.2–2.5 cm, cylindrical to clavate, white to pale yellowish (4C3), nearly glabrous, staining reddish brown when touched; context solid, white, staining reddish brown when hurt; basal mycelium white. Taste bitter.

Basidia 20–38 × 7–11 μm, clavate to narrowly clavate, 4-spored, hyaline in KOH. Basidiospores 5.5–7 × 3.5–4 μm (Q = 1.57–1.86, Q_m = 1.68 ± 0.11), subellipsoid to elongated and inequilateral in side view

Fig.22.14 a *Tylopilus griseolus* Yan C. Li & Zhu L. Yang. Photos by G. Wu (KUN-HKAS 99967, type).

with slight suprahilar depression, ellipsoid to elongated in ventral view, smooth, hyaline to yellowish in KOH, yellow to brownish yellow in Melzer's reagent. Hymenophoral trama boletoid composed of 4–8 μm wide filamentous hyphae, hyaline to yellowish in KOH, yellowish to yellow in Melzer's reagent. Cheilo- and pleurocystidia 30–62 × 12–20 μm, abundant, subfusoid-ventricose to subfusiform, hyaline to yellowish in KOH, thin-walled. Pileipellis a trichoderm to palisadoderm composed of 4–9 μm wide filamentous hyphae with cylindrical terminal cells 23–74 × 4–7 μm, yellowish to pale brownish in KOH and yellow to brownish in Melzer's reagent. Pileal trama composed of 3–10 μm wide interwoven hyphae, colorless to yellowish in KOH and yellow to yellowish brown in Melzer's reagent. Clamp connections absent in all tissues.

Habitat: Solitary on soil in tropical forests dominated by plants of the family Fagaceae.

Fig.22.14b *Tylopilus griseolus* Yan C. Li & Zhu L. Yang. Photos by F. Li 813 (KUN-HKAS 106432).

Known distribution: Known from central and southern China.

Specimens examined CHINA, GUANGDONG PROVINCE: Fengkai County, Heishiding Nature Reserve, alt. 250 m, 3 July 2012, F. Li 586 (KUN-HKAS 106354), the same location, 20 July 2012, F. Li 699 (KUN-HKAS 106455), 13 August 2012, F. Li 782 (KUN-HKAS 106429), 14 August 2012, F. Li 813 (KUN-HKAS 106432), 14 August 2013, F. Li 1446 (KUN-HKAS 82187) and 16 August 2013, F. Li 1466 (KUN-HKAS 82188).

Commentary: *Tylopilus griseolus* is characterized by the pale grayish, grayish or gray then olive-gray to dark olive-gray pileus, the white context staining indistinct reddish brown when bruised, the adnate to slightly decurrent hymenophore, the white to cream then yellowish to grayish yellow hymenophoral surface staining brown to reddish brown when bruised, the white to pale yellowish stipe, the ovoid to ellipsoid or amygdaloid basidiospores, and the trichoderm to palisadoderm pileipellis composed of 4–9 μm wide filamentous hyphae.

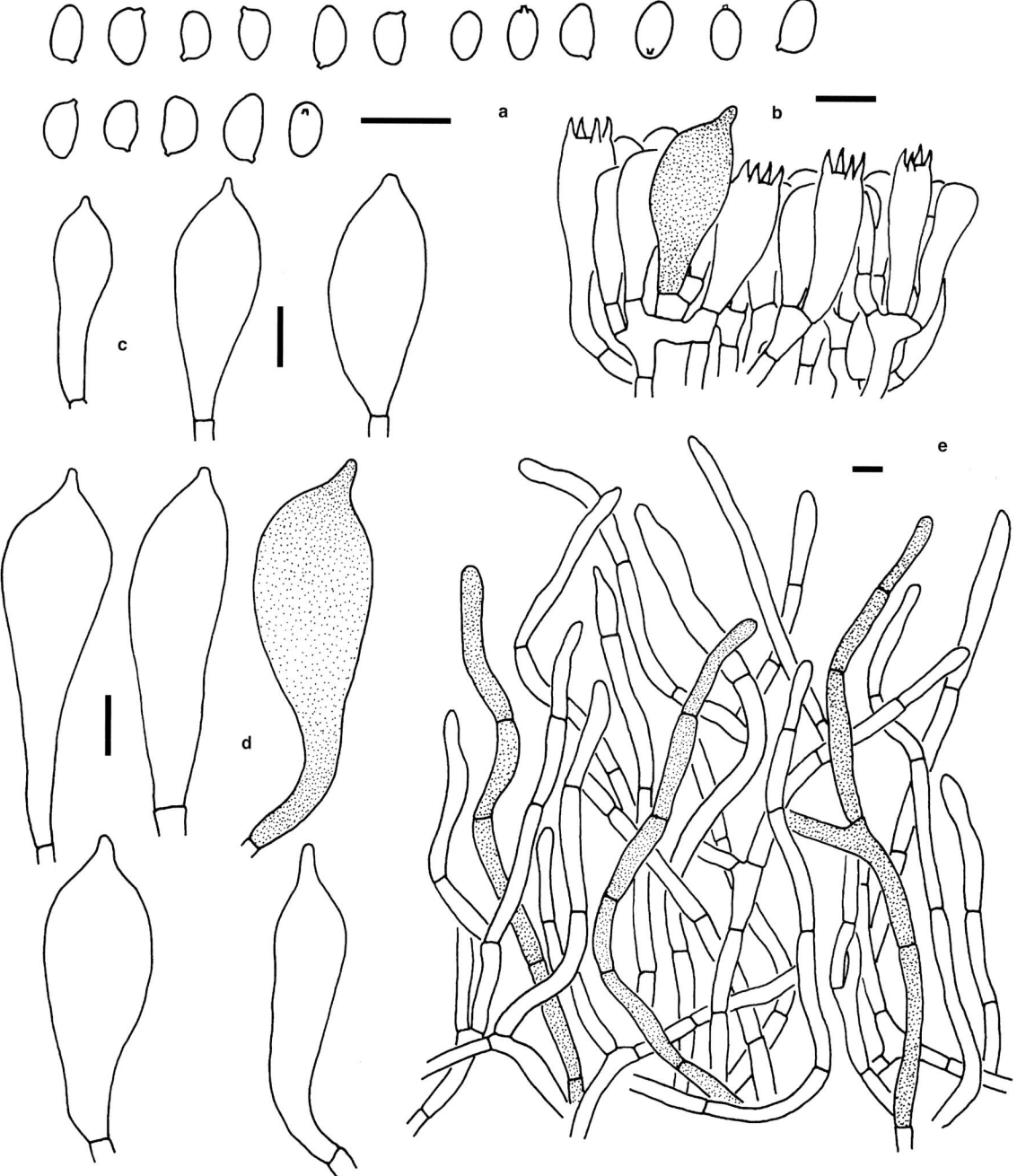

Fig. 22.14 c *Tylopilus griseolus* Yan C. Li & Zhu L. Yang (KUN-HKAS 99967, type).
a. Basidiospores; b. Basidia and cheilocystidium; c. Cheilocystidia; d. Pleurocystidia; e. Pileipellis. Scale bars = 10 μm.

Our molecular phylogenetic analysis (Fig. 4.1) indicates that *T. griseolus* is related to the species in *T. balloui* group and *T. virescens* (Har. Takah. & Taneyama) N.K. Zeng, H. Chai & Zhi Q. Liang. However, these species differ in the color of the basidiomata and the discoloration of the context when bruised.

22.15 *Tylopilus himalayanus* D. Chakr., K. Das & Vizzini, MycoKeys 33: 109 (2018)

Basidioma medium-sized to large. Pileus 5–13 cm in diam., hemispherical to subhemispherical when young, applanate to plano-convex at maturity, green to yellowish green or pale brownish green to olivaceous when young, always reddish brown to pink-brown at maturity or when bruised; surface dry, tomentose to fibrillose; context solid, whitish to pallid, staining reddish brown when bruised. Hymenophore adnate to slightly depressed around apex of stipe; surface white to yellowish when young, dingy pink to pinkish at maturity, staining reddish brown when bruised; pores subangular to roundish, 0.3–1 mm wide; tubes up to 6 mm long, concolorous or much paler than hymenophoral surface, staining a pale reddish brown tinge when bruised. Spore print grayish red to reddish. Stipe 4–11 × 1–1.8 cm, subcylindrical to clavate, always enlarged downwards, brown to reddish brown, sometimes with olivaceous tinge, nearly glabrous, staining reddish brown when touched; context solid, white, staining reddish brown when injured; basal mycelium whitish to pallid. Taste bitter.

Basidia 19–29 × 7–9 μm, clavate to broadly clavate, 4-spored, hyaline in KOH. Basidiospores 11–14 × 3.5–4.5 μm (Q = 2.56–3.29, Q_m = 2.86 ± 0.14), subcylindrical to subfusiform in side view with slight suprahilar depression, cylindrical to fusiform in ventral view, smooth, hyaline to yellowish in KOH, yellow to brownish yellow in Melzer's reagent. Hymenophoral trama boletoid composed of 4–7 μm wide filamentous hyphae, hyaline to yellowish in KOH, yellowish to yellow in Melzer's reagent. Cheilo- and pleurocystidia 39–67 × 8–17 μm, abundant, fusoid-ventricose to subfusiform, yellowish to brownish yellow in KOH, thin-walled. Pileipellis a palisadoderm composed of 4–13 μm wide vertically arranged

Fig. 22.15 a *Tylopilus himalayanus* D. Chakr., K. Das & Vizzini. Photos by Y.J. Hao (KUN-HKAS 107165).

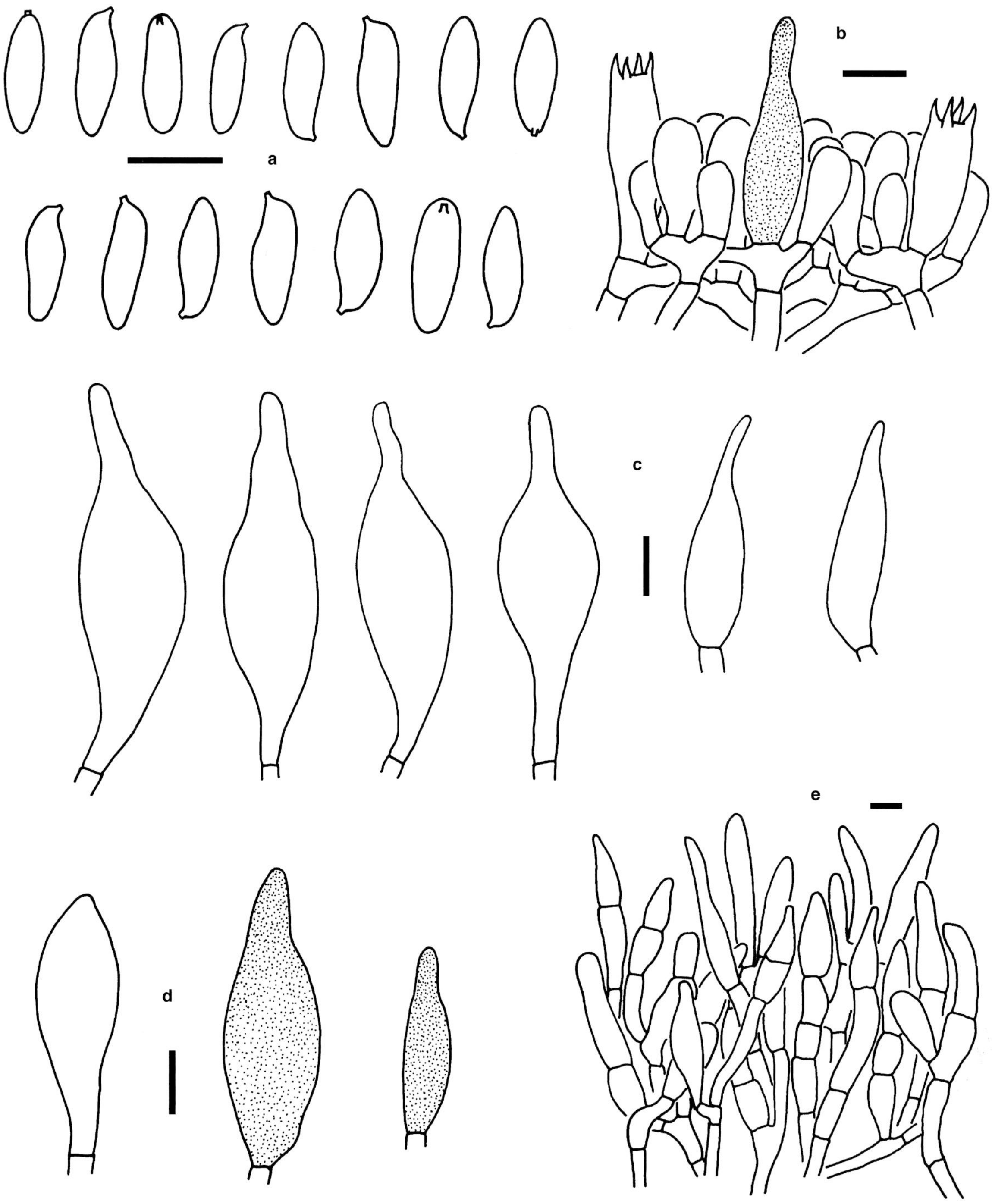

Fig. 22.15 b *Tylopilus himalayanus* D. Chakr., K. Das & Vizzini (KUN-HKAS 107165).
a. Basidiospores; b. Basidia and cheilocystidium; c. Pleurocystidia; d. Cheilocystidia; e. Pileipellis. Scale bars = 10 μm.

hyphae, pale yellowish to brownish yellow in KOH and yellow to brownish in Melzer's reagent; terminal cells 22–58 × 5–10 μm, pyriform to cystidioid. Pileal trama composed of 3–8 μm wide interwoven hyphae, hyaline to yellowish in KOH and yellowish to brownish in Melzer's reagent. Clamp connections absent in all tissues.

Habitat: Scattered to solitary on soil in temperate forests dominated by plants of the family Fagaceae, or in mixed forests dominated by plants of the families Fagaceae and Pinaceae.

Known distribution: Known from China and India.

Specimens examined: CHINA, HENAN PROVINCE: Luanchuan County, Laojun Mountain, alt. 2200 m, 13 August 2015, X.H. Wang 3659 (KUN-HKAS 89932). LIAONING PROVINCE: Benxi County, Xiamatang Town, Majia Village, alt. 430 m, 21 August 2015, P.M. Wang 385 (KUN-HKAS 93345), the same location and date, J. Li 220 (KUN-HKAS 91248); Anshan, Qianshan Forest Park, alt. 400 m, 24 August 2015, Q. Cai 1423 (KUN-HKAS 92042), P.M. Wang 419 (KUN-HKAS 93425), and J. Li 250 (KUN-HKAS 91278), the same location, 25 August 2015, P.M. Wang 425 (KUN-HKAS 93431). ANHUI PROVINCE: Jinzhai County, alt. 1100 m, 19 July 2017, Y.J. Hao 1445 (KUN-HKAS 107146), the same location, 20 July 2017, Y.J. Hao 1483 (KUN-HKAS 107165), the same location, 21 July 2017, Y.J. Hao 1499 (KUN-HKAS 107168) and Y.J. Hao 1522 (KUN-HKAS 107166).

Commentary: *Tylopilus himalayanus*, originally described by Chakraborty *et al.* (2018) from India, is characterized by the green to yellowish green or pale brownish green to olivaceous pileus staining reddish brown to pink-brown with age or injury, the white to pallid context staining reddish brown when bruised, the white to yellowish then dingy pink to pinkish hymenophore, the brown to reddish brown stipe, and the palisadoderm pileipellis composed of 4–13 μm wide vertically arranged hyphae. It is macroscopically similar to the species with an olivaceous pileus, viz. *T. alpinus*, *T. brunneirubens*, *T. olivaceobrunneus*, and *T. otsuensis* Hongo. However, *T. alpinus* differs from *T. himalayanus* in its distinctly reticulate stipe, trichoderm pileipellis, and distribution in alpine areas with altitude up to 3500 m (Wu *et al.* 2016a). *Tylopilus brunneirubens* differs in its distinctly reticulate stipe, relatively small basidiospores measuring 8.5–12.5 × 3.5–4.5 μm, and distribution in tropical to subtropical forests. *Tylopilus olivaceobrunneus* differs in its white context without discoloration when bruised, relatively small basidiospores measuring (8.5) 9–11 (12) × 3–3.5 μm, and trichoderm pileipellis composed of 4–10 μm wide interwoven filamentous hyphae. *Tylopilus otsuensis* differs in its dark green to olivaceous pileus and stipe, broadly ellipsoid to ovoid basidiospores measuring 5.5–6.5 × 4–5 μm, epithelium pileipellis made up of vertically arranged moniliform hyphae, and distribution in tropical forests (Hongo 1966, Wu *et al.* 2016a).

Our multi-locus phylogenetic analysis (Fig. 4.1) indicates that *T. himalayanus* is related to *T. rubrobrunneus*. However, *T. rubrobrunneus* has a dark vinaceous brown to chocolate-colored pileus, a white context staining olivaceous when injured, and a trichoderm pileipellis (Mazzer and Smith 1967).

22.16 *Tylopilus jiangxiensis* Kuan Zhao & Yan C. Li, Phytotaxa 434: 285 (2020)

Basidioma very small to small. Pileus 2–3.5 cm in diam., convex; surface olive-brown to olive-yellow, becoming yellowish brown to reddish brown when aged or injured, slightly darker in the center, margin always enrolled when immature, without discoloration when injured. Context whitish, becoming rufescent slowly when injured. Hymenophore adnate when young, and then slightly depressed around apex of stipe when mature; surface pallid to pinkish, becoming rufescent when bruised; pores 0.3–1 mm wide; tubes up to 4 mm long, concolorous or much paler than hymenophoral surface. Spore print grayish red to reddish. Stipe 3–7 × 0.4–0.7 cm, central, subcylindrical, solid; surface dry, concolorous with pileal surface but becoming lighter upward, nearly whitish at the apex; context whitish, without color change when hurt; basal mycelium white. Taste bitter.

Basidia 22–30 × 6–9 μm, clavate to narrowly clavate, 4-spored, hyaline to yellowish. Basidiospores 9–13 × 3.5–5 μm [Q = 2–3.43 (3.5), Q_m = 2.68 ± 0.36], subcylindrical to subfusiform in side view with slight suprahilar depression, cylindrical to fusiform in ventral view, smooth, hyaline to yellowish in KOH, yellow to brownish yellow in Melzer's reagent. Hymenophoral trama boletoid, hyphae cylindrical, 2–10 μm wide filamentous hyphae, hyaline to yellowish in KOH, yellowish to yellow in Melzer's reagent. Cheilo- and pleurocystidia 29–58 × 6–13 μm, subfusiform to fusiform, sometimes with a long pedicel, hyaline to yellowish in KOH, yellow to yellowish brown in Melzer's reagent. Pileipellis a trichoderm composed of 4–8 μm wide interwoven filamentous hyphae, yellowish to pale brownish in KOH and yellowish brown to brownish in Melzer's reagent; terminal cells 22–45× 4–8 μm, subcylindrical. Pileal trama composed of

Fig. 22.16 a *Tylopilus jiangxiensis* Kuan Zhao & Yan C. Li. Photos by G. Wu (KUN-HKAS 107152).

Fig. 22.16 b *Tylopilus jiangxiensis* Kuan Zhao & Yan C. Li. Photo by K. Zhao (KUN-HKAS 105252, type).

4–13 μm wide interwoven hyphae, hyaline to yellowish in KOH and yellowish to brownish in Melzer's reagent. Clamp connections absent in all tissues.

Habitat: Scattered on soil in subtropical mixed forests dominated by plants of the families Fagaceae and Pinaceae.

Known distribution: Currently known from southeastern China.

Specimens examined: CHINA, JIANGXI PROVINCE: Jiujiang, Lushan Mountain, alt. 950 m, 21 July 2017, G. Wu 2177 (KUN-HKAS 107152), the same location, 10 June 2019, K. Zhao 1221 (KUN-HKAS 105250), the same location, 13 June 2019, K. Zhao 1222 (KUN-HKAS 105251); Yichun, Sanzhualun National Forest Park, alt. 650 m, 17 June 2019, K. Zhao 1226 (KUN-HKAS 105252, type), and K. Zhao 1233 (KUN-HKAS 105253).

Commentary: *Tylopilus jiangxiensis* was originally described by Zhao *et al.* (2020) from southeastern China and is characterized by the yellowish brown to reddish brown pileus, and the pallid to pinkish hymenophore. In the original description, there was no discoloration in the context when bruised. However, additional collection from the type locality shows that the pileus olive-brown to olive-yellow when young and then becoming yellowish brown to reddish brown when aged or injured, and the context rufescent when injured.

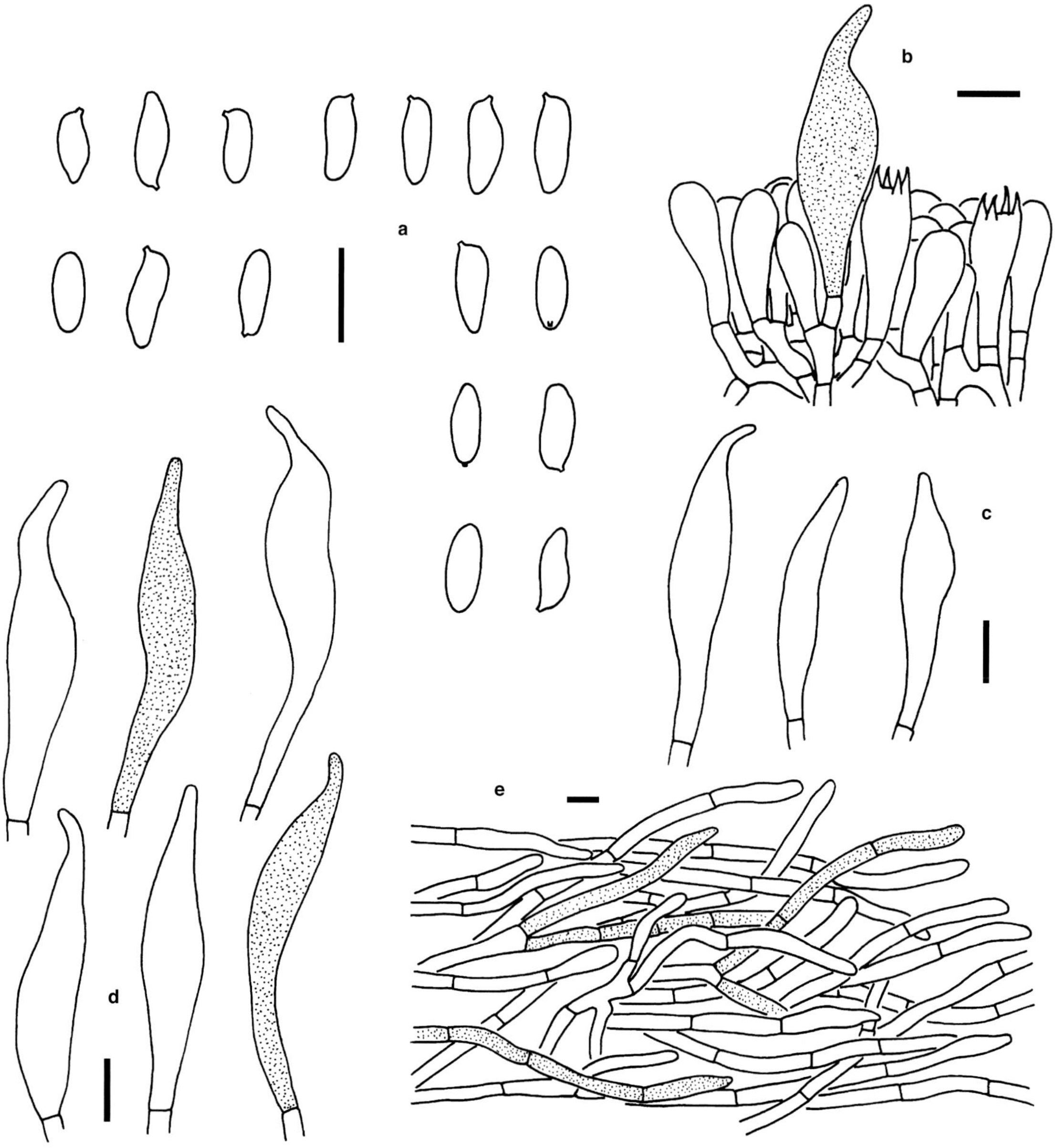

Fig. 22.16 c *Tylopilus jiangxiensis* Kuan Zhao & Yan C. Li (KUN-HKAS 105252, type).
a. Basidiospores; b. Basidia and cheilocystidium; c. Cheilocystidia; d. Pleurocystidia; e. Pileipellis. Scale bars = 10 μm.

Our multi-locus phylogenetic analysis (Fig. 4.1) indicates that *T. jiangxiensis* clusters with *T. felleus* as a monophyletic lineage but without statistical support. Moreover, *T. felleus* differs from *T. jiangxiensis* in its gray to grayish brown or grayish pileus, distinctly reticulate stipe over the upper 1/3 or 1/2 part, and relatively large basidiospores measuring 14–16 (17) × 4.5–5.5 μm (Wu *et al*. 2016a).

22.17 *Tylopilus neofelleus* Hongo, J Jap Bot 42: 154 (1967)

Synonym: *Tylopilus microsporus* S.Z. Fu *et al.*, Mycotaxon 96: 42 (2006).

Basidioma medium-sized to very large. Pileus 5–16 cm in diam., hemispherical to subhemispherical when young and then subhemispherical to plano-convex or applanate when mature, brownish to drab, always with purple tinge when young; surface dry, finely tomentose, without discoloration when touched; context solid, whitish, without discoloration when injured. Hymenophore adnate when young, adnate to depressed around apex of stipe when mature; surface purple to purplish pink when young, and then pink to pinkish when mature, without discoloration when bruised; pores subangular to roundish, 0.2–0.5 mm wide; tubes 5–10 mm long, white when young and pink to pinkish when mature, without discoloration when bruised. Spore print pinkish to reddish (Fig. 3.4 m). Stipe 5–16 × 1.5–4 cm, cylindrical, brownish to drab, always with purple tinges at apex of stipe when young, without discoloration when touched; surface almost glabrous, sometimes with indistinct reticulum restricted to the apex; context white, without discoloration when bruised; basal mycelium white. Taste and odor better.

Basidia 20–30 × 7–10 μm, clavate to narrowly clavate, 4-spored, hyaline in KOH. Basidiospores 8–9 (10) × 3–4 μm (Q = 2.25–2.81, Q_m = 2.6 ± 0.17), subcylindrical to subfusiform in side view with slight suprahilar depression, cylindrical to fusiform in ventral view, smooth, hyaline to yellowish in KOH, yellow to brownish yellow in Melzer's reagent. Hymenophoral trama boletoid composed of 3–11 μm wide filamentous hyphae, hyaline to yellowish in KOH, yellowish to yellow in Melzer's reagent. Cheilocystidia 35–50 × 5–9 μm, narrowly subfusiform to lanceolate, yellowish to pale brownish yellow in KOH, brown to dark brown in Melzer's reagent, thin-walled. Pleurocystidia 70–85 × 10–15 μm, subfusiform to broad subfusiform or broad fusoid-ventricose, concolorous with cheilocystidia, thin-walled. Pileipellis an intricate trichoderm composed of 5–7 μm wide interwoven filamentous hyphae, yellowish to pale brownish in KOH and yellowish brown to brownish in Melzer's reagent; terminal cells 20–40 × 5–8 μm, subcylindrical. Pileal trama composed of 7–15 μm wide interwoven hyphae, hyaline to yellowish in KOH and yellowish

Fig. 22.17 a *Tylopilus neofelleus* Hongo. Photo by Y.C. Li (**KUN-HKAS 59459**).

Fig. 22.17 b *Tylopilus neofelleus* Hongo. Photo by J.W. Liu (KUN-HKAS 107176).

to brownish in Melzer's reagent. Clamp connections absent in all tissues.

Habitat: Scattered to solitary on soil in tropical to subtropical forests dominated by plants of the family Pinaceae, or in mixed forests dominated by plants of the families Pinaceae and Fagaceae.

Known distribution: Known from China and Japan.

Specimens examined: CHINA, GUIZHOU PROVINCE: Daozhen County, Dashahe Nature Reserve, alt. 1500 m, 28 July 2010, X.F. Shi 389 (KUN-HKAS 107183). HUBEI PROVINCE: Shiyan, Dachuan Town, alt. 600 m, 1 July 2013, Y.J. Hao 923 (KUN-HKAS 80203). SICHUAN PROVINCE: Xichang, Puge County, Luoji Mountain, alt. 2000 m, 29 July 2012, T. Guo 503 (KUN-HKAS 761950); Xichang, Lushan Mountain, alt. 2300 m, 4 October 1983, M.S. Yuan 263 (KUN-HKAS 11967). YUNNAN PROVINCE: Kunming, Heping Village, alt. 1900 m, 10 August 2003, Q.B. Wang 85 (HMAS 79720, type, as *T. microsporus*); Jingdong County, Ailao Mountain, alt. 2450 m, 19 July 2006, Y.C. Li 565 (KUN-HKAS 50319); Jinghong County, Dadugang Town, alt. 1300 m, 21 July 2007, Y.C. Li 913 (KUN-HKAS 52600); Tengchong County, Gudong Town, Pojiao Village, alt. 1900 m, 20 July 2009, Y.C. Li 1712 (KUN-HKAS 59459); Yongping County, on the way from Yongping to Baoshan, at the landmark of 3295 km, alt. 2080 m, 3 July 2009 Y.C. Li 1913 (KUN-HKAS 59661); Lanping County, Hexi Town, alt. 2600 m, 16 August 2010, X.T. Zhu 180 (KUN-HKAS 68356); Menghai County, Xiding Town, alt. 700 m, 1 July 2014, Z.W. Ge 3568 (KUN-HKAS 84406); Lancang County, alt. 1000 m, 28 June 2017, J.W. Liu 530828MF0043 (KUN-HKAS 107176). CHINA, HUNAN PROVINCE: Yizhang County, Mangshan National Forest Park, alt. 1800 m, 2 September 2007, Y.C. Li 1066 (KUN-HKAS 53411), the same location, 2 August 2005, P. Zhang 691 (KUN-HKAS 55829). JAPAN, Otsu City, 2 September 1966, T. Hong 3301 (TNS-F-174771, type of *T. neofelleus*).

Commentary: *Tylopilus neofelleus* originally described from Japan (Hongo 1967) is characterized by the brownish to drab pileus always with purple tinge when young, the white context without discoloration when injured, the initially purple to purplish pink hymenophoral surface and then becoming pink to pinkish when mature, the brownish to drab stipe always with purple tinges at apex, the intricate trichoderm pileipellis, and the distribution in tropical to subtropical forests. This species is macroscopically similar to *T. felleus*. However, *T. felleus* differs from *T. neofelleus* in its gray to grayish brown or grayish white pileus without any purple tinge, large basidiospores measuring 14–16 (17) × 4.5–5.5 μm, and distribution throughout the holarctic region.

Our phylogenetic data (Fig. 4.1) indicate that sequence of *T. neofelleus* from type specimen forms a monophyletic clade. Species to which *T. neofelleus* is phylogenetically related remain as yet unknown.

22.18 *Tylopilus obscureviolaceus* Har. Takah., Mycoscience 45(6): 374 (2004)

Basidioma small to medium-sized. Pileus 3–7 cm in diam., hemispherical to subhemispherical when young, plano-convex to applanate when mature, pallid purplish, grayish purplish to drab purple, sometimes purplish pink, much darker in the center; surface dry, tomentose, without discoloration when touched; context solid, white, without discoloration when bruised. Hymenophore adnate when young, and adnate to slightly depressed around apex of stipe when mature; surface initially white to pallid and then pinkish to grayish pink, without discoloration when bruised; pores subangular to roundish, 0.3–0.5 mm wide; tubes 5–10 mm long, concolorous or much paler than hymenophoral surface, without discoloration when bruised. Spore print pinkish to reddish. Stipe 6–10 × 1.2–2.5 cm, subcylindrical to clavate, always enlarged downwards, concolorous with pileal surface, almost glabrous, without discoloration when touched; context solid, white, without discoloration when injured; basal mycelium white. Taste bitter, odor mild.

Basidia 18–24 × 8–10 μm, clavate, 4-spored, hyaline in KOH. Basidiospores 6.5–8 × 3.5–4.5 μm [Q = (1.44) 1.56–1.88 (2), Q_m = 1.73 ± 0.14], subellipsoid to elongated and inequilateral in side view with slight suprahilar depression, ellipsoid to elongated in ventral view, smooth, hyaline to yellowish in KOH, yellow to brownish yellow in Melzer's reagent. Hymenophoral trama boletoid composed of 4–11 μm wide filamentous hyphae, hyaline to yellowish in KOH, yellowish to yellow in Melzer's reagent. Cheilocystidia 26–54 × 4–6 μm, subfusiform to fusoid-ventricose or lanceolate, hyaline to yellowish or yellowish brown, thin-walled. Pleurocystidia 34–76 × 9–16 μm, fusoid-ventricose to subfusiform, usually with long and subacute apex, hyaline to yellowish or yellowish brown, thin-walled. Pileipellis a trichoderm to palisadoderm composed of 5–11 μm wide interwoven hyphae with clavate to cylindrical terminal cells 13–50 × 5–11 μm, yellowish to pale brownish in KOH and yellow to brownish in Melzer's reagent. Pileal trama composed of 3–8 μm wide interwoven hyphae, colorless to yellowish in KOH and yellow to yellowish brown in Melzer's reagent. Clamp connections absent in all tissues.

Habitat: Scattered on soil in tropical forests dominated by plants of the family Fagaceae.

Known distribution: Known from China and Japan.

Specimens examined: CHINA, HAINAN PROVINCE: Baisha County, Yinggeling National Nature Reserve, alt. 1300 m, 28 July 2009, N.K. Zeng 372 (KUN-HKAS 107161); Qiongzhong County, Limu Mountain, alt. 850 m, 3 August 2010, N.K. Zeng 813 (KUN-HKAS 59842); Ledong County, Jianfengling, alt. 870 m, 4 August 2009, N.K. Zeng 444 (KUN-HKAS 107163). GUANGDONG PROVINCE: Fengkai County, Heishiding Nature Reserve, alt. 250 m, 4 July 2012, F. Li 583 (KUN-HKAS 106452) and F. Li 592 (KUN-HKAS 106453), the same location, 2 June 2013, Y.J. Hao 835 (KUN-HKAS 80115).

Commentary: *Tylopilus obscureviolaceus*, originally described from Japan, is characterized by the pallid purplish, grayish purplish to drab purple pileus, the white to pallid then pinkish to purplish pink hymenophore, the white context without discoloration when injured, the trichoderm to palisadoderm pileipellis, and the distribution in tropical forests (Takahashi 2004).

Tylopilus obscureviolaceus is phylogenetically related and macroscopically similar to *T. argillaceus*, *T. atroviolaceobrunneus*, and *T. purpureorubens*. However, *T. argillaceus* is characterized by the reddish brown to castaneous or salmon pileus, the white context becoming reddish when injured, and the trichoderm pileipellis composed of 3–6.5 μm wide interwoven filamentous hyphae (see our description above). *Tylopilus atroviolaceobrunneus* differs in its dark colored pileus, reddish to reddish brown discoloration of the basidiomata when injured, big basidiospores measuring 10–14 × 4–5 μm (Wu *et al.*

2016a; see our description above). *Tylopilus purpureorubens* differs in its reddish brown or rufescent discoloration of the basidiomata on injury, relatively long basidiospores measuring 7.5–9.5 × 3.5–4.5 μm (see our description below). *Tylopilus obscureviolaceus* is also macroscopically similar to *T. violatinctus*. However, the pileus of *T. violatinctus* turns bright rusty violet to deep dark violet when bruised. Moreover, the basidiospores of *T. violatinctus* are clearly slender than those of *T. obscureviolaceus* with a *Q* value of 2–2.8 (Baroni and Both 1998). *Tylopilus obscureviolaceus* is documented and illustrated here for the first time in China.

Fig. 22.18 a *Tylopilus obscureviolaceus* Har. Takah. Photos by Y.J. Hao (KUN-HKAS 80115).

Chapter 22 *Tylopilus* P. Karst.   271

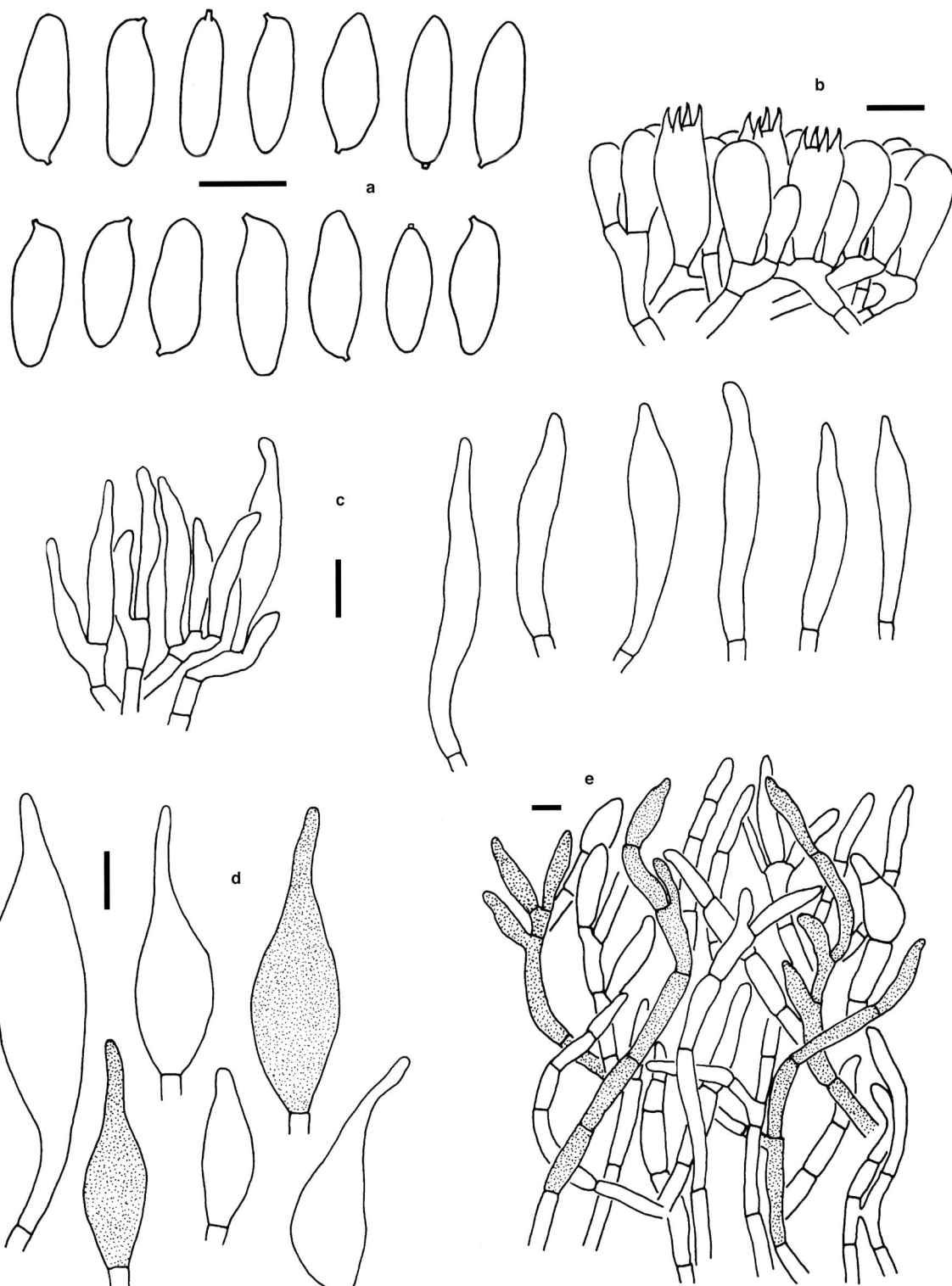

Fig. 22.18 b *Tylopilus obscureviolaceus* Har. Takah. **(KUN-HKAS 80115).**
a. Basidiospores; b. Basidia; c. Cheilocystidia; d. Pleurocystidia; e. Pileipellis. Scale bars = 10 μm.

22.19 *Tylopilus olivaceobrunneus* Yan C. Li & Zhu L. Yang, The Boletes of China: *Tylopilus* s.l. 319 (2021)

MycoBank: MB 834752

Etymology: The epithet "*olivaceobrunneus*" refers to the color of the basidiomata.

Type: CHINA, ANHUI PROVINCE: Jinzhai County, alt. 940 m, 19 July 2017, Y. J. Hao 1433 (KUN-HKAS 107145, GenBank Acc. No.: MT154722 for nrLSU).

Basidioma small to medium-sized. Pileus 3–8 cm in diam., hemispherical to subhemispherical or convex, yellowish olivaceous (2D3–5) with brownish (6E4–5) to pale brown (5D4–5) tinge; surface dry, covered with fine tomentose to fibrillose squamules; context white (1A1), slowly staining reddish tinge when bruised. Hymenophore adnate to slightly depressed around apex of stipe; surface white (11A1) when young, pale grayish red (11B2) to pinkish (11A2) at maturity, slowly staining reddish tinge when bruised; pores subangular to roundish, up to 2 mm wide; tubes 4–7 mm long, concolorous or much paler than hymenophoral surface, slowly staining reddish tinge when bruised. Spore print pinkish to reddish. Stipe 4–9 × 0.8–2.1 cm, subcylindrical to clavate, brownish gray (8E2) to dark brownish gray (9E2) or dark brown (9F4–5), without discoloration when touched; context solid, white, without discoloration when hurt; basal mycelium white. Taste bitter.

Basidia 21–58 × 6–10 μm, clavate to narrowly clavate, 4-spored, hyaline in KOH. Basidiospores [40/1/1] (8.5) 9–11 (12) × 3–3.5 μm [Q = (2.42) 2.57–3.33 (4), Q_m = 3.01 ± 0.23], subcylindrical to subfusiform in side view with slight suprahilar depression, cylindrical to fusiform in ventral view, smooth, hyaline to yellowish in KOH, yellow to brownish yellow in Melzer's reagent. Hymenophoral trama boletoid, hyphae cylindrical, 3.5–10 μm wide, hyaline to yellowish in KOH, yellowish to yellow in Melzer's reagent. Cheilo- and pleurocystidia 32–50 × 6–9 μm, subfusiform to fusoid-ventricose with a spindly apical part, brown to yellowish brown in KOH, with brown to dark brown encrustations, somewhat thick-walled (up to 1 μm thick). Pileipellis a palisadoderm composed of 4–10 μm wide interwoven filamentous hyphae with subcylindrical terminal cells 18–55 × 6–10 μm, pale yellowish to brownish yellow in KOH and yellow to brownish in Melzer's reagent. Pileal trama composed of 4–12 μm wide interwoven hyphae, colorless to yellowish in KOH and yellow to yellowish brown in Melzer's reagent. Clamp connections absent in all tissues.

Habitat: Scattered on soil in subtropical forests dominated by plants of the family Fagaceae.

Known distribution: Currently known from eastern China.

Commentary: *Tylopilus olivaceobrunneus* is characterized by the yellowish olivaceous pileus with brownish to pale brown tinge, the white then pale grayish red to pinkish hymenophore, the white context without discoloration or slowly staining reddish tinge when bruised, the brownish gray to dark brownish gray or dark brown stipe, the narrowly oblong to subfusiform basidiospores (8.5) 9–11 (12) × 3–3.5 μm with Q value mostly 2.57–3.33, and the palisadoderm pileipellis composed of 4–10 μm wide somewhat vertically arranged hyphae. It is macroscopically similar to the species with an olivaceous pileus, viz. *T. alpinus*, *T. brunneirubens*, *T. otsuensis*, and *T. himalayanus*. However, *T. alpinus* differs in its white context that stains brownish red when injured, distinctly reticulate stipe, relatively long basidiospores measuring 13–14.5 (15.5) × (3.5) 4–5 μm, and distribution in alpine forests with altitude up to 3500 m (Wu *et al.*

Chapter 22 *Tylopilus* P. Karst. 273

Fig. 22.19 a *Tylopilus olivaceobrunneus* Yan C. Li & Zhu L. Yang. Photos by Y. J. Hao (KUN-HKAS 107145, type).

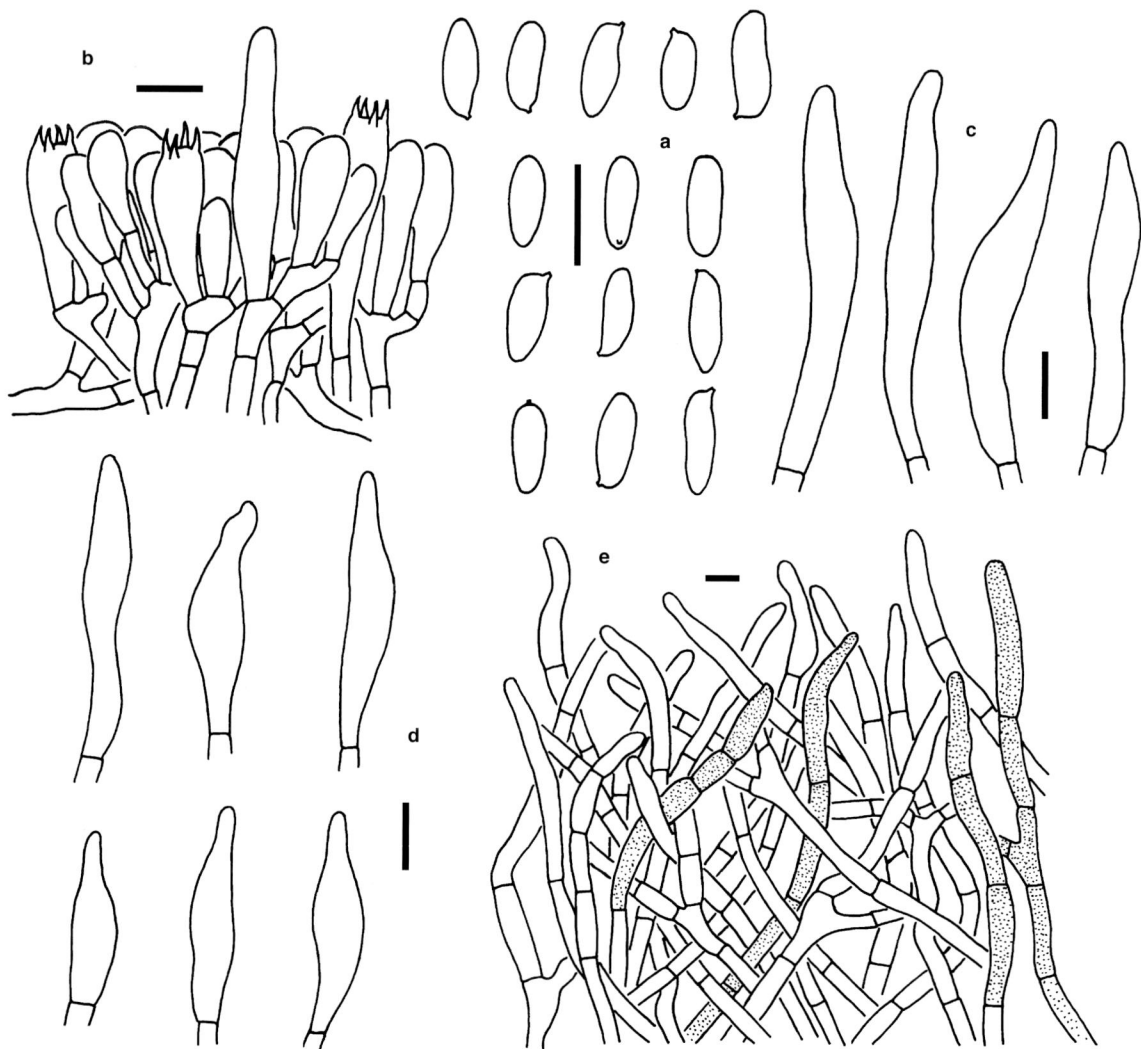

Fig. 22.19 b *Tylopilus olivaceobrunneus* Yan C. Li & Zhu L. Yang (KUN-HKAS 107145, type).
a. Basidiospores; b. Basidia and cheilocystidium; c. Pleurocystidia; d. Cheilocystidia; e. Pileipellis. Scale bars = 10 μm.

2016a; see our description above). *Tylopilus brunneirubens* differs in its white context that stains grayish red to brownish red or rust-brown when injured, distinctly reticulate stipe, and distribution in tropical to subtropical forests. *Tylopilus otsuensis* differs in its dark green to olivaceous pileus and stipe, white context that stains reddish brown when bruised, broadly ellipsoid to ovoid basidiospores measuring 5.5–6.5 × 4–5 μm, epithelium pileipellis made up of vertically arranged moniliform hyphae, and distribution in tropical forests (Hongo 1966, Wu *et al.* 2016a). *Tylopilus himalayanus* differs from *T. olivaceobrunneus* in its white to pallid context that stains distinct reddish brown when bruised, large basidiospores measuring 11–14 × 3.5–4.5 μm, palisadoderm pileipellis composed of 4–13 μm wide vertically arranged hyphae, and distribution in temperate to subalpine forests. *Tylopilus olivaceobrunneus* is phylogenetically related to *T. castanoides*. However, *T. castanoides* differs in its dark reddish brown to chestnut-brown or brown pileus and stipe, and trichoderm pileipellis composed of 5–8 μm wide interwoven hyphae (Takahashi 2002).

22.20 *Tylopilus otsuensis* Hongo, Mem Shiga Univ 16: 60 (1966)

Synonyms: *Boletus olivaceirubens* Corner, *Boletus* in Malaysia 178 (1972); *Tylopilus olivaceirubens* (Corner) E. Horak, Malay Fores Rec 51: 50 (2011).

Basidioma small to large. Pileus 3.5–9.5 cm in diam., hemispherical when young, subhemispherical to applanate when mature, dark green to grayish green or olivaceous; surface dry, nearly glabrous to finely tomentose or fibrillose; context solid, whitish to pallid, staining reddish brown when bruised. Hymenophore adnate when young, adnate to slightly depressed around apex of stipe when mature; surface white to pallid when young, pinkish, sometimes with olivaceous tinge when mature, staining ferruginous or brownish red when bruised; pores subangular to roundish, 0.3–0.5 mm wide; tubes up to 3 mm long, concolorous or much paler than hymenophoral surface, staining pale reddish brown or ferruginous when bruised. Spore print pinkish to reddish. Stipe 9.5–11 × 1–3 cm, subcylindrical, always enlarged downwards, concolorous with pileal surface, staining reddish brown when touched; surface nearly glabrous or sometimes covered with pulverous squamules; context solid, white to pallid, staining reddish brown when injured; basal mycelium white. Taste bitter.

Fig. 22.20 *Tylopilus otsuensis* Hongo. Photos by G. Wu (KUN-HKAS 99778).

Basidia 18–25 × 6.5–9 μm, clavate to narrowly clavate, 4-spored, sometimes 2-spored, hyaline in KOH. Basidiospores 5.5–6.5 × 4–5 μm [Q = (1) 1.2–1.5 (1.63), Q_m = 1.36 ± 0.11], subglobose to subellipsoid and inequilateral in side view, subglobose to broadly ellipsoid or ellipsoid in ventral view, smooth (Fig. 3.3 b), hyaline to yellowish in KOH, yellow to brownish yellow in Melzer's reagent. Hymenophoral trama boletoid composed of 3–8 μm wide filamentous hyphae. Cheilo- and pleurocystidia 45–55 × 9–11 μm, abundant, fusoid-ventricose to subfusiform, hyaline to yellowish in KOH, thin-walled. Pileipellis an epithelium composed of 1–2 inflated concatenate cells, yellowish to pale brownish in KOH and yellowish brown to brownish in Melzer's reagent; terminal cells 21–45 × 9–16 μm, cystidioid or pyriform. Pileal trama composed of 3–7 μm wide interwoven hyphae, hyaline to yellowish in KOH and yellowish to brownish in Melzer's reagent. Clamp connections absent in all tissues.

Habitat: Scattered or solitary on soil in tropical forests dominated by plants of the family Fagaceae.

Known distribution: Known from central and southwestern China, Japan and Malaysia.

Specimens examined: CHINA, YUNNAN PROVINCE: Jinghong County, Dadugang Town, alt. 1400 m, 7 July 2006, Y.C. Li 486 (KUN-HKAS 50212); Maguan County, Dulong Town, Nanjia Village, alt. 1200 m, 28 July 2016, G. Wu 1659 (KUN-HKAS 99778). HUNAN PROVINCE: Yizhang County, Mangshan National Forest Park, alt. 1800 m, 2 September 2007, Y.C. Li 1056 (KUN-HKAS 53401).

Commentary: *Tylopilus otsuensis* originally described from Japan by Hongo (1966) is characterized by the dark green to olivaceous basidioma, the white context staining reddish brown when bruised, the initially pallid to whitish and then pinkish hymenophore sometimes with greenish tinge, the subglobose to subellipsoid or broadly ellipsoid basidiospores, the epithelium pileipellis with terminal cells pyriform to cystidioid, and the distribution in tropical forests. This species is macroscopically similar to *T. griseiviridus*. However, *T. griseiviridus* can be easily distinguished by its very small to small basidioma, relatively large basidiospores measuring 7–11 × (2.5) 3–4 μm, small hymenial cystidia measuring 30–39 × 5–7 μm, and palisadoderm pileipellis composed of 5–9.5 μm wide vertically arranged hyphae (see our description above).

In our multi-locus phylogenetic analysis (Fig. 4.1), *T. otsuensis* clusters with *T. subotsuensis*. However, *T. subotsuensis* has a white context without discoloration when injured, large and slender basidiospores measuring 7.5–9.5 (10.5) × 3–3.5 μm, and hymeniderm pileipellis composed of a layer of cystidioid cells.

22.21 *Tylopilus phaeoruber* Yan C. Li & Zhu L. Yang, The Boletes of China: *Tylopilus* s.l. 324 (2021)

MycoBank: MB 834754

Etymology: The epithet "*phaeoruber*" refers to the color of the pileus.

Type: CHINA, YUNNAN PROVINCE: Tengchong County, alt. 1700 m, 11 August 2011, G. Wu 611 (KUN-HKAS 74925, GenBank Acc. No.: KF112473 for nrLSU, KF112222 for *tef1-α*, KF112577 for *rpb1*, KF112739 for *rpb2*).

Basidioma medium-sized. Pileus 5–7 cm in diam., hemispherical to subhemispherical; surface pale reddish brown (8D3–4) to reddish gray (9C2–3) or dull red (10C2–3), but deep reddish brown (10E3–4) to grayish red (11D3–4) in the center

Fig. 22.21 a *Tylopilus phaeoruber* Yan C. Li & Zhu L. Yang. Photos by G. Wu (KUN-HKAS 74925, type).

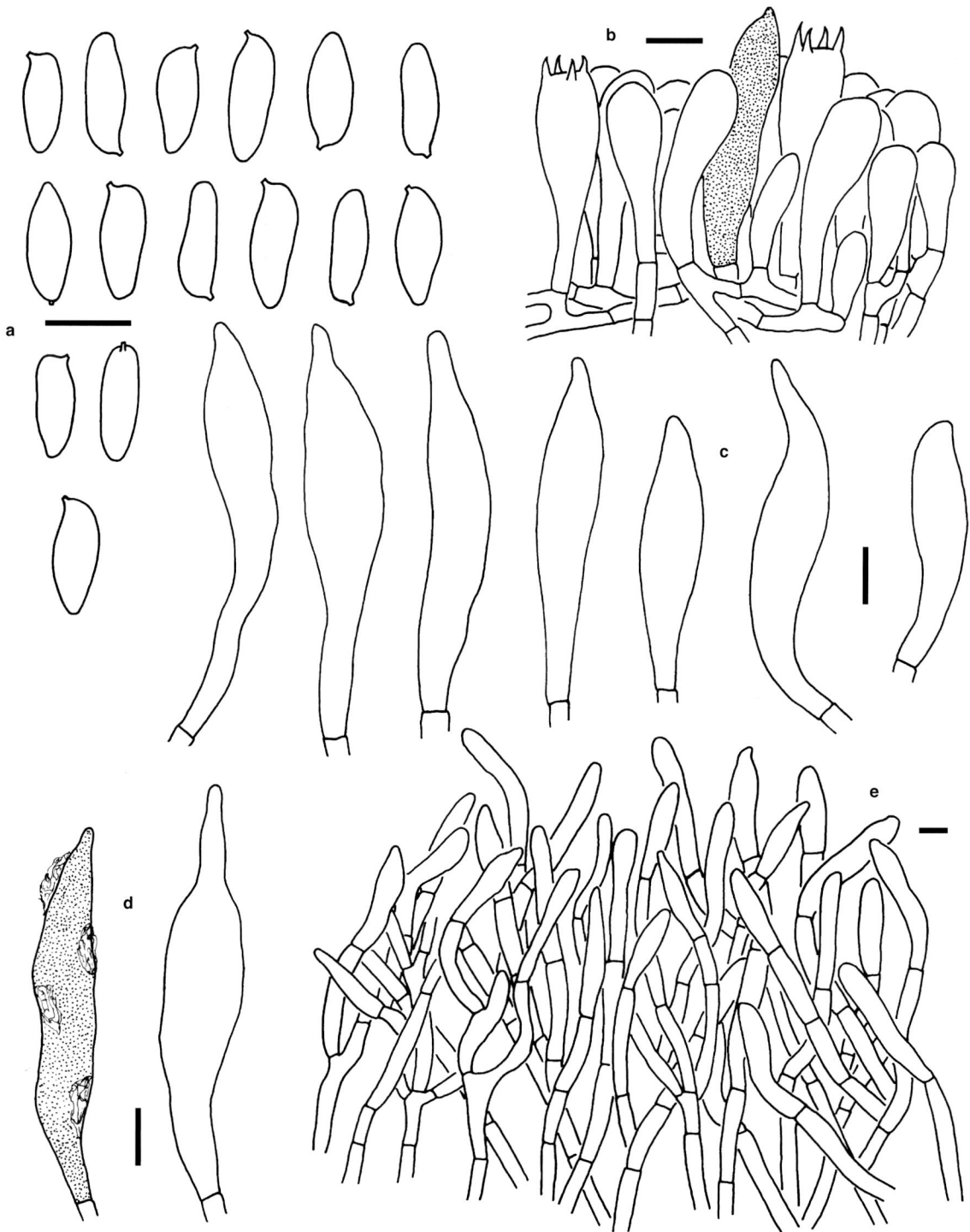

Fig. 22.21 b *Tylopilus phaeoruber* Yan C. Li & Zhu L. Yang (KUN-HKAS 74925, type).
a. Basidiospores; b. Basidia and cheilocystidium; c. Cheilocystidia; d. Pleurocystidia; e. Pileipellis. Scale bars = 10 μm.

and pale brownish (7E6–7) towards margin; surface dry with tomentose to fibrillose squamules, without discoloration when touched; context solid, white (1A1), without discoloration when bruised. Hymenophore adnate to slightly depressed around apex of stipe; surface white (11A1) when young, pale pinkish (11A2) when mature, without color change or slowly staining brownish when bruised; pores subangular to roundish, 0.3–0.5 mm wide; tubes 5–8 mm long, concolorous or much paler than hymenophoral surface, without color change or slowly staining brownish when bruised. Spore print reddish (12B2) to pinkish (13B2). Stipe 4 –8 × 1.2–2.2 cm, subcylindrical or clavate, always enlarged downwards, almost glabrous, concolorous or a little paler than pileal surface, without color change when touched; context solid, white at apex but reddish brown to red-brown downwards, without discoloration when injured; basal mycelium white. Taste bitter, odor mild.

Basidia 34–42 × 8–12 µm, clavate, 4-spored, sometimes 2-spored, hyaline in KOH. Basidiospores 12.5–14.5 (15.5) × 4.5–5.5 µm [Q = (2.55) 2.6–3, Q_m = 2.78 ± 0.13], subcylindrical to subfusiform in side view with slight suprahilar depression, cylindrical to fusiform in ventral view, smooth, hyaline to yellowish in KOH, yellow to brownish yellow in Melzer's reagent. Hymenophoral trama boletoid composed of 3–10 µm wide filamentous hyphae, hyaline to yellowish in KOH, yellowish to yellow in Melzer's reagent. Cheilo- and pleurocystidia 48–72 × 9–14 µm, subfusiform to fusoid-ventricose, sometimes with subacute apex and long pedicel, yellow to brownish yellow in KOH, thin-walled. Pileipellis a palisadoderm composed of 4–10 µm wide interwoven hyphae, pale yellowish to brownish yellow in KOH and yellow to brownish in Melzer's reagent; terminal cells 31–65 × 7–10 µm, pyriform to clavate or cystidioid. Pileal trama composed of 3–9 µm wide interwoven hyphae, colorless to yellowish in KOH and yellowish to pale yellow in Melzer's reagent. Clamp connections absent in all tissues.

Habitat: Solitary on soil in tropical forests dominated by plants of the family Fagaceae.

Known distribution: Currently known from southwestern China.

Commentary: *Tylopilus phaeoruber* is characterized by the reddish gray to pale reddish brown then dark reddish brown to grayish red pileus and stipe, the white then pinkish hymenophore staining brownish when injured, the white context without discoloration when bruised, and the palisadoderm pileipellis composed of 4–10 µm wide interwoven hyphae. This species is macroscopically similar to *T. castanoides* Har. Takah. However, *T. castanoides* differs from *T. phaeoruber* in its white context without color change or gradually staining rufescent when injured, relatively small basidiospores measuring 9–11(13) × 3.5–4 µm, small pleurocystidia measuring 30–60 × 6–10 µm, and white to dull pinkish hymenophore without color change when injured (Takahashi 2002).

22.22 *Tylopilus plumbeoviolaceoides* T.H. Li, B. Song & Y.H. Shen, Mycosystema 21: 3 (2002)

Basidioma small to large. Pileus 3.5–10 cm in diam., hemispherical to convex when young, becoming broadly convex to plano-convex or applanate when mature; vinaceous to dark violaceous-brown or purple-chestnut, sometimes paler to vinaceous buff, usually paler towards the margin; surface subviscid to viscid when wet, subtomentose to fibrillose; context white, staining pinkish to purplish when bruised. Hymenophore adnate when young,

Fig. 22.22 *Tylopilus plumbeoviolaceoides* T.H. Li, B. Song & Y.H. Shen. Photos by T.H. Li & C.Y. Deng (HMIGD 26167).

adnate to slightly depressed around apex of stipe when mature; surface initially white and then pinkish to pale purplish when mature, becoming pinkish to reddish brown when bruised; pores subangular to roundish, up to 1.5 mm wide; tubes 7–10 mm long, concolorous or a little paler than hymenophoral surface, staining pinkish to reddish brown when bruised. Stipe 6–9 × 1.2–2.5 cm, subcylindrical to clavate, always enlarged downwards, concolorous or paler than pileal surface, staining reddish brown when touched; context white, becoming reddish brown when hurt; basal mycelium white. Taste bitter, odor mild.

Basidia 18–25 × 7–9 µm, clavate, 4-spored, hyaline in KOH. Basidiospores 8.5–12 × 3–4 µm (Q = 2.25–3.67, Q_m = 2.85 ± 0.21), subcylindrical to subfusiform in side view with slight suprahilar depression, cylindrical to fusiform in ventral view, smooth, hyaline to yellowish in KOH, yellow to brownish yellow in Melzer's reagent. Hymenophoral trama boletoid composed of 3.5–8 µm wide filamentous hyphae, hyaline to yellowish in KOH, yellowish to yellow in Melzer's reagent. Cheilo- and pleurocystidia 35–45 × 7–11 µm, abundant, fusiform to fusoid-ventricose, yellow to brownish yellow in KOH, thin-walled. Pileipellis an ixotrichoderm to trichoderm composed of gelatinous 3.5–5 µm wide interwoven filamentous hyphae, pale yellowish to brownish yellow in KOH and yellow to brownish in Melzer's reagent; terminal cells 25–70 × 5–8 µm, subcylindrical. Clamp connections absent in all tissues.

Habitat: Scattered, gregarious, or subcaespitose in tropical forests dominated by plants of the family Fagaceae.

Known distribution: Currently known from southern China.

Specimens examined: CHINA, GUANGDONG PROVINCE: Guangzhou, Tianluhu Forest Park, alt. 230 m, 5 April 2003, W.Q. Deng (HMIGD 21040), the same location, 10 April 2009, T.H. Li and C.Y. Deng (HMIGD 26167).

Commentary: *Tylopilus plumbeoviolaceoides* originally described from southern China by Li *et al.* (2002) is characterized by the violaceous-brown or purple-chestnut basidioma, the white context staining pinkish to purplish when bruised, the white to pinkish or pale purplish hymenophore staining pinkish to reddish brown when bruised, ixotrichoderm to trichoderm pileipellis composed of gelatinous 3.5–5 µm wide interwoven filamentous hyphae. Phylogenetically, *T. plumbeoviolaceoides* is related to *T. plumbeoviolaceus* and *T. atripurpureus*. Indeed, they are morphologically similar in that they share a violaceous to purple pileus. However, *T. plumbeoviolaceus* has a hymeniform pileipellis which is composed of a layer of cystidioid cells measuring 14.3–45.5 × 2.5–7.8 µm, relatively small and narrow cheilocystidia measuring 16.9–22 × 5.2–7.8 µm, and pleurocystidia measuring 26.3–39 × 5.2–7.8 µm (Snell and Dick, 1941; Smith and Thiers, 1971). *Tylopilus atripurpureus* has a dark purple to blackish purple pileus, a pinkish to purplish hymenophore without color change when injured, a white to pallid context without discoloration when injured, a palisadoderm pileipellis composed of 5–12 µm wide vertically arranged hyphae (Corner 1972; Horak 2011; Wu *et al*. 2016a).

22.23 *Tylopilus pseudoalpinus* Yan C. Li & Zhu L. Yang, The Boletes of China: *Tylopilus* s.l. 329 (2021)

MycoBank: MB 837928

Etymology: The epithet "*pseudoalpinus*" refers to its similarity to *T. alpinus*.

Type: CHINA, YUNNAN PROVINCE: Longling County, Gaoligongshan National Forest Park, near Nankang Station, alt. 2100 m, 17 June 2014, L.H. Han 306 (KUN-HKAS 84602, GenBank Acc. No.: MW114855 for nrLSU, MW165268 for *tef1-α*).

Basidioma small to medium-sized. Pileus 4–6 cm in diam., hemispherical to plano-convex, grayish ruby (12D5–6) to grayish red (10D6) or brownish red (10D6), but pale red (9B3–4) to grayish red (8C4–5) towards margin; surface covered with concolorous filamentous squamules, dry, minutely cracked with age, staining reddish brown (9E7–8) when touched; context solid at first and then spongy, white, staining grayish red (10D6) to brownish red (10D6) when bruised. Hymenophore adnate when young, depressed around apex of stipe when mature; surface whitish (10A1) to dull gray (10B1), staining grayish red (9C4–5) to reddish brown (9D7–8) when bruised; pores subangular to roundish, 0.5–1 mm wide; tubes 6–10 mm long, concolorous or a little paler than hymenophoral surface, staining grayish red (9C4–5) reddish brown (9D7–8) when bruised. Stipe 3–8 × 1.2–1.5 cm, clavate, always enlarged towards base, dingy white to cream, staining grayish ruby (12D5–6) to grayish red (10D6) or brownish red (10D6) when touched or aged; surface entirely with distinct reticulum; context solid to spongy, whitish to pallid, staining a grayish red to reddish brown tinge when hurt; basal mycelium whitish to pallid. Taste and odor mild.

Basidia 19–27 × 7–9 μm, clavate to narrowly clavate, yellowish to pale brownish or colorless in KOH, 4-spored. Basidiospores 9–11.5 × 3.5–4.5 μm [Q = (2.22) 2.25–2.88 (3), Q_m = 2.56 ± 0.18], subcylindrical to subfusiform in side view with slight suprahilar depression, cylindrical to fusiform in ventral view, smooth, hyaline to yellowish in KOH, yellow to brownish yellow in Melzer's reagent. Hymenophoral trama boletoid composed of 3–11 μm wide filamentous hyphae, hyaline to yellowish in KOH, yellowish to yellow in Melzer's reagent. Cheilo- and pleurocystidia remarkably numerous, 27–57 × 9–16 μm, subfusiform to subfusoid-ventricose, thin-walled, brown to yellow-brown in KOH. Pileipellis a trichoderm to subcutis, composed of 3–6 μm wide filamentous hyphae, yellowish to pale brownish in KOH and yellowish brown to brownish in Melzer's reagent; terminal cells 29–59 × 3–5 μm, subcylindrical. Pileal trama composed of 3–10 μm wide interwoven hyphae, colorless to yellowish in KOH and yellow to yellowish brown in Melzer's reagent. Clamp connections absent in all tissues.

Habitat: Solitary on soil in subtropical forests dominated by plants of the family Fagaceae.

Known distribution: Currently known from southwestern China.

Commentary: *Tylopilus pseudoalpinus* is characterized by the grayish ruby to grayish red or brownish red pileus, the whitish to dull gray hymenophore, the dingy white to cream stipe which is entirely reticulate, the trichoderm to subcutis pileipellis, and the distribution in subtropical forests.

Chapter 22 *Tylopilus* P. Karst. 283

In our multi-locus phylogenetic analysis (Fig. 4.1), *T. pseudoalpinus* clusters, with high support, with *T. alpinus* and *T. brunneirubens*. Indeed, they are morphologically similar in that they share a brown to brownish red pileus in aged basidiomata, a white to grayish context staining pale red to grayish red when

Fig. 22.23 a *Tylopilus pseudoalpinus* Yan C. Li & Zhu L. Yang. Photos by L.H. Han (KUN-HKAS 84602, type).

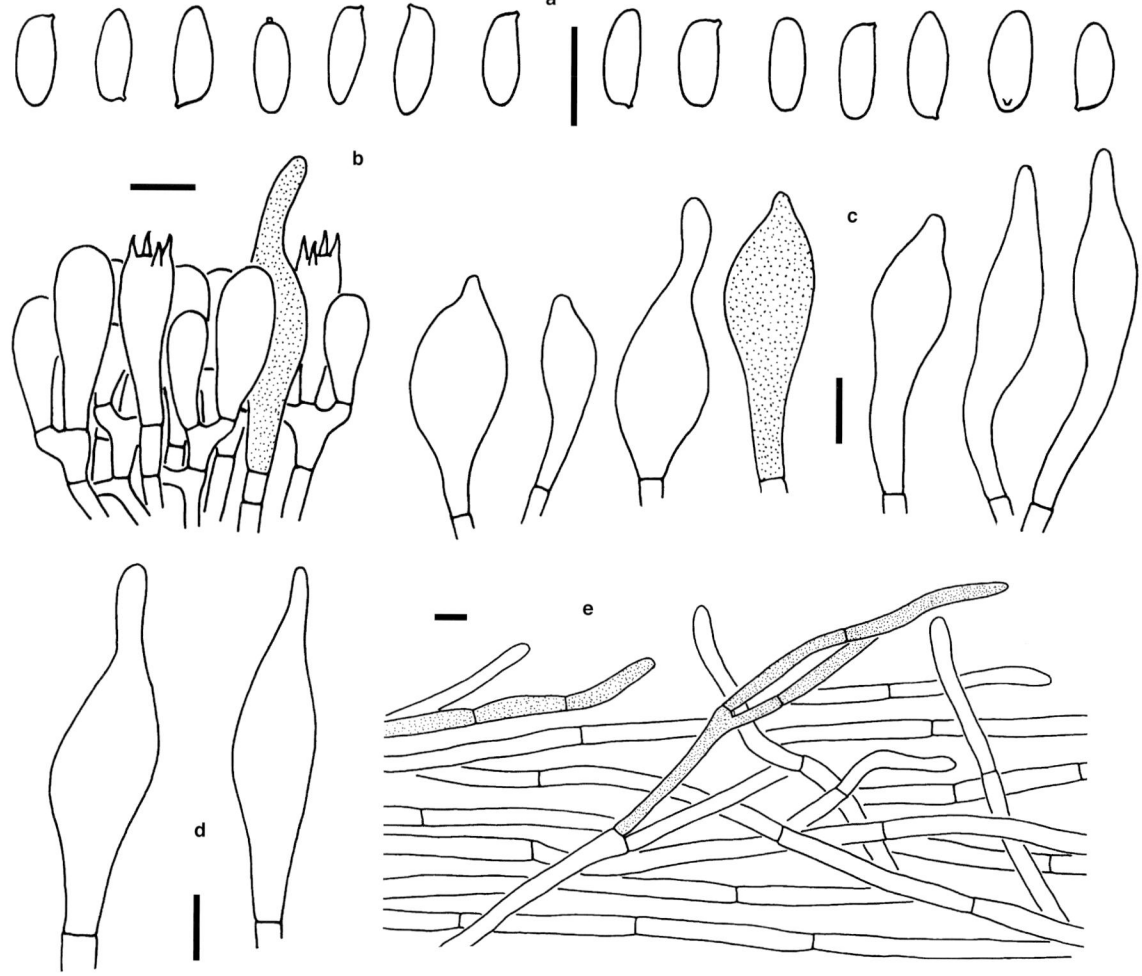

Fig. 22.23 b *Tylopilus pseudoalpinus* Yan C. Li & Zhu L. Yang (KUN-HKAS 84602, type).
a. Basidiospores; b. Basidia and pleurocystidium; c. Cheilocystidia; d. Pleurocystidia; e. Pileipellis. Scale bars = 10 μm.

bruised, and a reticulate stipe. However, *T. alpinus* differs from *T. pseudoalpinus* in the dark olive to olive-green then olive-brown pileus, the initially whitish and then pinkish hymenophore, the whitish to grayish or pale orange stipe which is reticulate on upper half, the relatively large basidiospores 13–15.5 × 3.5–5 μm, the large pleurocystidia 62–87 × 13–16 μm, and the distribution in subalpine to alpine forests. *Tylopilus brunneirubens* can be distinguished from *T. pseudoalpinus* by the olive-brown to brown then dark brown to chestnut-brown pileus, the distinct reticulum on the upper half of the stipe, and the distribution in subtropical to tropical forests.

22.24 *Tylopilus pseudoballoui* D. Chakr., K. Das & Vizzini, MycoKeys 33: 109 (2018)

Basidioma medium-sized. Pileus 5–9 cm in diam., hemispherical when young, subhemispherical to applanate when mature, yellow to orange-yellow when young, and then orange to orange-brown when mature; surface viscid when wet, shiny when dry, nearly glabrous, without discoloration when injured; context whitish to cream, without discoloration when bruised. Hymenophore adnate when young, adnate to slightly decurrent when mature; surface initially white to pallid and then pinkish to pale pink, without

Fig. 22.24 *Tylopilus pseudoballoui* D. Chakr., K. Das & Vizzini. Photos by Y.C. Li (KUN-HKAS 51151).

discoloration when bruised; pores subangular to roundish, 0.3–0.5 mm wide; tubes 4–5 mm long, concolorous or a little paler than hymenophoral surface, without discoloration when bruised. Spore print reddish to pinkish. Stipe 7–9 × 1.2–1.6 cm, clavate, yellow to orange-yellow, or orange to orange-brown, without discoloration when injured; context whitish to pallid, without discoloration when hurt; basal mycelium whitish to pallid. Taste and odor mild.

Basidia 20–35 × 7–10 μm, clavate to narrowly clavate, 4-spored, hyaline in KOH. Basidiospores 6–8 × 4–4.5 μm (Q = 1.5–1.95, Q_m = 1.73 ± 0.15), subellipsoid to elongated and inequilateral in side view with slight suprahilar depression, ellipsoid to elongated in ventral view (Fig. 3.3 a), smooth, hyaline to yellowish in KOH, yellow to brownish yellow in Melzer's reagent. Hymenophoral trama boletoid composed of 4–11 μm wide filamentous hyphae. Cheilo- and pleurocystidia 30–45 × 8–12 μm, abundant, fusiform to subfusiform, yellow to orange-yellow in KOH, thin-walled. Pileipellis an intricate trichoderm composed of 3.5–5 μm wide interwoven filamentous hyphae with subcylindrical terminal cells 20–35 × 4–5 μm, yellowish to pale brownish in KOH and yellow to brownish in Melzer's reagent. Clamp connections absent in all tissues.

Habitat: Solitary on soil in tropical to subtropical forests dominated by plants of the family Fagaceae.

Known distribution: Widely distributed in India and southwestern China.

Specimens examined: CHINA, YUNNAN PROVINCE: Kunming, Qiongzhu Temple, alt. 2100 m, 21 September 2006, Y.C. Li 714 (KUN-HKAS 51151); Kunming, Heilongtan Park, alt. 1900 m, 6 September 2000, F.Q. Yu 387 (KUN-HKAS 38693); Lancang County, Fazhanhe Town, alt. 1200 m, 28 June 2017, J.W. Liu 635 (KUN-HKAS 107175); Maguan County, Renhe Town, Gesa Village, alt. 1500 m, 15 October 2017, J. Wang 324 (KUN-HKAS 107174).

Commentary: *Tylopilus pseudoballoui*, originally described by Chakraborty *et al.* (2018) from India, is characterized by the orange-red to orange-brown pileus, the adnate hymenophore, the white context, the ellipsoid to elongated basidiospores, and the intricate trichoderm pileipellis. Just as many species endemic to East Asia were mislabeled as North America names (Wu *et al.* 2016a; Zeng *et al.* 2016), *T. pseudoballoui* was once misidentified as "*Tylopilus balloui*" by Mao (2000) and Ying and Zang (1994). However, *T. balloui* differs from *T. pseudoballoui* in its dry pileus (viscid in *T. pseudoballoui*), white to dingy white pores and context staining pinkish tan on exposure, and distribution in North America (Smith and Thiers 1971; Wolfe 1981; Both 1993; Bessette *et al.* 2016; Osmundson and Halling 2010).

In our multi-locus phylogenetic analysis (Fig. 4.1), *T. pseudoballoui* is closely related to *T. aurantiacus* and *T. rubrotinctus*. For comparisons of these species see the commentary on *T. aurantiacus*.

22.25 *Tylopilus purpureorubens* Yan C. Li & Zhu L. Yang, The Boletes of China: *Tylopilus* s.l. 334 (2021)

MycoBank: MB 837931

Etymology: The epithet "*purpureorubens*" refers to the rufescent discoloration of the purple basidiomata.

Type: CHINA, ZHEJIANG PROVINCE, Qingyuan County, Baishanzu National Nature Reserve, alt. 1400 m, 11 September 2013, G. Wu 1230 (KUN-HKAS 80605, GenBank Acc. No.: MT154727 for nrLSU, MT110344 for *tef1-α*, MT110420 for *rpb2*).

Basidioma small to medium-sized. Pileus 3.5–8 cm in diam., subhemispherical or convex when young, convex to plano-convex or applanate when mature, purplish gray (13D2–3) to purple-gray (14E2) when young, pale purplish gray (13C2), reddish gray (12D2) to dark brownish gray (11E2–3) when mature; margin much paler, gray (14D1) to reddish gray (12D2); surface dry, tomentose, becoming reddish brown (8D5–6) when touched; context solid, whitish (12A1), staining reddish brown when bruised. Hymenophore adnate when young, adnate to slightly depressed around apex of stipe; surface white (10A1) to pallid when young, becoming pinkish (11A2) when mature, staining reddish brown when bruised; pores subangular to roundish, 0.5–1 mm wide; tubes 5–10 mm long, concolorous or a little paler than hymenophoral surface, becoming reddish brown when bruised. Spore print reddish (12B2) to pinkish (13B2). Stipe 4–8 × 1.2–2.5 cm,

Fig. 22.25 a *Tylopilus purpureorubens* Yan C. Li & Zhu L. Yang. Photos by Y.J. Hao (KUN-HKAS 80106).

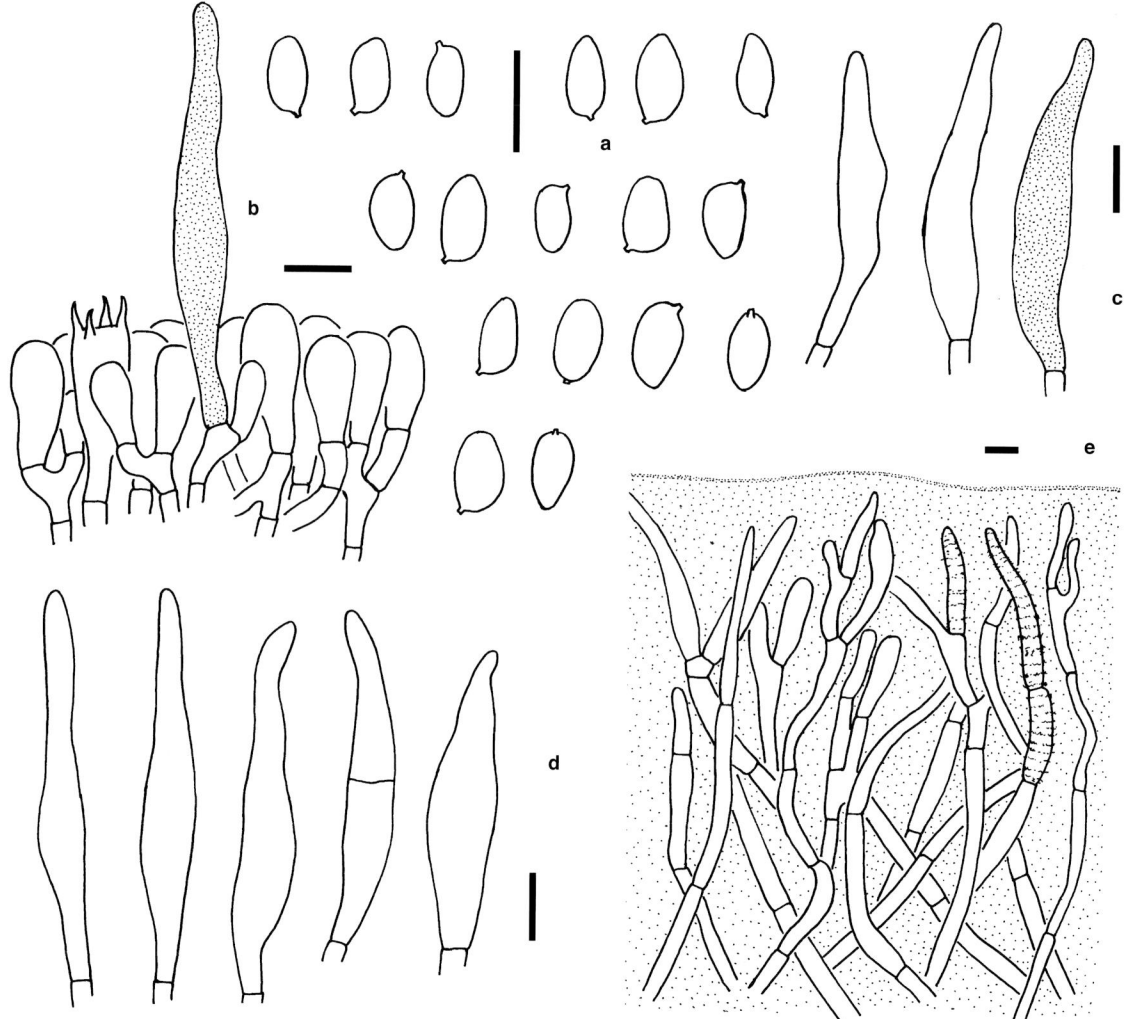

Fig. 22.25 b *Tylopilus purpureorubens* Yan C. Li & Zhu L. Yang (KUN-HKAS 80106).
a. Basidiospores; b. Basidia and pleurocystidium; c. Cheilocystidia; d. Pleurocystidia; e. Pileipellis. Scale bars = 10 μm.

subcylindrical to clavate, always enlarged downwards, pale grayish brown (9D3–4) to pale reddish brown (8D4–5), almost glabrous, becoming reddish brown when touched; context solid, white, becoming reddish brown when injured; basal mycelium pallid. Taste bitter, odor mild.

Basidia 20–27 × 9–10 μm, clavate to narrowly clavate, 4-spored, sometimes 2-spored, hyaline in KOH. Basidiospores [140/7/7] 7–9.5 × 3.5–4.5 μm [Q = 1.67–2.29 (2.43), Q_m = 1.94 ± 0.17], elongated to subcylindrical in side view, elongated to cylindrical in ventral view, smooth, hyaline to yellowish in KOH, yellow to brownish yellow in Melzer's reagent. Hymenophoral trama boletoid composed of 4–10 μm wide filamentous hyphae. Cheilo- and pleurocystidia abundant, 48–66 × 9–15 μm, subfusiform to fusoid-ventricose, often spindly in the upper parts, sometimes septate, brown to brownish yellow in KOH, yellow-brown to dark brown in Melzer's reagent, thin-walled. Pileipellis a trichoderm to palisadoderm composed of 4.5–9 μm wide interwoven filamentous hyphae with subcylindrical terminal cells 16–43 × 4.5–7 μm, pale yellowish to brownish yellow in KOH and yellow to brownish in Melzer's reagent. Pileal trama composed of 4–7 μm wide interwoven hyphae, colorless to yellowish in KOH and yellowish to pale yellow in Melzer's reagent. Clamp connections absent in all tissues.

Habitat: Scattered to solitary on soil in tropical forests dominated by plants of the family Fagaceae.

Known distribution: Currently known from central, southeastern and southern China.

Specimens examined: CHINA, HUNAN PROVINCE: Yizhang County, Mangshan National Forest Park, alt. 1800 m, 5 September 2007, Y.C. Li 1097 (KUN-HKAS 53459). HAINAN PROVINCE: Ledong County, Jianfengling, alt. 850 m, 28 August 2009, N.K. Zeng 637 (KUN-HKAS 59837). GUANGDONG PROVINCE: Fengkai County, Heishiding Nature Reserve, alt. 250 m, 3 July 2012, F. Li 579 (KUN-HKAS 106451), the same location, 4 July 2012, F. Li 594 (KUN-HKAS 106356) and F. Li 605 (KUN-HKAS 106454), the same location, 1 June 2013, Y.J. Hao 826 (KUN-HKAS 80106).

Commentary: *Tylopilus purpureorubens* is characterized by the purplish gray to purple-gray pileus becoming pale purplish gray, reddish gray or dark brownish gray when aged, the white context staining reddish brown when bruised, the white to pallid then pinkish hymenophore, the nearly glabrous pale

Fig. 22.25 c *Tylopilus purpureorubens* Yan C. Li & Zhu L. Yang. Photo by G. Wu (KUN-HKAS 80605, type)

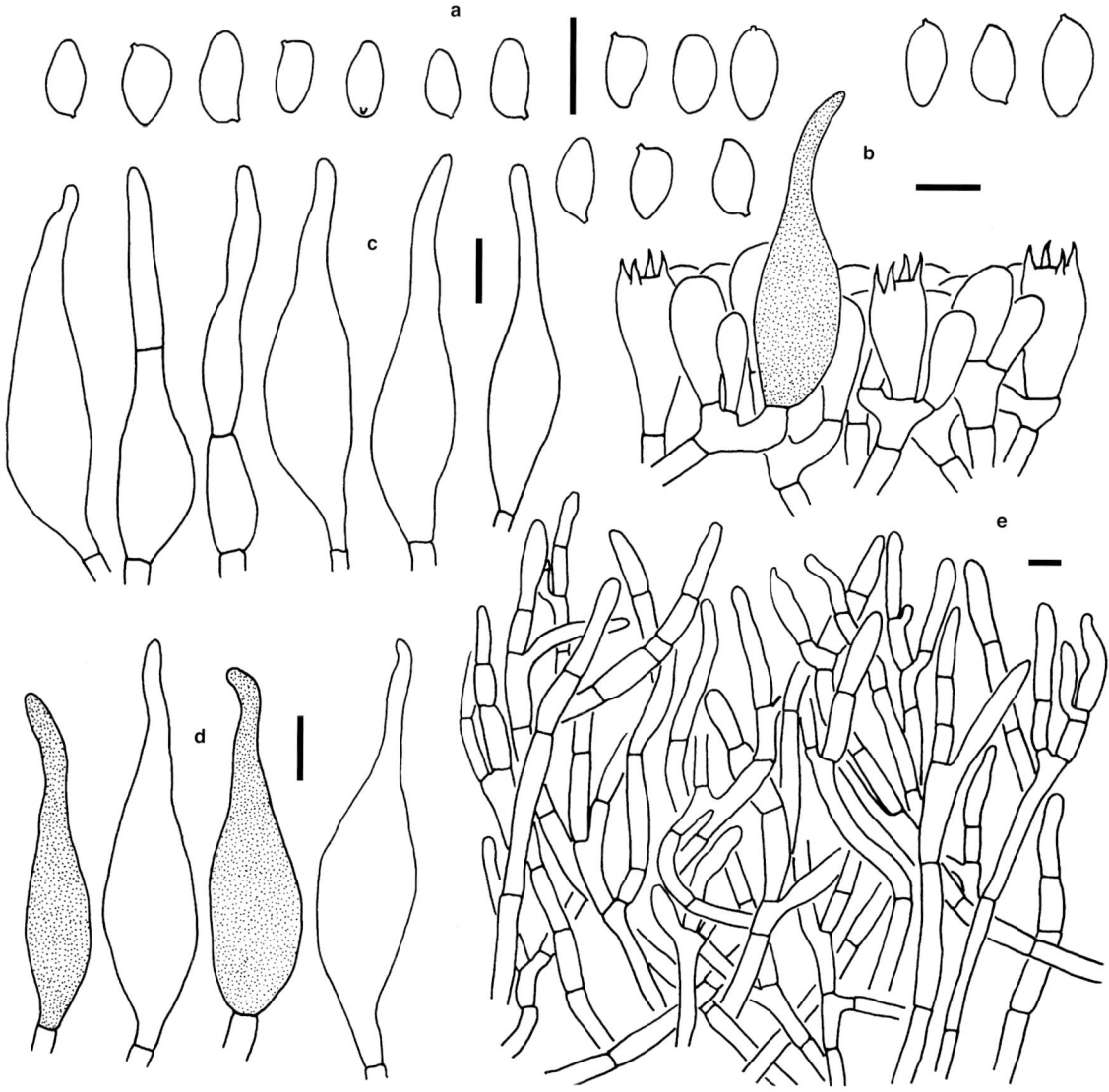

Fig. 22.25 d *Tylopilus purpureorubens* Yan C. Li & Zhu L. Yang (KUN-HKAS 80605, type).
a. Basidiospores; b. Basidia and cheilocystidium; c. Cheilocystidia; d. Pleurocystidia; e. Pileipellis. Scale bars = 10 μm.

grayish brown to pale reddish brown stipe, and the trichoderm to palisadoderm pileipellis composed of 4.5–9 μm wide interwoven hyphae.

Tylopilus purpureorubens is phylogenetically related and macroscopically similar to *T. argillaceus*. Moreover, these two species are sympatric in their distributions. However, *T. argillaceus* has a dark reddish brown to castaneous or salmon basidioma which is without any purple tinge, and trichoderm pileipellis composed of 3–6 μm wide narrow hyphae.

22.26 *Tylopilus rubrotinctus* Yan C. Li & Zhu L. Yang, The Boletes of China: *Tylopilus* s.l. 339 (2021)

MycoBank: MB 834756

Etymology: The epithet "*rubrotinctus*" refers to the color of the basidiomata.

Type: CHINA, GUANGDONG PROVINCE: Fengkai County, Heishiding Nature Reserve, alt. 250 m, 2 June 2013, K. Zhao 259 (KUN-HKAS 80684, GenBank Acc. No.: MT154733 for nrLSU, MW165264 for *tef1-α*, MW165283 for *rpb2*).

Basidioma small to medium-sized. Pileus 3–7 cm in diam., subhemispherical to plano-convex or applanate; surface grayish red (7B3–4) to pale red (7A3–4) or reddish orange (6B2–3) in the center, pale orange (6A2–3) to orangish (5A3–4) towards margin, without discoloration when touched; context white to cream, without discoloration when bruised. Hymenophore adnate when young, and depressed around apex of stipe when mature; surface whitish (4A1) when young, pinkish (4A2) when mature, without discoloration when bruised; pores subangular to roundish, up to 2.5 mm wide; tubes 5–10 mm long, concolorous or a little paler than hymenophoral surface, without discoloration when bruised. Spore print reddish (12B2) to pinkish (13B2). Stipe 5–8 × 1–1.2 cm, subcylindrical to clavate, pale yellow or yellowish (4A3–4), without color change when touched; context white to cream, without color change when hurt; basal mycelium pallid. Taste and odor mild.

Basidia 22–30 × 6–8 μm, clavate to narrowly clavate, 4-spored, sometimes 2-spored, hyaline in KOH. Basidiospores [80/4/4] 5–6.5 × 3.5–4.5 μm [Q = (1.11) 1.33–1.71, Q_m = 1.54 ± 0.15], subellipsoid to elongated and inequilateral in side view with slight suprahilar depression, ellipsoid to elongated in ventral view, smooth, hyaline to yellowish in KOH, yellow to brownish yellow in Melzer's reagent. Hymenophoral trama boletoid composed of 2.5–8 μm wide filamentous hyphae. Cheilo- and pleurocystidia 41–80 × 6–16 μm, abundant, fusiform to fusoid-ventricose, yellow to orange-yellow in KOH, thin-walled. Pileipellis a trichoderm to palisadoderm composed of 5–9 μm wide interwoven hyphae with subcylindrical terminal cells 16–60 × 5–9 μm, yellowish to pale brownish in KOH and yellowish brown to brownish in Melzer's

Fig. 22.26 a *Tylopilus rubrotinctus* Yan C. Li & Zhu L. Yang. Photos by F. Li (KUN-HKAS 106346).

Fig. 22.26 b *Tylopilus rubrotinctus* Yan C. Li & Zhu L. Yang. Photo by K. Zhao (KUN-HKAS 80684, type).

reagent. Clamp connections absent in all tissues.

Habitat: Scattered to solitary on soil in tropical forests dominated by plants of the family Fagaceae.

Known distribution: Currently known from southern China.

Additional specimens examined: CHINA, GUANGDONG PROVINCE: Fengkai County, Heishiding Nature Reserve, alt. 250 m, 25 June 2012, F. Li 557 (KUN-HKAS 106346), the same location, 12 August 2012 F. Li 843 (KUN-HKAS 106435), and 10 December 2012, F. Li 1101 (KUN-HKAS 78271).

Commentary: *Tylopilus rubrotinctus* is characterized by the grayish red, pale red, or reddish orange pileus, the white context without discoloration when bruised, the whitish then pinkish hymenophore, the pale yellow or yellowish stipe, the broadly ellipsoid to ovoid basidiospores, and the trichoderm to palisadoderm pileipellis composed of 5–9 μm wide interwoven hyphae.

Tylopilus rubrotinctus is phylogenetically related and morphologically similar to *T. aurantiacus* and *T. pseudoballoui* (Fig. 4.1). However, these species differ in the color of the pileus and stipe, the structure of the pileipellis, the size of the basidiospores and pleurocystidia, and the habitat (see our commentary under *T. aurantiacus* above).

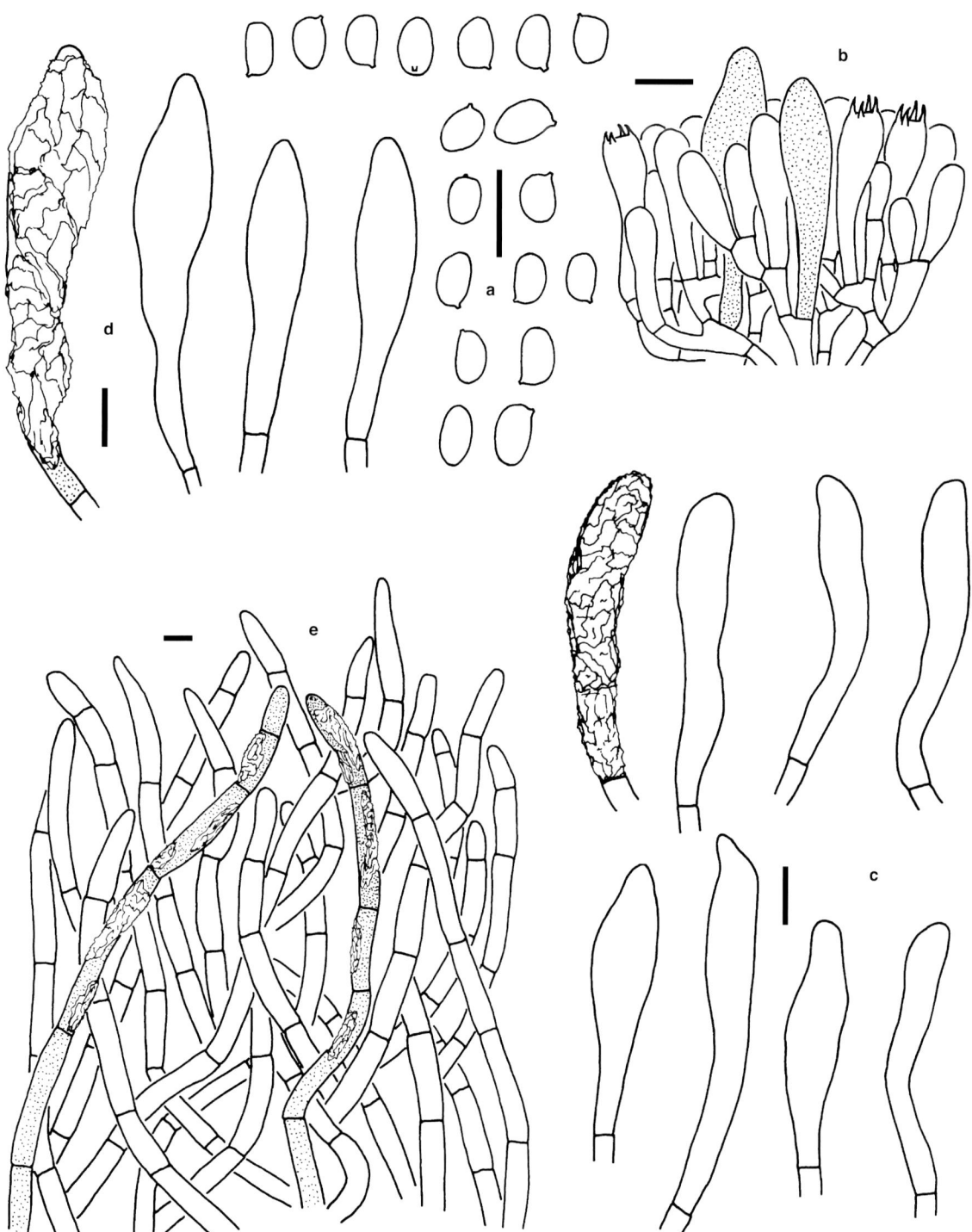

Fig. 22.26 c *Tylopilus rubrotinctus* Yan C. Li & Zhu L. Yang (KUN-HKAS 80684, type).
a. Basidiospores; b. Basidia and cheilocystidia; c. Cheilocystidia; d. Pleurocystidia; e. Pileipellis. Scale bars = 10 μm.

22.27 *Tylopilus rufobrunneus* Yan C. Li & Zhu L. Yang, The Boletes of China: *Tylopilus* s.l. 343 (2021)

MycoBank: MB 834757

Etymology: The epithet "*rufobrunneus*" refers to the color of the basidiomata.

Type: CHINA, GUANGDONG PROVINCE: Fengkai County, Heishiding Nature Reserve, alt. 250 m, 25 May 2012, F. Li 389 (KUN-HKAS 106331, GenBank Acc. No.: MT154714 for nrLSU, MT110337 for *tef1-α*, MT110380 for *rpb1*, MT110413 for *rpb2*).

Basidioma medium-sized to large. Pileus 5–12 cm in diam., hemispherical to subhemispherical when young, subhemispherical to applanate when mature, brown (6D7–8) to brownish orange (6C7) when young, brownish (6D5–6) to brownish yellow (5C7–8) or pale brownish yellow (7D3–4) when mature, slightly darker in the center; tubes becoming reddish brown when bruised. Spore print reddish (12B2) to pinkish (13B2). Stipe cylindrical, 4–7 × 0.8–1.2 cm, concolorous with pileal surface, upper 1/3–1/2 part with distinct reticulum, staining reddish brown when touched; context whitish to pallid, staining reddish brown when hurt; basal mycelium white to cream (1A2) or yellowish (1A2). Taste and odor mild.

Basidia 25–30 × 7–11 μm, clavate, 4-spored, hyaline in KOH. Basidiospores 10.5–13.5 × 3.5–4 μm [Q = 2.88–3.57 (3.71), Q_m = 3.09 ± 0.13], subcylindrical to subfusiform in side view with slight suprahilar depression, cylindrical to fusiform in ventral view, smooth, hyaline to yellowish in KOH, yellow to brownish yellow in Melzer's reagent. Hymenophoral trama boletoid composed of 3–7 μm wide filamentous hyphae. Cheilo- and pleurocystidia 50–84 × 13–22 μm, subfusiform to fusoid-ventricose, brownish to yellow-brown in KOH, thin-walled. Pileipellis a trichoderm composed of 4–6.5 μm wide filamentous hyphae with clavate to cylindrical terminal cells 16–101 × 4–6 μm, yellowish to pale brownish in KOH and yellow to brownish in Melzer's reagent. Pileal trama composed of 3–8 μm wide interwoven hyphae, hyaline to yellowish in KOH and yellowish to brownish in Melzer's reagent. Clamp connections absent in all tissues.

Habitat: Scattered to solitary on soil in tropical forests dominated by plants of the family Fagaceae.

Known distribution: Currently known from southern China.

Additional specimen examined: CHINA, HAINAN PROVINCE: Baisha County, Yinggeling National Nature Reserve, alt. 1300 m, 28 July 209, N.K. Zeng 370 (KUN-HKAS 106466).

Commentary: *Tylopilus rufobrunneus* is characterized by the brown, brownish, brownish orange or brownish yellow pileus, the white to pallid context staining reddish brown when injured, the white then pinkish to reddish gray hymenophore staining reddish brown when injured, the distinctly reticulate stipe over the upper 1/3–1/2 part, and the trichoderm pileipellis composed of 4–6.5 μm wide filamentous hyphae.

Tylopilus rufobrunneus formed a distinct lineage in our phylogenetic study (Fig. 4.1). It can be easily confused with species with reticulate stipes, viz. *T. alpinus*, *T. brunneirubens*, *T. felleus*, and *T. violaceobrunneus*. However, *T. alpinus*, *T. felleus*, and *T. violaceobrunneus* have distributions in temperate or subalpine forests. Moreover, *T. alpinus* has an olive-green to olive-brown pileus in younger basidioma,

Chapter 22 *Tylopilus* P. Karst. 295

Fig. 22.27 a *Tylopilus rufobrunneus* Yan C. Li & Zhu L. Yang. Photos by F. Li (KUN-HKAS 106331, type).

Fig. 22.27 b *Tylopilus rufobrunneus* Yan C. Li & Zhu L. Yang. Photos by N.K. Zeng (KUN-HKAS 106466).

a grayish to pale orange stipe, and large basidiospores 13–14.5 (15.5) × (3.5) 4–5 μm (Wu *et al*. 2016a; see our description above). *Tylopilus felleus* has a gray to grayish brown or grayish white pileus, a white to pallid context without discoloration when bruised and large basidiospores 14–16 (17) × 4.5–5.5 μm (Wu *et al*. 2016a; see our description above). *Tylopilus violaceobrunneus* has a brownish violet or reddish brown to light brown pileus always with magenta tinge, a white context without color change when injured, and the small hymenial cystidia 25–50 × 6–10 μm (Wu *et al*. 2016a; see our description below). *Tylopilus brunneirubens* is not only similar in macroscopic appearance, but also has a sympatric distribution with *T. rufobrunneus*. However, *T. brunneirubens* has an olive-brown to blackish olive pileus when young and then becomes brown to dark brown or chestnut-brown with age or injury, and relatively short basidiospores measuring 8.5–11 (12.5) × 3.5–4.5 μm (Wu *et al*. 2016a; see our description above).

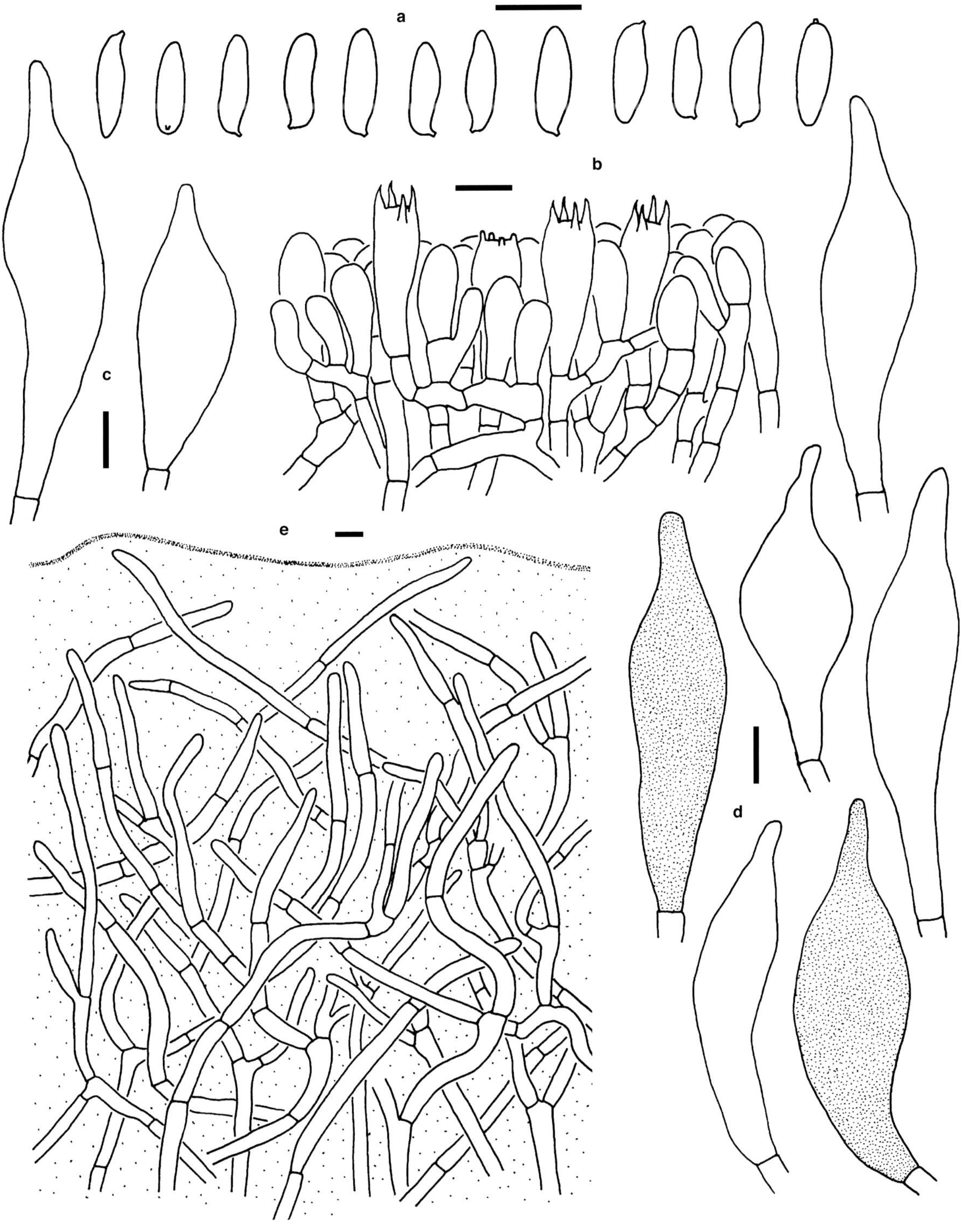

Fig. 22.27 c *Tylopilus rufobrunneus* Yan C. Li & Zhu L. Yang (KUN-HKAS 106331, type).
a. Basidiospores; b. Basidia; c. Cheilocystidia; d. Pleurocystidia; e. Pileipellis. Scale bars = 10 μm.

22.28 *Tylopilus subotsuensis* T.H.G. Pham, A.V. Alexandrova & O.V. Morozova, Persoonia 45: 397 (2020)

Basidioma small. Pileus 3–6 cm in diam., hemispherical to subhemispherical; surface grayish green (2C3–4) to gray-olivaceous (3C3–4) always with greenish tinge when young, dark brown (7F5–6), brown (7E4–5), grayish brown (7D2–3) or brownish (6D3–4) when mature, dry, with fine tomentose to fibrillose squamules, without discoloration when hurt; context solid, white, without discoloration when injured. Hymenophore adnate; surface whitish (10B1) to dingy white (9B1) when young, pinkish (8A2) when mature, without discoloration when bruised; pores subangular to roundish, 0.3–1 mm wide; tubes 3–7 mm long, concolorous or a little paler than hymenophoral surface, without discoloration when bruised. Spore print reddish (12B2) to pinkish (13B2). Stipe 4–8 × 1–2.5 cm, subcylindrical to clavate, sometimes enlarged downwards, pale grayish green (1C3–4) to grayish green (2C3–4) or olivaceous (2D4–5), without discoloration when touched; surface almost glabrous; context solid, white, without discoloration when bruised; basal mycelium white. Taste bitter. Odor indistinct.

Fig. 22.28 a *Tylopilus subotsuensis* T.H.G. Pham, A.V. Alexandrova & O.V. Morozova. Photos by N.K. Zeng (KUN-HKAS 107180)

Fig. 22.28 b *Tylopilus subotsuensis* T.H.G. Pham, A.V. Alexandrova & O.V. Morozova. Photos by N.K. Zeng (KUN-HKAS 106464).

Basidia 22–30 × 6–9 μm, clavate to narrowly clavate, 4-spored, rarely 2-spored, hyaline in KOH. Basidiospores [60/3/2] 6.5–9.5 (10.5) × (2.5) 3–3.5 μm [Q = (1.86) 2–3 (3.33), Q_m = 2.54 ± 0.17], subfusiform to subcylindrical in side view, cylindrical to fusiform in ventral view, smooth, yellowish to brownish yellow in KOH, yellow to yellow-brown in Melzer's reagent. Hymenophoral trama boletoid composed of 2.5–8 μm wide filamentous hyphae, hyaline to yellowish in KOH, yellowish to yellow in Melzer's reagent. Cheilo- and pleurocystidia 46–67 × 7–12 μm, narrowly lanceolate, subfusiform or fusoid-ventricose to narrowly fusoid-ventricose, or subfusiform with subacute apex, yellowish to pale brownish yellow in KOH, thin-walled. Pileipellis a hymeniderm composed of a layer of cystidioid cells (20–32 × 5–12 μm), yellowish to pale brownish in KOH and yellow to brownish in Melzer's reagent. Pileal trama composed of 5–15 μm wide interwoven hyphae, colorless to yellowish in KOH and yellowish to pale yellow in Melzer's reagent. Clamp connections absent in all tissues.

Habitat: Solitary on soil in tropical forests dominated by plants of the family Pinaceae.

Known distribution: Currently known from southern China and Vietnam.

Specimens examined: CHINA, HAINAN PROVINCE: Qiongzhong County, Limu Mountain, alt. 750 m, 6 May 2009, N.K. Zeng 119 (KUN-HKAS 106464, GenBank Acc. No.: MT154724 and MT154725 for nrLSU, MT110342 for *tef1-α,* MT110383 for *rpb1*, MT110418 for *rpb2*); Qiongzhong County, Limu Mountain, alt. 850 m, 3 August 2010, N.K. Zeng 818 (KUN-HKAS 107180 GenBank Acc. No.: MW114847 and MW114846 for nrLSU).

Commentary: *Tylopilus subotsuensis* originally described from Vietnam (Crous *et al.* 2020) is characterized by the grayish green to gray-olivaceous pileus and stipe, the white context without

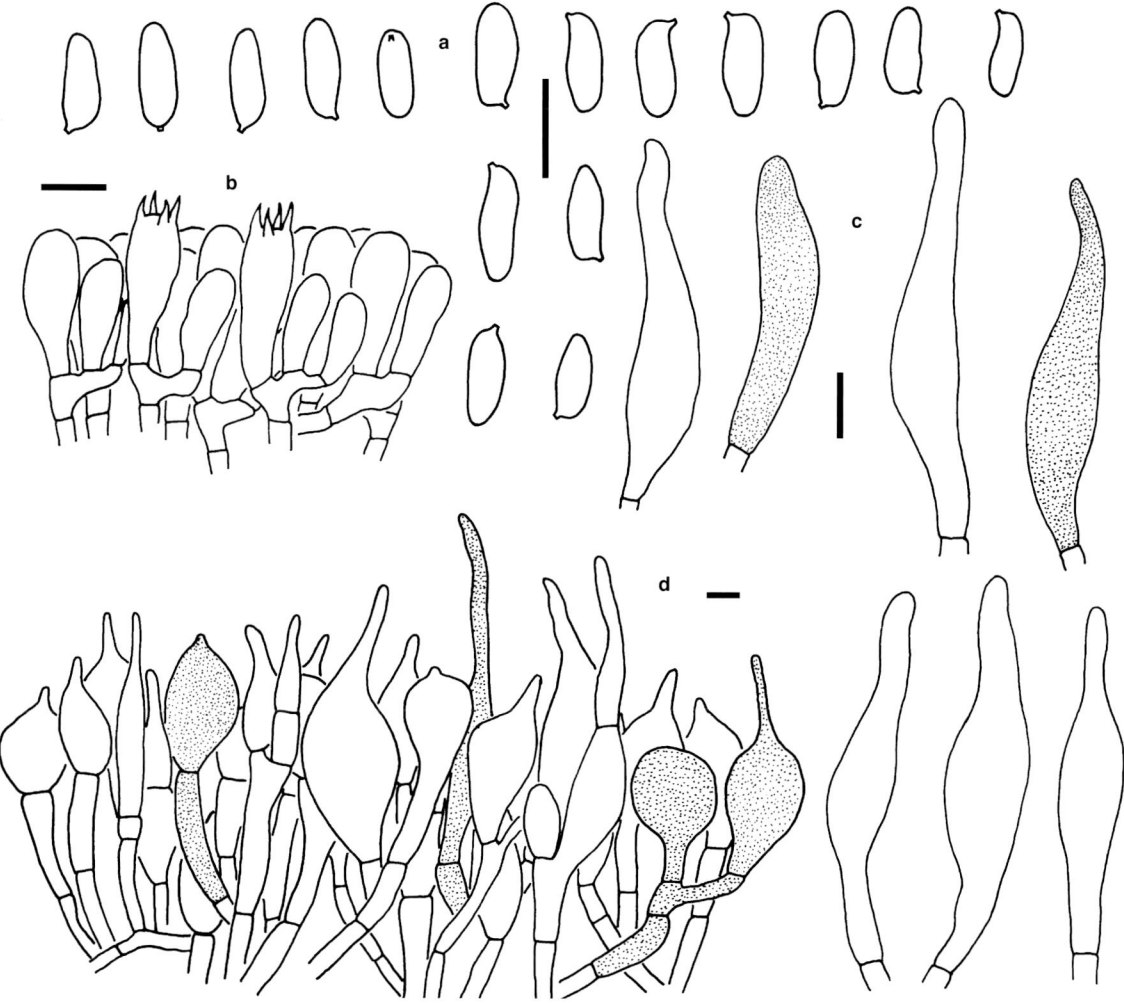

Fig. 22.28 c *Tylopilus subotsuensis* T.H.G. Pham, A.V. Alexandrova & O.V. Morozova (KUN-HKAS 106464). a. Basidiospores; b. Basidia; c. Cheilo- and pleurocystidia; d. Pileipellis. Scale bars = 10 μm.

discoloration when injured, the whitish to dingy white then pinkish hymenophore, the oblong to fusiform basidiospores, and the hymeniderm pileipellis.

Tylopilus subotsuensis is phylogenetically related and macroscopically similar to *T. otsuensis* Hongo. However, *T. otsuensis* differs from *T. subotsuensis* in its dark green to olivaceous pileus and stipe, white context staining reddish brown when bruised, broadly ellipsoid to ovoid basidiospores measuring 5.5–6.5 × 4–5 μm, and epithelium pileipellis made up of erectly arranged moniliform hyphae (Hongo 1966; Wu et al. 2016a).

22.29 *Tylopilus vinaceipallidus* (Corner) Watling & E. Turnbull, Mycologia 91: 663 (1999)

Basionym: *Boletus vinaceipallidus* Corner, *Boletus* in Malaysia 171 (1972).

Basidioma small to large. Pileus 3–12.5 cm in diam., hemispherical when young, subhemispherical to applanate or plano-convex when mature, dark reddish brown to castaneous, slightly darker in the center; surface viscid when wet, nearly glabrous to finely tomentose; context solid, white, without discoloration when bruised. Hymenophore adnate when young, adnate to depressed around apex of stipe; surface white to pinkish when young, pink to grayish pink when mature or aged, without discoloration when bruised; pores subangular to angular or roundish, 0.3–1 mm wide; tubes 6–15 mm long, pinkish to grayish pink, without discoloration when bruised. Spore print pinkish to reddish. Stipe 6.5–8 × 1–2.5 cm, subcylindrical to clavate, concolorous with pileal surface; surface covered with white farinose squamules, without discoloration when touched; context white, without discoloration when hurt; basal mycelium white to pallid. Taste bitter.

Basidia 25–50 × 7–13 μm, clavate to narrowly clavate, 4-spored, hyaline in KOH. Basidiospores (8) 10–12 (13.5) ×3.5–4.5 μm (Q = 2.51–3.17, Q_m = 2.78 ± 0.19), subcylindrical to subfusiform in side view with slight suprahilar depression, cylindrical to fusiform in ventral view, smooth, hyaline to yellowish in KOH, yellow to brownish yellow in Melzer's reagent. Hymenophoral trama boletoid composed of 3–15 μm wide filamentous hyphae. Cheilo- and pleurocystidia 30–46 × 8–10 μm, subfusiform to fusoid-ventricose, hyaline to yellowish in KOH, thin-walled. Pileipellis an intricate trichoderm composed of 4–6 μm wide interwoven filamentous hyphae with cylindrical terminal cells 17–58 × 3.5–6 μm, pale yellowish to brownish yellow in KOH and yellow to brownish in Melzer's reagent. Clamp connections absent in all tissues.

Habitat: Scattered to solitary on soil in tropical forests dominated by plants of the family Fagaceae.

Known distribution: Currently known from China and Malaysia.

Fig. 22.29 a *Tylopilus vinaceipallidus* (Corner) Watling & E. Turnbull. Photos by Y.J. Hao (KUN-HKAS 82858).

Specimens examined: CHINA, GUIZHOU PROVINCE: Suiyang County, Kuankuoshui National Nature Reserve, alt. 1000 m, 8 July 1993, X.L. Wu 3942 (KUN-HKAS 29233). YUNNAN PROVINCE: Jinghong County, Dadugang Town, alt. 1400 m, 7 July 2006, Y.C. Li 456 (KUN-HKAS 50210); Mengla County, Menglun Town, alt. 600 m, 9 July 2006, Y.C. Li 476 (KUN-HKAS 50230); Lancang County, Huimin Town, Baixiang Mountain, alt. 1250 m, 25 June 2017, J.W. Liu 613 (KUN-HKAS 107170). GUANGDONG PROVINCE: Fengkai County, Heishiding Nature Reserve, alt. 250 m, 23 May 2012, F. Li 353 (KUN-HKAS 90184). SICHUAN PROVINCE: Hejiang County, Hutou Town, alt. 260 m, 16 June 2014, Y.J. Hao 1067 (KUN-HKAS 82858).

Fig. 22.29 b *Tylopilus vinaceipallidus* (Corner) Watling & E. Turnbull. Photos by J.W. Liu (KUN-HKAS 107170).

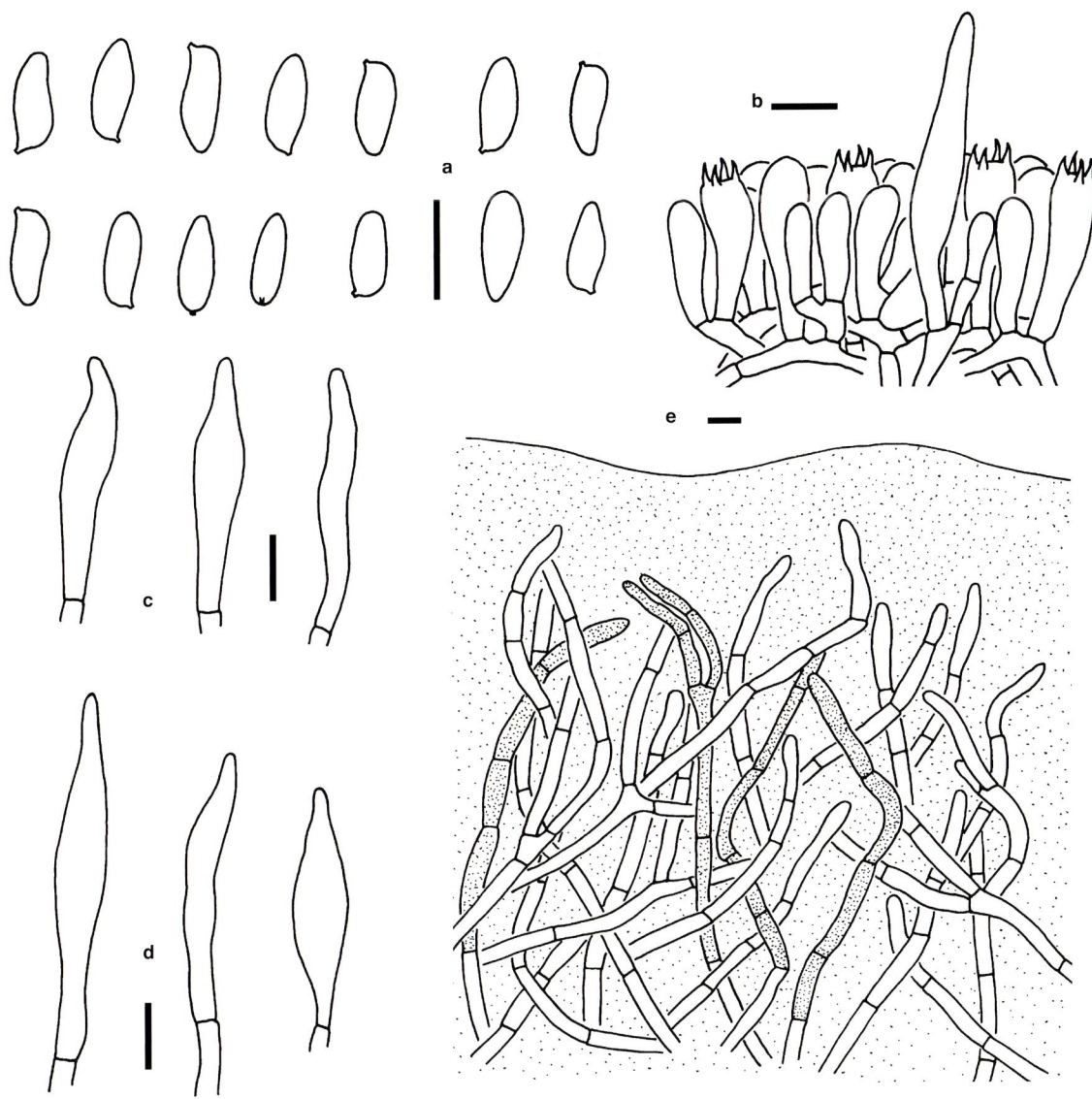

Fig. 22.29 c *Tylopilus vinaceipallidus* **(Corner) Watling & E. Turnbull (KUN-HKAS 107170).**
a. Basidiospores; b. Basidia and pleurocystidium; c. Cheilocystidia; d. Pleurocystidia; e. Pileipellis. Scale bars = 10 μm.

Commentary: *Tylopilus vinaceipallidus* originally described from Singapore by Corner (1972) is characterized by the dark reddish brown to castaneous basidioma, the white to pinkish or pink hymenophore, the white context without discoloration when bruised, the intricate trichoderm pileipellis, and the distribution in tropical forests. This species is phylogenetically related and morphologically similar to *T. phaeoruber* (Fig. 4.1). However, *T. phaeoruber* differs from *T. vinaceipallidus* in its white then pinkish hymenophore staining brownish when injured, relatively large basidiospores measuring 12.5–14.5 (15.5) × 4.5–5.5 μm, relatively large hymenial cystidia measuring 48–72 × 9–14 μm, palisadoderm pileipellis composed of 4–10 μm wide vertically arranged hyphae.

22.30 *Tylopilus violaceobrunneus* Yan C. Li & Zhu L. Yang, Fungal Divers 81: 163 (2016)

Basidioma small to medium-sized. Pileus 4–8 cm in diam., hemispherical when young, subhemispherical to plano-convex or applanate, light brown to reddish brown with magenta to brownish violet tinge; surface dry, finely tomentose without discoloration when touched; context white, without discoloration when bruised. Hymenophore adnate when young, adnate to slightly depressed around apex of stipe when mature; surface whitish to white when young, pinkish to dingy pinkish when mature, without color change when bruised; pores round to angular, 0.3–0.5 mm wide; tubes 10–15 mm long, concolorous or a little paler than hymenophoral surface, without discoloration when bruised. Spore print pinkish to reddish. Stipe 6–8 × 1.5–2.5 cm, subcylindrical to clavate, concolorous with pileal surface, but grayish magenta to brownish violet downwards; surface reticulate over the upper part, without discoloration when touched; context solid, white, without discoloration when injured; basal mycelium pallid to white. Taste bitter, odor mild.

Basidia 26–39 × 8–11 μm, clavate to narrowly clavate, 4-spored, hyaline in KOH. Basidiospores 10–12 × 3–3.5 (4) μm (Q = 2.86–3.83, Q_m = 3.32 ± 0.3), subcylindrical to subfusiform in side view with slight suprahilar depression, cylindrical to fusiform in ventral view, smooth, hyaline to yellowish in KOH, yellow to brownish yellow in Melzer's reagent. Hymenophoral trama boletoid composed of 4.5–11 μm wide filamentous hyphae. Cheilo- and pleurocystidia 25–50 × 6–10 μm, subfusiform to fusoid-ventricose with a long pedicel, yellow to brownish yellow in KOH, thin-walled. Pileipellis a palisadoderm composed of 3.5–9 μm wide nearly vertically arranged hyphae with clavate to cystidioid terminal cells 29–64 × 4–7 μm, pale yellowish to brownish yellow in KOH and yellow to brownish in Melzer's reagent. Pileal trama composed of 5–10 μm wide interwoven hyphae, colorless to yellowish in KOH and yellow to yellowish brown in Melzer's reagent. Clamp connections absent in all tissues.

Habitat: Scattered to solitary on soil in temperate mixed forests of *Pinus tabuliformis* and *Quercus* sp.

Known distribution: Currently known from eastern and southwestern China.

Fig. 22.30 a *Tylopilus violaceobrunneus* Yan C. Li & Zhu L. Yang. Photos by Y.C. Li (KUN-HKAS 89443, type).

Fig. 22.30 b *Tylopilus violaceobrunneus* Y.an C. Li & Zhu L. Yang. Photos by P. Qiao (KUNHKAS 62677)

Specimens examined: CHINA, SHANDONG PROVINCE: Tai'an, Taishan Mountain, alt. 1350 m, 4 August 2012, Y.C. Li 2800 (KUN-HKAS 89443, type). YUNNAN PROVINCE: Deqin County, Xiaruo Town, alt. 2400 m, 18 September 2010, P. Qiao HBB2010-D-49 (KUN-HKAS 62677).

Commentary: *Tylopilus violaceobrunneus* originally described from China by Wu *et al.* (2016a) is characterized by the light brown to reddish brown pileus always with magenta to brownish violet tinge, the white pileal context without discoloration when bruised, the initially white to pallid and then pinkish hymenophore, the distinct reticulum over the upper part of stipe, the palisadoderm pileipellis composed of 3.5–9 μm wide nearly vertically arranged hyphae, and the distribution in temperate forests in northern China. This species is macroscopically similar to *Boletus violaceofuscus* W. F. Chiu. However, *B. violaceofuscus* differs from *T. violaceobrunneus* in its white stuffed hymenophoral pores when young, yellow to dingy yellow hymenophoral surface when mature, relatively big basidiospores measuring 12–14 × 5–6 μm (Cui *et al.* 2016).

In our multi-locus phylogenetic analysis (Fig. 4.1), *T. violaceobrunneus* clusters together with *T. felleus*. However, *T. felleus* differs from *T. violaceobrunneus* in its grayish to gray or grayish brown pileus, relatively large basidiospores measuring 14–16 (17) × 4.5–5.5 μm, large hymenial cystidia measuring 50–70 × 10–18 μm, and the intricate trichoderm pileipellis composed of 4.5–7 μm wide interwoven filamentous hyphae.

22.31 *Tylopilus violaceorubrus* Yan C. Li & Zhu L. Yang, The Boletes of China: *Tylopilus* s.l. 356 (2021)

MycoBank: MB 837924

Etymology: *violaceorubrus*, from *violaceo* = violaceous, and *rubrus* = red, referring to its violaceous basidioma staining red when injured.

Type: CHINA, YUNNAN PROVINCE: Malipo County, Yangwan Town, Banpo Village, alt. 1100 m, 21 June 2017, G. Wu 1956 (KUN-HKAS 107154, GenBank Acc. No.: MT154729 for nrLSU, MW165286 for *rpb2*).

Basidioma small to medium-sized. Pileus 4–7 cm in diam., subhemispherical to hemispherical when young and subhemispherical to applanate with age; surface dry, finely tomentose, entirely dark ruby (12F7–8) to violet-brown (11F5–6) or gray-red (11D3–4) to gray-ruby (12D3–5). Context white to pallid, without color change or slowly staining rufescent or reddish (7B4) when injured. Hymenophore adnate; surface initially white and then dull pinkish (10B1) when mature, slowly becoming reddish brown when bruised; pores subangular to roundish, 0.3–1 mm wide; tubes 3–5 mm long, concolorous or a little paler

Fig. 22.31 a *Tylopilus violaceorubrus* Yan C. Li & Zhu L. Yang. Photo by J.W. Liu (KUN-HKAS 107169).

Fig. 22.31 b *Tylopilus violaceorubrus* Yan C. Li & Zhu L. Yang. Photo by G. Wu (KUN-HKAS 107154, type).

than hymenophoral surface, slowly staining reddish brown or rufescent (5C6–7) when bruised. Spore print grayish red (11B2) to reddish (12B2). Stipe 3–7 × 1.5–2 cm, clavate to subcylindrical, enlarged downwards, concolorous with pileal surface, but much paler towards base, staining a reddish tinge when injured; surface subglabrous or fibrillose to finely tomentose; context white to pallid, slowly staining a reddish tinge when bruised; basal mycelium white to pallid. Taste bitter, odor mild.

Basidia 24–32 × 7–10 μm, clavate, 4-spored, hyaline in KOH. Basidiospores 8–10.5 × 3.5–4.5 μm [Q = (1.78) 1.89–2.5, Q_m = 2.25 ± 0.17], elongated to subcylindrical and inequilateral in side view with slight suprahilar depression, elongated to cylindrical in ventral view, smooth, hyaline to yellowish in KOH, yellow to brownish yellow in Melzer's reagent. Hymenophoral trama boletoid composed of 3–8 μm wide filamentous hyphae. Cheilocystidia 34–62 × 6–10 μm, fusiform to subfusoid-ventricose, yellowish to yellow-brown in KOH, brown to dark yellow-brown or olivaceous brown in Melzer's reagent, thin-walled. Pleurocystidia 46–85 × 8–15 μm, relatively bigger than cheilocystidia, fusiform to subfusiform,

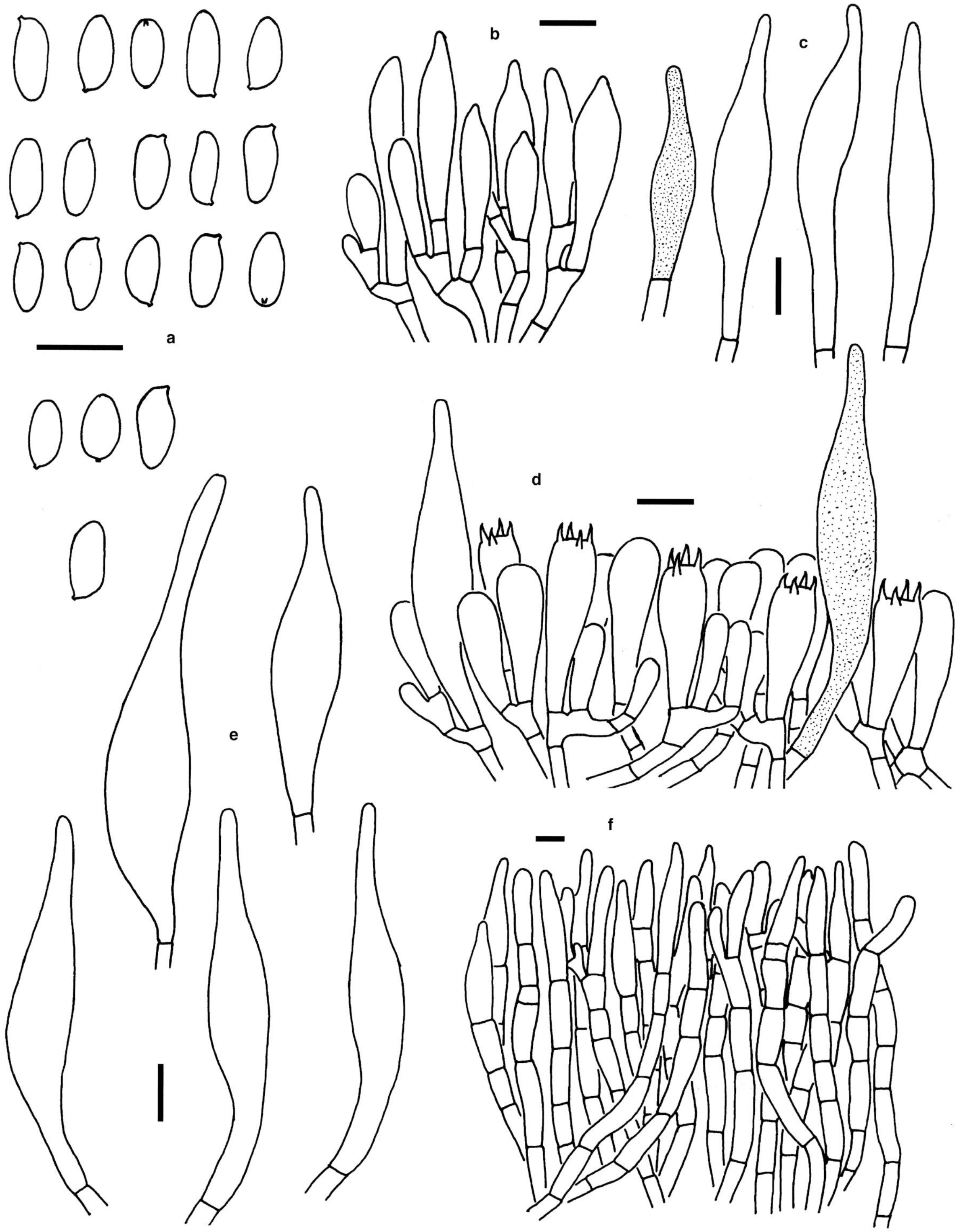

Fig. 22.31 c *Tylopilus violaceorubrus* Yan C. Li & Zhu L. Yang (KUN-HKAS 107169).
a. Basidiospores; b. Basidia and cheilocystidia; c. Cheilocystidia; d. Basidia and pleurocystidia; e. Pleurocystidia; f. Pileipellis. Scale bars = 10 μm.

concolorous with cheilocystidia, thin-walled. Pileipellis a palisadoderm composed of 6–10 μm wide vertically arranged hyphae, yellowish brown to brownish in KOH and yellow-brown to dark brown in Melzer's reagent; terminal cells 26–48 × 7–10 μm, pyriform to clavate or cystidioid. Pileal trama composed of 5–8 μm wide interwoven hyphae, pale yellowish to brownish yellow in KOH and yellow to brownish in Melzer's reagent. Clamp connections absent in all tissues.

Habitat: Gregarious to scattered on soil in tropical forests dominated by plants of the family Fagaceae.

Known distribution: Currently known from southwestern China.

Additional specimen examined: CHINA, YUNNAN PROVINCE: Lancang County, Huimin Town, Baixiang Mountain, alt. 1000 m, 26 June 2017, J.W. Liu 615 (KUN-HKAS 107169).

Commentary: *Tylopilus violaceorubrus* is characterized by the dark ruby to violet-brown or gray-red to gray-ruby pileus and stipe, the white to cream context without color change or slowly staining indistinct reddish when injured, the initially white to pallid and then pinkish hymenophore staining a reddish tinge when injured, the palisadoderm pileipellis and the bitter taste. This species is phylogenetically related and morphologically similar to *T. atroviolaceobrunneus* (Fig. 4.1). However, these two species differ from each other in the color of the basidiomata and the size of the basidiospores (see our commentary on *T. atroviolaceobrunneus*).

22.32 *Tylopilus virescens* (Har. Takah. & Taneyama) N.K. Zeng, H. Chai & Zhi Q. Liang, MycoKeys 46: 82 (2019)

Basionym: *Boletus virescens* Har. Takah. & Taneyama, Fungal Flora Southwestern Japan: Agarics and boletes 45 (2016), Figs. 36–40.

Synonym: *Tylopilus callainus* N.K. Zeng, Zhi Q. Liang & M.S. Su, Phytotaxa 343 (3): 271 (2018), Figs. 2, 3.

Basidioma small to medium-sized. Pileus 3–6 cm in diam., subhemispherical to hemispherical, then expanding to convex or applanate; surface dry, rugulose, densely tomentose, yellowish brown to brown when young, but softly fading into yellowish to whitish towards margin when mature, tinged with greenish blue when bruised; context whitish to cream, staining greenish blue slowly when injured. Hymenophore adnate when young, adnate to slightly decurrent when mature; surface pallid to whitish or cream to yellowish when young, pinkish to yellowish pink or purplish pink when mature, staining greenish blue when injured, pores angular, 0.5–3 mm wide; tubes 3–6 mm long, concolorous with hymenophoral surface, staining greenish blue when injured. Spore print pinkish to reddish. Stipe 2.5–6 × 0.5–0.8 cm, clavate to

Fig. 22.32 a *Tylopilus virescens* (Har. Takah. & Taneyama) N.K. Zeng, H. Chai & Zhi Q. Liang. Photo by N.K. Zeng (FHMU 3147).

Fig. 22.32 b *Tylopilus virescens* (Har. Takah. & Taneyama) N.K. Zeng, H. Chai & Zhi Q. Liang. Photos by L.K. Jia (KUN-HKAS 107310).

subcylindrical, solid, usually flexuous; surface dry, cream to yellowish upwards, pale brown, brown to dark brown downwards, always with a greenish blue ring at apex of stipe; context cream to whitish upwards, staining greenish blue slowly when injured, pale brown to reddish brown downwards, without discoloration when injured; basal mycelium white. Taste bitter, odor indistinct.

Basidia 25–38 × 8–12 μm, clavate, 4-spored, hyaline to yellowish. Basidiospores 7–11 × 4–5 μm [Q = (1.55) 1.6–2 (2.3), Q_m = 1.85 ± 0.14], subellipsoid to elongated or subcylindrical and inequilateral in side view with slight suprahilar depression, ellipsoid to elongated or cylindrical in ventral view, smooth, somewhat thick-walled (up to 0.5 μm), hyaline to yellowish in KOH, yellow to brownish yellow in Melzer's reagent. Hymenophoral trama boletoid composed of 3–8 μm wide filamentous hyphae, hyaline to yellowish in KOH. Cheilocystidia 23–35 × 5–9 μm, rare, ventricose, fusiform or subfusiform, yellowish to brownish yellow. Pleurocystidia 25–45 × 6–10 μm, ventricose, fusiform or subfusiform, yellowish to brownish yellow. Pileipellis a palisadoderm composed of 6–14 μm wide vertically arranged hyphae with clavate terminal cells 25–56 × 10–13 μm, yellowish to pale brownish in KOH and yellowish brown to brownish in Melzer's reagent. Pileal trama made up of 6–10 μm wide filamentous hyphae, colorless to yellowish in KOH and yellow to yellowish brown in Melzer's reagent. Clamp connections absent in all tissues.

Habitat: Solitary or gregarious on the ground in tropical forests dominated by plants of the family Fagaceae.

Known distribution: Currently known from southern and southwestern China and Japan.

Specimens examined: CHINA, HAINAN PROVINCE: Ledong County, Jianfengling, alt. 850 m, 5 July 2021, M.X. Li 181 (KUN-HKAS 121979); Changjiang County, Bawangling National Nature Reserve, alt. 650 m, 23 May 2019, N.K. Zeng 4052 (FHMU 3147). YUNNAN PROVINCE: Mengla County, Xishuangbanna National Natural Reserve, alt. 1000 m, 27 August 2019, L.K. Jia 476 (KUN-HKAS 107310).

Commentary: *Tylopilus virescens* is characterized by the rugulose yellowish brown to brown pileus which is rimose and cracked into small squamules on a yellowish to whitish background, the whitish to cream or yellowish and then pinkish to yellowish pink or purplish pink hymenophore, the greenish blue ring at apex of stipe, the greenish blue discoloration when bruised and the palisadoderm pileipellis. Such traits are very similar to those in *P. orientifumosipes*. However, *P. orientifumosipes* Yan C. Li & Zhu L. Yang has a dark brown to brown or reddish brown pileus and never whitish to yellowish towards margin, an epithelium pileipellis composed of inflated (16–21 μm wide) concatenated cells and large abundant hymenial cystidia 55–70 × 14–19 μm.

Tylopilus virescens was originally described as *Boletus virescens* Har. Takah. & Taneyama from Japan (Terashima *et al*. 2016). This species was then transferred to *Tylopilus* due to its color of hymenophore and basidiospores by Liang *et al*. (2018). Our multi-locus phylogenetic analysis (Fig. 4.1) suggests that the species nests into the *Tylopilus* clade and is related to the species of *T. balloui* complex. However, species of *T. balloui* complex differ from *T. virescens* in the discoloration when injured, the color of the pileus, the morphology of basidiospores and the structure of the pileipellis.

Chapter 23

Veloporphyrellus L.D. Gómez & Singer

Veloporphyrellus L.D. Gómez & Singer, Brenesia 22: 293 (1984)

Type species: *Veloporphyrellus pantoleucus* L.D. Gómez & Singer, Brenesia 22: 293. 1984.

Diagnosis: This genus differs from the other genera in Boletaceae in its extended pileal margin and often embracing the apex of the stipe in younger basidiomata, white to pallid context without color change when hurt, initially white to pinkish then pink to purplish pink hymenophore, trichoderm pileipellis, smooth or mixed of smooth and warty basidiospores under SEM in some species.

Pileus hemispherical when young then subhemispherical to convex or plano-convex when mature; surface dry, subtomentose, tomentose or matted tomentose; pileal margin extended and embracing the stipe in younger basidiomata then breaking into pieces and hanging on the pileal margin when mature; context white, without discoloration when injured. Hymenophore adnate when young, adnate to slightly depressed around apex of stipe when mature; surface white to pinkish or pink when young, becoming pink to purplish pink when mature, without discoloration when bruised; pores subangular to roundish; tubes concolorous or much paler than hymenophoral surface. Stipe subglabrous, fibrillose or farinose; basal mycelium whitish. Basidiospores subfusiform, fusiform or cylindrical; surface smooth or in some species both with smooth and warty basidiospores. Hymenial cystidia abundant, subfusiform, ventricose, or fusoid-ventricose. Pileipellis a trichoderm composed of filamentous interwoven hyphae. Clamp connections absent in all tissues.

Commentary: *Veloporphyrellus* shares extended pileal margin and the same colored hymenophore with *Austroboletus*. However, *Austroboletus* differs from *Veloporphyrellus* in the lightly to heavily ornamented basidiospores under light microscopy and never with smooth basidiospores and the distinctly reticulate stipe (Gómez and Singer 1984; Wolfe 1979a; Singer 1986; Fulgenzi *et al.* 2010). Currently, five species have been recognized from China, including one new species.

Key to the species of the genus *Veloporphyrellus* in China

1. Pileus brown, chestnut-brown or reddish brown; margin with distinct membranous veil remnants; surface covered with concolorous floccose squamules; species with smooth basidiospores under SEM ·········· 2
1. Pileus brownish red to blackish red; margin extended but without membranous veil remnants; surface nearly glabrous or tomentose; species with both smooth and warty basidiospores under SEM ·········· 4
2. Species with a distribution in temperate to subtropical forests with altitudes ranging from 1500 m to 2000 m or subalpine to alpine forests with altitudes ranging from 3100 m to 3600 m; stipe surface orange-yellow to orange-brown or reddish brown ·········· 3
2. Species with a distribution in tropical forests; stipe surface white ·········· *V. velatus*
3. Species with a distribution in subalpine to alpine areas with high altitudes ranging from 3100 m to 3600 m; basidiospores 15.5–19.5 × 4.5–6.5 μm ·········· *V. alpinus*
3. Species with a distribution in temperate to subtropical areas with low altitude ranging from 1500 m to 2000 m; basidiospores small 12–16 × 4–5.5 μm ·········· *V. pseudovelatus*
4. Species with a distribution in temperate to subalpine forests; pileus orange-brown to brownish red; basidiospores 12–16.5 × 5.5–6.5 μm ·········· *V. gracilioides*
4. Species with a distribution in subtropical forests; pileus red to blackish red; basidiospores small 10–12 × 4.5–5 μm ·········· *V. castaneus*

23.1 *Veloporphyrellus alpinus* Yan C. Li & Zhu L. Yang, Mycologia 106: 293 (2014)

Basidioma very small to medium-sized. Pileus 1.8–7 cm in diam., conical to subconical when young, subconical to plano-convex when mature, always with a sharp umbo; surface dry, brown, cocoa-brown, chestnut-brown or dark reddish brown, densely covered with concolorous matted squamules; margin extended and embracing the apex of stipe in younger basidiomata, then breaking into pieces and hanging on the pileal margin in aged ones; context solid when young, spongy when mature, white to pallid, without discoloration when bruised. Hymenophore adnate when young, depressed around apex of stipe when mature; surface pallid to pale pinkish when young, pink to purplish pink when mature, without discoloration when bruised; pores subangular to roundish, up to 1 mm wide; tubes up to 6 mm long, concolorous or a little paler than hymenophoral surface, without discoloration when bruised. Spore print dull red to grayish red. Stipe 5.5–6.5 × 0.4–0.7 cm, clavate to subcylindrical, always enlarged downwards, nearly glabrous or finely fibrillose, yellowish orange to grayish orange upwards and brown to reddish brown downwards, context white, without discoloration when bruised; basal mycelium white. Taste and odor mild.

Basidia 34–39 × 8.5–12 μm, clavate, hyaline to light yellowish in KOH, thin-walled, 4-spored. Basidiospores (15.5) 16–19 (19.5) × (4.5) 5–6 (6.5) μm [Q = (2.77) 2.82–3.56 (3.6), Q_m = 3.13 ± 0.18], subcylindrical to subfusiform in side view with slight suprahilar depression, cylindrical to fusiform in

Fig. 23.1 a *Veloporphyrellus alpinus* Yan C. Li & Zhu L. Yang. Photos by B. Feng (KUN-HKAS 57490, type).

Fig. 23.1 b *Veloporphyrellus alpinus* Yan C. Li & Zhu L. Yang. Photo by X.T. Zhu (KUN-HKAS 68301).

ventral view, smooth under SEM, somewhat thick-walled (up to 0.5 μm thick), subhyaline to light olivaceous in KOH and yellow to yellowish brown in Melzer's reagent. Hymenophoral trama bilateral composed of hyphae up to 10 μm wide, hyaline to yellowish in KOH, yellowish to yellow in Melzer's reagent. Cheilocystidia 33–81 × 6–9 μm, subcylindrical to finger-like, always with 1–2 septa, thin-walled, hyaline to yellowish in KOH, yellow to yellowish brown in Melzer's reagent. Pleurocystidia 47–69 × 5.5–9 μm, fusiform to subfusiform or subfusoid-ventricose, thin-walled, hyaline to yellowish in KOH, yellow to yellowish brown in Melzer's reagent. Pileipellis a trichoderm composed of 4.5–7 μm wide more or less vertically arranged interwoven hyphae, colorless to yellowish in KOH and yellow to brownish yellow in Melzer's reagent; terminal cells 27–69 × 4.5–6.5 μm, subcylindrical to clavate. Pileal trama made up of 5–10 μm wide filamentous hyphae, colorless to yellowish in KOH and yellowish to pale yellow in Melzer's reagent. Clamp connections absent in all tissues.

Habitat: Solitary on soil in subalpine to alpine mixed forests dominated by plants of the families Fagaceae and Pinaceae.

Known distribution: Currently known from southeastern and southwestern China.

Specimens examined: CHINA, YUNNAN PROVINCE: Dali, Cangshan National Forest Park, alt. 3600 m, 12 August 2010, X.T. Zhu 125 (KUN-HKAS 68301); Yulong County, Shitou Town, alt. 3100 m, 2 September 2009, B. Feng 761 (KUN-HKAS 57490, type). TAIWAN PROVINCE: Hehuan Mountain, alt. 3200 m, 15 September 2012, B. Feng 1266 (KUN-HKAS 63669).

Commentary: *Veloporphyrellus alpinus*, originally described by Li *et al.* (2014a) from China, is characterized by the conical to subconical or plano-convex pileus, the relatively large basidiospores measuring (15.5) 16–19 (19.5) × (4.5) 5–6 (6.5) μm, and the distribution in subalpine to alpine forests with high altitude ranging from 3100 m to 3600 m. This species is phylogenetically related and macroscopically similar to *V. pseudovelatus* Yan C. Li & Zhu L. Yang. However, *V. pseudovelatus* differs from *V. alpinus* in its hemispherical to subhemispherical then convex to plano-convex pileus, small basidiospores measuring (12) 12.5–15 (16) × 4–5 (5.5) μm, and distribution in temperate to subtropical forests with low altitude ranging from 1500 m to 2000 m (see our description of *V. pseudovelatus* below).

23.2 *Veloporphyrellus castaneus* Yan C. Li & Zhu L. Yang, The Boletes of China: *Tylopilus* s.l. 367 (2021)

MycoBank: MB 834758

Etymology: The epithet "*castaneus*" refers to the color of the basidiomata.

Type: CHINA, GUIZHOU PROVINCE: Suiyang County, Kuankuoshui National Nature Reserve, alt. 1500 m, 24 July 2010, X.F. Shi 368 (KUN-HKAS 107147, GenBank Acc. No.: MT154764 for nrLSU, MT110366 for *tef1-α*, MT110441 for *rpb2*).

Basidioma small. Pileus 3–5 cm in diam., hemispherical to subhemispherical, grayish brown (10E2–3), reddish brown (8F6–8) to chestnut-brown (9F5–6) when young, then dark brownish red (10E7–8) to dark brown (9F5–6) when mature; surface dry, nearly glabrous, margin extended; context white, without discoloration when bruised. Hymenophore adnate when young, adnate to depressed around apex of stipe when mature; surface white (11A1) to pinkish (11A2) when young, pink (12A2) to purplish pink (12A3)

Fig. 23.2 a *Veloporphyrellus castaneus* Yan C. Li & Zhu L. Yang. Photos by Z.H. Chen (MHHNU 30640).

318　The Boletes of China: *Tylopilus* s.l.

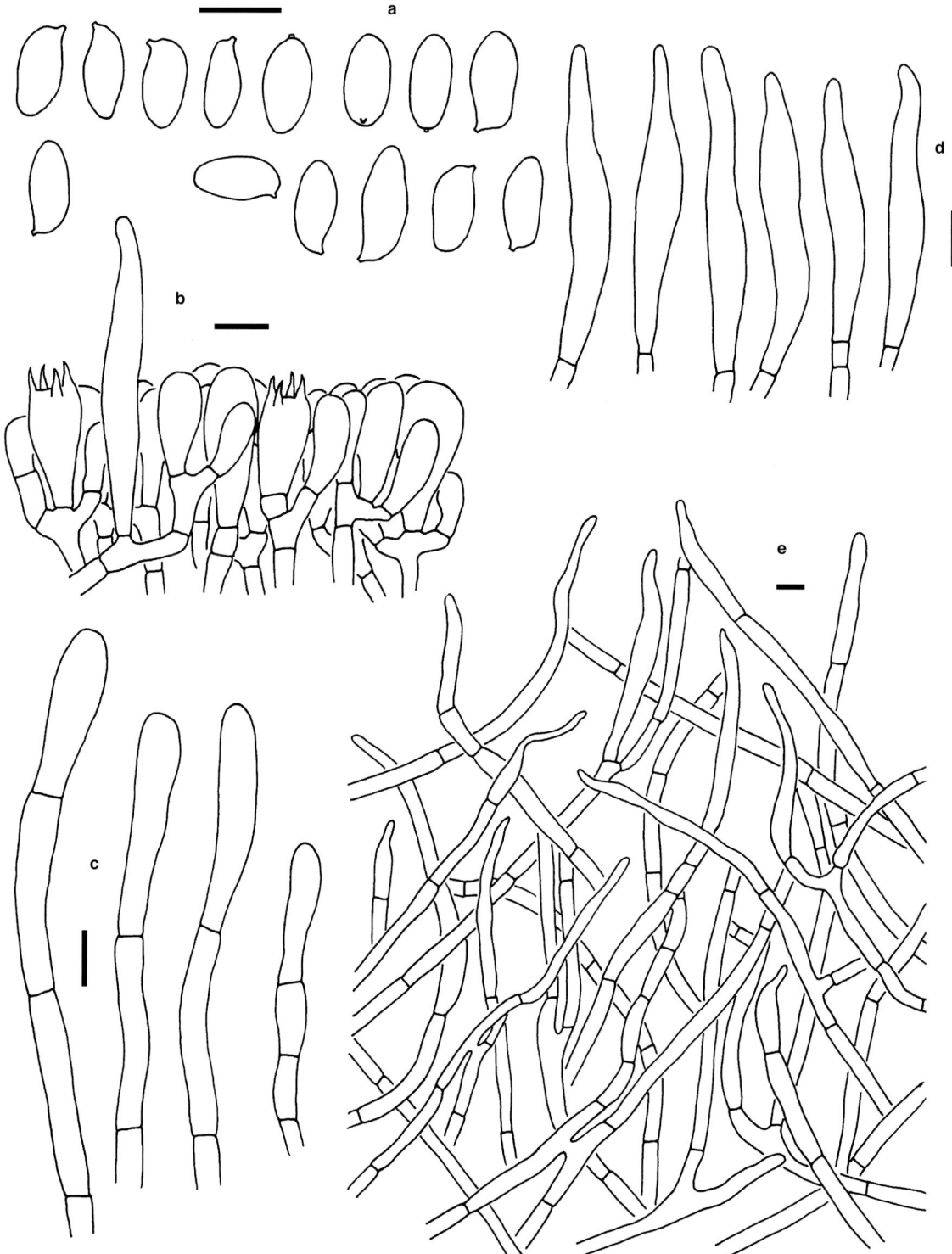

Fig. 23.2 b *Veloporphyrellus castaneus* Yan C. Li & Zhu L. Yang (KUN-HKAS 107147, type).
a. Basidiospores; b. Basidia and pleurocystidium; c. Cheilocystidia; d. Pleurocystidia; e. Pileipellis. Scale bars = 10 μm.

when mature, without discoloration when injured; pores angular to roundish, 0.3–1 mm wide; tubes 5–10 mm long, concolorous or a little paler than hymenophoral surface, without discoloration when injured. Spore print dull red (9C4) to grayish red (9C5). Stipe 4–8 × 1.2–1.8 cm, subcylindrical to clavate, often enlarged downwards, concolorous with pileal surface, occasionally covered with white powder-elements, without discoloration when touched; context white, without discoloration when injured; basal mycelium white. Taste and odor mild.

Basidia 18–28 × 9–11 μm, clavate to broadly clavate, 4-spored, hyaline in KOH. Basidiospores (10) 10.5–12 × 4.5–5 μm (Q = 2.1–2.44, Q_m = 2.3 ± 0.1), subcylindrical to subfusiform in side view with slight suprahilar depression, cylindrical to fusiform in ventral view, with smooth and warty surface under SEM (Fig. 3.3 i), somewhat thick-walled (up to 1 μm thick), hyaline to yellowish in KOH and yellow to yellowish brown in Melzer's reagent. Hymenophoral trama boletoid composed of 4–10 μm wide filamentous hyphae, hyaline to yellowish in KOH, yellowish to yellow in Melzer's reagent. Cheilocystidia 45–120 × 4–7 μm, finger-like, usually with 1–2 septa or secondary septa, hyaline to yellowish in KOH and yellow to brownish in Melzer's reagent; terminal cells 25–40 × 5–7 μm, thin-walled. Pleurocystidia 48–60 × 5.5–7 μm, lanceolate to subfusiform or subfusoid-ventricose, yellowish to light brownish in KOH and yellow to brownish in Melzer's reagent, thin-walled. Pileipellis a trichoderm composed of 4–8 μm wide interwoven hyphae, hyaline to yellowish in KOH and yellowish to brownish in Melzer's reagent; surface often covered with yellowish to yellow encrustation in KOH; terminal cells 26–105 × 5–10 μm, subcylindrical. Pileal trama composed of 4–7 μm wide interwoven hyphae, colorless to yellowish in KOH and yellowish to yellow in Melzer's reagent. Clamp connections absent in all tissues.

Habitat: Scattered on soil in subtropical forests dominated by plants of the family Fagaceae.

Known distribution: Currently known from central and southwestern China.

Additional specimen examined: CHINA, HUNAN PROVINCE: Dong'an County, Shunhuangshan National Nature Reserve, alt. 900 m, 14 September 2010, Z.H. Chen 30640 (MHHNU 30640).

Commentary: *Veloporphyrellus castaneus* is characterized by the reddish brown to chestnut-brown then dark brownish red to dark brown pileus, the somewhat extended pileal margin, the white context without discoloration when bruised, the initially white to pinkish then pink to purplish pink hymenophore, the trichoderm pileipellis composed of 4–8 μm wide interwoven filamentous hyphae, and the distribution in subtropical forests.

Veloporphyrellus castaneus is phylogenetically related and morphologically similar to *V. gracilioides* Yan C. Li & Zhu L. Yang (Fig. 4.1). However, *V. gracilioides* differs from *V. castaneus* in its large basidiospores measuring (12) 14–15 (16.5) × (5.5) 6–6.5 μm, and distribution in temperate to subalpine forests.

23.3 *Veloporphyrellus gracilioides* Yan C. Li & Zhu L. Yang, Fungal Divers 81: 165 (2016)

Basidioma small to medium-sized. Pileus 3–5.5 cm in diam., hemispherical to subhemispherical then applanate or plano-convex; surface dry, reddish brown to castaneous when young, brown to orange-brown when mature, sometimes with orange-yellow tinge towards margin, usually covered with concolorous fibrillose squamules, margin somewhat extended; context white, without discoloration when bruised. Hymenophore adnate when young, depressed around apex of stipe when mature; surface pinkish to pink when young, becoming purplish to ocher-purple when mature, without discoloration when injured; pores angular to roundish, 0.3–1 mm wide; tubes 6–10 mm long, concolorous or a little paler than hymenophoral surface, without discoloration when injured. Spore print dull red to grayish red. Stipe 4–8 × 0.8–1 cm, clavate, often enlarged downwards, concolorous with pileal surface, covered with white farinose squamules, without discoloration when touched; context white, without discoloration when injured; basal mycelium white. Taste and odor mild.

Basidia 25–40 × 9–12 μm, clavate to narrowly clavate, 4-spored, hyaline in KOH. Basidiospores (12) 14–15 (16.5) × (5.5) 6–6.5 μm (Q = 2.23–2.33, Q_m = 2.28 ± 0.05), subfusiform and inequilateral in side view with distinct suprahilar depression, cylindrical to fusiform in ventral view, with smooth and warty surface under SEM, somewhat thick-walled (up to 1 μm thick), hyaline to yellowish in KOH, yellow to brownish yellow in Melzer's reagent. Hymenophoral trama boletoid composed of 3–15 μm wide filamentous hyphae, hyaline to yellowish in KOH, yellowish to yellow in Melzer's reagent. Cheilocystidia 75–121 × 4–9.5 μm, clavate to subcylindrical, always with 1–2 septa, with clavate to subcylindrical terminal cells 29–64 × 4.5–9.5 μm, colorless to yellowish in KOH, yellowish to yellow in Melzer's reagent, thin-walled. Pleurocystidia 34–51 × 6.5–10 μm, fusiform, subfusiform, subfusoid-mucronate or subfusoid-

Fig. 23.3 a *Veloporphyrellus gracilioides* Yan C. Li & Zhu L. Yang. Photos by Q. Cai (KUN-HKAS 92025).

Fig. 23.3 b *Veloporphyrellus gracilioides* Yan C. Li & Zhu L. Yang. Photos by J. Li (KUN-HKAS 91262).

ventricose, yellowish to pale brownish yellow in KOH, yellow to brownish yellow in Melzer's reagent, thin-walled. Pileipellis a trichoderm composed of 4–6 μm wide interwoven filamentous hyphae, hyaline to yellowish or pale brownish yellow in KOH and yellow to yellow-brown in Melzer's reagent, covered with yellowish to yellow incrustation in KOH; terminal cells 50–80 × 5–7 μm, subcylindrical. Pileal trama composed of 7–11 μm wide filamentous hyphae, colorless to yellowish in KOH and yellowish to yellow in Melzer's reagent. Clamp connections absent in all tissues.

Habitat: Scattered on soil in temperate to subalpine forests dominated by plants of the families Fagaceae and Pinaceae.

Specimens examined: CHINA, SICHUAN PROVINCE: Danba County, Donggu Town, Dongma Village, alt. 3000 m, 24 July 2007, Z.W. Ge 1504 (KUN-HKAS 53590, type). LIAONING PROVINCE: Benxi County, Xiamatang Town, Majia Village, alt. 320 m, 21 August 2015, J. Li 206 (KUN-HKAS 91234), the same location, 22 August 2015, J. Li 234 (KUN-HKAS 91262) and Q. Cai 1405 (KUN-HKAS 92025).

Known distribution: Currently known from northeastern and southwestern China.

Commentary: *Veloporphyrellus gracilioides* was described by Wu *et al*. (2016a) from China. In the protologue, the units of length and width of hymenial cystidia were mistakenly used millimeters. However, the units should be microns. Here we correct the units in the description above.

Veloporphyrellus gracilioides is phylogenetically related and morphologically similar to *V. castaneus* (Fig. 4.1). However, *V. castaneus* differs in its small basidiospores measuring (10) 10.5–12 × 4.5–5 μm, and distribution in subtropical forests.

23.4 *Veloporphyrellus pseudovelatus* Yan C. Li & Zhu L. Yang, Mycologia 106: 301 (2014)

Basidioma very small to small. Pileus 2–5 cm in diam., hemispherical to subhemispherical when young, convex to plano-convex when mature; surface rimose when dry or mature, always cracked into large cocoa-brown to chestnut-brown floccose squamules on a whitish to light brownish background; margin extended and embracing apex of stipe in younger basidiomata, then breaking into pieces and hanging on pileal margin when mature; context white, without discoloration when injured. Hymenophore adnate when young, depressed around apex of stipe when mature; surface initially white, then pinkish to flesh colored when mature, without discoloration when injured; pores angular to roundish, about 0.3–1 mm wide; tubes 3–6 mm long, concolorous or a little paler than hymenophoral surface, without discoloration when injured. Spore print dull red to grayish red (Fig. 3.4 n). Stipe 3–9 × 0.5–1 cm, subcylindrical to clavate, always enlarged downwards; surface pale chestnut-brown to reddish brown, nearly glabrous or with fibrillose squamules; context white, without discoloration when injured; basal mycelium white to pallid. Taste and odor indistinct.

Fig. 23.4 a *Veloporphyrellus pseudovelatus* Yan C. Li & Zhu L. Yang. Photo by Z.L. Yang (KUN-HKAS 52258, type).

Fig. 23.4 b *Veloporphyrellus pseudovelatus* Yan C. Li & Zhu L. Yang. Photos by Y.C. Li (KUN-HKAS 59444).

Basidia 23–32 × 8–12 µm, clavate, hyaline to light yellowish in KOH and yellowish in Melzer's reagent, 4-spored. Basidiospores (12) 12.5–15 (16) × 4–5 (5.5) µm [Q = (2.45) 2.6–3.38 (3.63), Q_m = 2.94 ± 0.18], subcylindrical to subfusiform in side view with slight suprahilar depression, cylindrical to fusiform in ventral view, smooth under SEM, somewhat thick-walled (up to 0.5 µm thick), olive-brown to yellowish brown in KOH, brown to ochreous-brown in Melzer' reagent. Cheilocystidia 37–68 × 5–10 µm, abundant, clavate to subcylindrical or finger-like, often with 1–2 septa, thin-walled, hyaline to light yellowish in KOH and yellowish to yellow in Melzer's reagent. Pleurocystidia 50–69 × 6–9 µm, fusoid, fusoid-mucronate or fusoid-ventricose, thin-walled, pale yellowish to yellowish in KOH and yellow to yellowish brown in Melzer's reagent. Pileipellis a trichoderm composed of 3–6 µm wide interwoven filamentous hyphae, yellowish to pale brownish in KOH and yellow-brown to brownish in Melzer's reagent; terminal cells 20–61 × 4–6 µm, subcylindrical. Pileal trama composed of 6–15 µm wide hyphae, colorless to yellowish in KOH and yellowish to light brownish in Melzer's reagent. Clamp connections absent in all tissues.

Habitat: Solitary to scattered in temperate to subtropical forests dominated by plants of the family Pinaceae or in mixed forests dominated by plants of the families Fagaceae and Pinaceae.

Known distribution: Known from southwestern China.

Specimens examined: CHINA, YUNNAN PROVINCE: Kunming, Jindian, alt. 2000 m, 1 August 2007, Z.L. Yang 4941 (KUN-HKAS 52258, type); Kunming, Xishan Mountain, alt. 2050 m, 10 August 2007, Y.C. Li 986 (KUN-HKAS 52673); Nanhua County, Wild Mushroom Market, altitude unknown,

25 August 2007, Z.L. Yang 4927 (KUN-HKAS 52244), the same location, 2 August 2009, Y.C. Li 1947 (KUN-HKAS 59695); Tengchong County, on the way from Tengchong to Longling, alt. 2010 m, 19 July 2009, Y.C. Li 1697 (KUN-HKAS 59444); Baoshan, Daxishan Mountain, alt. 1900 m, 9 August 2010, L.P. Tang 1212 (KUN-HKAS 63032), the same location, 10 August 2010, L.P. Tang 1219 (KUN-HKAS 63039); Baoshan, Longyang District, alt. 1700 m, 10 August 2010, Y.J. Hao 205 (KUN-HKAS 69189), the same location and date, Q. Cai 340 (KUN-HKAS 67903), the same location, 13 August 2010, G. Wu 662 (KUN-HKAS 74976); Tengchong County, Houqiao Town, Yangjiatian Village, alt. 1650 m, 10 August 2011, Y.J. Hao 442 (KUN-HKAS 71551), the same location and date, Q. Zhao 1258 (KUN-HKAS 74211); Lanping County, Jinding Town, Qingmen Village, alt. 2300 m, 15 August 2011, G. Wu 708 (KUN-HKAS 75022) and G. Wu 712 (KUN-HKAS 75026); Kunming, Yeya Lake, alt. 1950 m, 18 August 2012, Y.C. Li 2815 (KUN-HKAS 63670), the same location and date, B. Feng 1237 (KUN-HKAS 82379); Jingdong County, Ailao Mountain, alt. 2450 m, 24 July 2013, J. Qin 713 (KUN-HKAS 81151), the same location, 5 August 2013, L.H. Han 209 (KUN-HKAS 80439); Lancang County, Fofang, alt. 860 m, 20 August 2016, Y.J. Hao LC-29 (KUN-HKAS 96866).

Commentary: *Veloporphyrellus pseudovelatus*, originally described by Li *et al.* (2014a) from China, is characterized by the hemispherical to subhemispherical then convex to plano-convex pileus, the relatively small basidiospores measuring (12) 12.5–15 (16) × 4–5 (5.5) μm, and the distribution in temperate to subtropical forests with altitudes ranging from 860 m to 2450 m. This species is phylogenetically related and macroscopically similar to *V. alpinus*. However, *V. alpinus* differs from *V. pseudovelatus* in its sharp umbonate pileus, large basidiospores measuring (15.5) 16–19 (19.5) × (4.5) 5–6 (6.5) μm, and distribution in subalpine to alpine forests with high altitudes ranging from 3100 m to 3600 m (see our description of *V. alpinus* above).

23.5 *Veloporphyrellus velatus* (Rostr.) Yan C. Li & Zhu L. Yang, Mycologia 106: 303 (2014)

Basionym: *Suillus velatus* Rostr., Bot Tidsskr 24: 357 (1902).

Synonyms: *Boletus velatus* (Rostr.) Sacc. & D. Sacc., Syll fung 17: 97 (1905); *Tylopilus velatus* (Rostr.) F.L. Tai, Syll Fung Sinicorum 758 (1979).

Basidioma very small to small. Pileus 2–4 cm wide, subhemispherical to convex; surface densely covered with brown to cocoa-brown or chestnut-brown to dark reddish brown squamules on a white background, dry, inviscid when wet; margin extended with white membranous veil remnants; context white, without discoloration when injured. Hymenophore adnate when young, depressed around apex of stipe when mature; surface pallid to pale pinkish when young, becoming pinkish to pink when mature, without discoloration when injured; pores subangular to roundish, up to 1 mm wide; tubes up to 10 mm long, concolorous or a little paler than hymenophoral surface, without discoloration when injured. Spore print dull red to grayish red. Stipe 7.2 × 0.6–0.8 cm, clavate, glabrous, white to bright white; basal mycelium white; context solid, white to bright white, without discoloration when bruised; basal mycelium white to pallid. Taste and odor indistinct.

Basidia 25–32.5 × 10–12.5 μm, broadly clavate to clavate, hyaline to light yellowish in KOH, thin-walled, 4-spored, occasionally 2-spored. Basidiospores 11–12.5 (13) × (4) 4.5–5 μm [Q = (2.4) 2.44–2.67 (2.75), Q_m = 2.56 ± 0.12], subcylindrical to subfusiform in side view with slight suprahilar depression, cylindrical to fusiform in ventral view, smooth under SEM, somewhat thick-walled (up to 0.5 μm thick), subhyaline to light olivaceous in KOH and yellowish brown in Melzer's reagent. Pleurocystidia 61–75 × 8.5–12 μm, fusiform to subfusiform or subfusoid-mucronate to ventricose-mucronate with a short pedicel, sometimes narrowly mucronate, rostrate, thin-walled, hyaline in KOH and yellowish to yellow in Melzer's reagent. Cheilocystidia 25–38 × 7–11 μm, broadly clavate to subfusiform or ventricose, thin-walled, often with 1–2 septa, hyaline in KOH and yellowish to yellow in Melzer's reagent. Hymenophoral trama bilateral composed of broad (up to 9.5 μm wide) hyaline hyphae. Pileipellis a trichoderm composed of 3–7 μm wide loosely interwoven hyphae, colorless to yellowish in KOH and yellowish to pale yellow in Melzer's reagent; terminal cells 15–57.5 × 3–5.5 μm, subcylindrical. Pileal trama made up of 5–11 μm wide filamentous hyphae, colorless to yellowish in KOH and yellowish to pale yellow in Melzer's reagent. Clamp connections absent in all tissues.

Habitat: Solitary on the ground in tropical forests dominated by plants of the families Fagaceae and Pinaceae.

Known distribution: Currently known from Thailand and southern China.

Specimen examined: CHINA, HAINAN PROVINCE: Wuzhishan County, Wuzhishan National Nature Reserve, alt. 1200 m, 31 Jul 2010, N.K. Zeng 763 (KUN-HKAS 63668).

Commentary: *Veloporphyrellus velatus*, originally described from Thailand then reported from China, is well characterized by the dense tomentose, brown to cocoa-brown, chestnut-brown or dark reddish brown squamules on the pileus, the white membranous veil remnants hanging on the pileal margin, the pinkish to flesh colored hymenophore, the white glabrous stipe and the distribution in tropical forests (Gómez and

Singer 1984; Li *et al.* 2014a).

Phylogenetically, *V. velatus* clusters with *V. alpinus* and *V. pseudovelatus*, with high statistical support based on our multi-locus phylogenetic analysis (Fig. 4.1). However, *V. alpinus* and *V. pseudovelatus* differ from *V. velatus* in their distributions in temperate to subtropical or subalpine to alpine forests, and the orange-yellow to orange-brown or reddish brown stipe surface (see our descriptions of *V. alpinus* and *V. pseudovelatus* above). For detailed descriptions, comparisons with similar species, line drawings and images of *V. velatus* see Li *et al.* (2014a).

Fig. 23.5 *Veloporphyrellus velatus* (Rostr.) Yan C. Li & Zhu L. Yang. Photos by N.K. Zeng (KUN-HKAS 63668).

Chapter 24
Zangia Yan C. Li & Zhu L. Yang

Zangia Yan C. Li & Zhu L. Yang, Fungal Divers 49: 129 (2011)

Type species: *Zangia roseola* (W.F. Chiu) Yan C. Li & Zhu L. Yang ≡ *Boletus roseolus* W.F. Chiu, Mycologia 40(2): 208 (1948).

Diagnosis: The genus is different from other genera in the Boletaceae in its rugose to roughened pileus, initially white to pallid and then pinkish to pink hymenophore, pink to pinkish brown spore deposit, pink scabrous squamules over the stipe surface, white to cream pileal context staining blue asymmetrically when injured, golden yellow to bright yellow base of stipe, ixohyphoepithelium pileipellis, and smooth basidiospores.

Pileus subhemispherical when young, subhemispherical to convex or plano-convex; surface viscid when wet, rugose and pulverescent when young; context white to pallid, asymmetrically bluish when injured. Hymenophore adnate to depressed around apex of stipe; surface white to pinkish when young, becoming pinkish to pink or purplish when mature, without discoloration when injured. Spore print pinkish, pink to purplish pink or pale purple. Stipe central, clavate to subcylindrical, always enlarged downwards, pallid to whitish but golden yellow to chrome-yellow at base, staining asymmetric blue when injured; surface covered with red to pinkish red scabrous squamules; context white to pallid in the upper part, but cream to yellowish downwards and golden yellow to bright yellow at base, asymmetrically bluish when injured; mycelium on the base golden yellow to chrome-yellow. Pileipellis an ixohyphoepithelium with outer layer consisting of filamentous hyphae and inner layer made up of subglobose concatenated cells arising from radially arranged filamentous hyphae. Hymenial cystidia subcylindrical, clavate to subfusiform. Basidiospores smooth, elongated to cylindrical or fusiform, pinkish, light olivaceous to nearly colorless. Clamp connections absent in all tissues.

Commentary: This genus was originally described from China by Li *et al.* (2011), and harbored six species. The genus is endemic to southwestern China.

Key to the species of the genus *Zangia* in China

1. Species with a distribution in subtropical or temperate to subalpine forests; pileus without pale yellow to lemon-yellow tinge ··· 2
1. Species with a distribution in tropical forests; pileus lemon-yellow to pale yellow ········ *Z. citrina*
2. Pileus with a red, bright red or purple-red tinge ··· 3
2. Pileus with olivaceous tinge, without red, bright red or purple-red tinge ································ 4
3. Species with a distribution in temperate to subalpine forests; basidioma small to medium-sized; pileus dark red to red or brownish red; distributed in temperate to subalpine forests ···················· ··· *Z. erythrocephala*
3. Species with a distribution in subtropical forests; basidioma very small to small; pileus purple-red, purplish red, carnelian, or dull red; distributed in subtropical forests ···················· *Z. roseola*
4. Species with a distribution in subtropical forests; pileus reddish olivaceous or purplish brown to reddish brown with olivaceous tinge; basidiospores relatively narrow, not more than 6.5 μm wide ··· *Z. olivaceobrunnea*
4. Species with a distribution in temperate, subalpine or alpine forests; pileus honey-yellowish or yellowish brown with olivaceous tinge, or greenish brown to olive-green or brownish olivaceous; basidiospores relatively broad up to 7.5 μm wide ··· 5
5. Pileus orange-yellow to honey-yellowish or yellowish brown, with olivaceous tinge; hymenial cystidia relatively large measuring 61–75 × 8–12 μm ··· *Z. chlorinosma*
5. Pileus greenish brown to olive-green or brownish olivaceous; hymenial cystidia small measuring 32–61 × 6–12 μm ·· *Z. olivacea*

24.1 *Zangia chlorinosma* (Wolfe & Bougher) Yan C. Li & Zhu L. Yang, Fungal Divers 49: 129 (2011)

Basionym: *Tylopilus chlorinosmus* Wolfe et Bougher, Aust Syst Bot 6(3): 207 (1993).

Basidioma medium-sized. Pileus 5–8 cm in diam., hemispherical when young, subhemispherical to convex or applanate when mature, brownish olivaceous, or orange-yellow to honey-yellow or yellowish brown with olivaceous tinge; surface dry, slightly viscid when wet, rugose; context whitish to cream, asymmetrically bluish when bruised. Hymenophore adnate when young, depressed around apex of stipe when mature; surface white to pinkish when young, pink to brownish pink when mature, without color change when touched; pores angular to roundish, about 0.5–1 mm wide, tubes up to 16 mm long, pinkish to dingy pink, without color change when bruised. Spore print pinkish to pink. Stipe 4–10 × 0.7–1.2 cm, clavate, enlarged downwards, cream to yellowish at apical part, pink to vinaceous pink in middle part, chrome-yellow or golden yellow at base; surface covered with concolorous scabrous squamules, indistinctly reticulate on the lower part, asymmetrically bluish when injured; mycelium on the base golden yellow to chrome-yellow; context cream to yellowish, but yellowish to yellow downwards and golden yellow to chrome-yellow at base, asymmetrically bluish when injured. Taste and odor mild.

Basidia 20–42 × 9–14 μm, clavate, 4-spored, hyaline in KOH. Basidiospores (12) 13–15 (17) × (5.5) 6–7 (7.5) μm [$Q =$ (1.86) 2.07–2.5 (2.67), $Q_m = 2.27 \pm 0.16$], elongated to subcylindrical and inequilateral in side view with slight suprahilar depression, elongated to cylindrical in ventral view, smooth, somewhat thick-walled (0.5–1 μm thick), hyaline to pale olivaceous brown in KOH, yellow to yellowish brown in Melzer's reagent. Pleuro- and cheilocystidia 61–75 × 8–12 μm, subfusiform to lanceolate or ventricose to ventricose-mucronate with a long pedicel, thin-walled, yellowish to pale brownish in KOH and yellow to brownish in Melzer's reagent. Caulocystidia forming

Fig. 24.1 a *Zangia chlorinosma* (Wolfe et Bougher) Yan C. Li & Zhu L. Yang. Photos by J.W. Liu (KUN-HKAS 97884).

Fig. 24.1 b *Zangia chlorinosma* (Wolfe et Bougher) Yan C. Li & Zhu L. Yang. Photo by X.T. Zhu (KUN-HKAS 68198).

the scabrous squamules over the surface of stipe, ventricose, fusoid-ventricose, clavate, or lanceolate. Pileipellis an ixohyphoepithelium: outer layer consisting of 3.5–8 μm wide filamentous hyphae with subcylindrical terminal cells 8–31 × 4–7 μm, yellowish to pale brownish in KOH and yellowish brown to brownish in Melzer's reagent, inner layer composed of inflated (up to 30 μm wide) concatenated cells arising from 3–4 μm wide radially arranged filamentous hyphae, hyaline to yellowish in KOH and yellowish to brownish in Melzer's reagent. Pileal trama composed of 3–15 μm wide hyphae, colorless to yellowish in KOH and yellowish to pale yellow in Melzer's reagent. Clamp connections absent in all tissues.

Habitat: Solitary to scattered on soil in temperate to subalpine mixed forests dominated by plants of the families Pinaceae and Fagaceae.

Known distribution: Currently known from southwestern China.

Specimens examined: CHINA, YUNNAN PROVINCE: Yulong County, Heibaishui River, alt. 3200 m, 4 August 2005, Z.L. Yang 4531 (KUN-HKAS 48695), the same location, 12 July 2010, X.T. Zhu 22 (KUN-HKAS 68198). SICHUAN PROVINCE: Kangding County, Pusharong Town, Lianhua Lake, alt. 3100 m, 7 September 2016, J.W. Liu KD91 (KUN-HKAS 97884).

Commentary: *Zangia chlorinosma*, originally described by Wolfe and Bougher (1993) as *Tylopilus chlorinosmus* Wolfe & Bougher from southwestern China, is characterized by the brownish olivaceous to olivaceous yellow or yellowish brown pileus, the initially white to pallid and then pinkish to pink or purplish pink hymenophore, the bluish color change when injured, the broad basidiospores up to 7.5 μm wide, and the ixohyphoepithelium pileipellis. This species is macroscopically similar to *Z. olivacea* Yan C. Li & Zhu L. Yang. Indeed, they share somewhat olivaceous pileus. However, *Z. olivacea* has an olive-green pileus when young and then becoming greenish brown to brownish olivaceous when mature without honey-yellow to orange-yellow or yellowish brown tinge, and relatively short hymenial cystidia measuring 32–61 × 6–12 μm (Li *et al.* 2011).

In our multi-locus phylogenetic analysis (Fig. 4.1), *Z. chlorinosma* is related to *Z. roseola* and *Z. erythrocephala*. However, *Z. roseola* differs from *Z. chlorinosma* in its very small to small sized basidioma, purple-red to purplish red or carnelian to dull red pileus, and distribution in subtropical forests (Li *et al.* 2011). *Zangia erythrocephala* has a red to dark red or brownish red pileus, and relatively narrow basidiospores which are mostly 5.5–6.5 μm wide (Li *et al.* 2011).

24.2 *Zangia citrina* Yan C. Li & Zhu L. Yang, Fungal Divers 49: 132 (2011)

Basidioma very small to small. Pileus 2–5 cm in diam., subhemispherical to convex or convexo-applanate, lemon-yellow to yellowish or pale yellow, margin much paler in color; surface viscid when wet, nearly glabrous to rugose; context whitish to pallid white, asymmetrically bluish when bruised. Hymenophore adnate when young, adnate to depressed around apex of stipe when mature; surface white to pinkish when young, becoming pink or grayish pink when mature, without color change when injured; pores angular to roundish, up to 1 mm wide; tubes 4–10 mm long, concolorous or a little paler than hymenophoral surface, without discoloration when injured. Spore print pinkish to pink. Stipe 4–7 × 0.4–0.7 cm, clavate to subcylindrical, enlarged downwards, yellow to yellowish at apical part, yellow to lemon-yellow downwards and golden yellow at base; surface covered with pink to pinkish red scabrous squamules; context yellowish to yellow at apical part, but yellow to bright yellow downwards and golden yellow or chrome-yellow at base, without color change or asymmetrically bluish in younger basidioma when bruised or injured; mycelium on the base golden yellow or chrome-yellow. Taste and odor mild.

Basidia 18–37 × 10–15 μm, clavate, 4-spored, hyaline in KOH. Basidiospores 10–14 × 4–5.5 μm [Q = (2) 2.18–2.86 (3), Q_m = 2.53±0.19], elongated to subcylindrical and inequilateral in side view with slight suprahilar depression, elongated to cylindrical in ventral view, smooth, somewhat thick-walled (up to 0.5 μm thick), hyaline to dull olivaceous in KOH, yellow to yellowish brown in Melzer's reagent. Pleuro- and cheilocystidia 21–61 × 8–13 μm, subfusiform to ventricose or ventricose-rostrate, thin-walled, yellowish to pale yellowish brown in KOH, yellow to pale yellow-brown in Melzer's reagent. Caulocystidia forming the scabrous squamules over the surface of stipe, subfusiform to subfusoid-mucronate or clavate, sometimes with 1–2 fibrillous or flagelliform terminal cells. Pileipellis an ixohyphoepithelium: outer layer composed of 4–8 μm wide filamentous hyphae with clavate to cylindrical terminal cells 20–76 × 4–8 μm, inner layer

Fig. 24.2 a *Zangia citrina* Yan C. Li & Zhu L. Yang. Photos by Y.C. Li (KUN-HKAS 52684, type).

made up of inflated (up to 25 µm wide) concatenated cells, arising from 2.5–4.5 µm wide radially arranged filamentous hyphae, hyaline to yellowish in KOH and yellow to brownish in Melzer's reagent. Pileal trama composed of 3–11 µm wide interwoven filamentous hyphae, colorless to yellowish in KOH and yellowish to pale yellow in Melzer's reagent. Clamp connections absent in all tissues.

Habitat: Solitary to scattered, in tropical forests dominated by plants of the family Fagaceae.

Fig. 24.2 b *Zangia citrina* Yan C. Li & Zhu L. Yang. Photos by J. Li (KUN-HKAS 85919).

Known distribution: Currently known from central, southern, southwestern, and southeastern China.

Specimens examined: CHINA, FUJIAN PROVINCE: Sanming, Sanyuan National Forest Park, alt. 260 m, 24 August 2007, Y.C. Li 990 (KUN-HKAS 52677) and Y.C. Li 997 (KUN-HKAS 52684, type), the same location, 25 August 2007, Y.C. Li 1000 (KUN-HKAS 53345) and Y.C. Li 1001 (KUN-HKAS 53346), the same location, 27 August 2007, Y.C. Li 1039 (KUN-HKAS 53384). GUANGDONG PROVINCE: Fengkai County, Heishiding Nature Reserve, alt. 250 m, 5 July 2012, F. Li 619 (KUN-HKAS 106417), the same location, 24 August 2012, F. Li 927 (KUN-HKAS 106439). HUBEI PROVINCE: Xingshan County, Muyu Town, alt. 1800 m, 12 August 2015, X. He 24 (KUN-HKAS 91197). HUNAN PROVINCE: Liuyang County, Daweishan National Forest Park, alt. 1000 m, 10 July 2003, H.C. Wang 296 (KUN-HKAS 42457); Yizhang County, Mangshan National Forest Park, 29 September 1981, Y.C. Zong and X.L. Mao 52 (HMAS 42680). YUNNAN PROVINCE: Guangnan County, Nanping Town, alt. 1350 m, 1 August 2014, J. Li 95 (KUN-HKAS 85919).

Commentary: *Zangia citrina* originally described from southeastern China by Li *et al*. (2011) is characterized by the distinct lemon-yellow to pale yellow or yellowish pileus, the initially white and then pinkish to pink hymenophore, the white to pallid white context of pileus, the yellow to bright yellow context of stipe, the asymmetrically bluish color change when bruised, the yellow to lemon-yellow stipe asymmetrically bluish when injured. This species can be easily confused with *Fistulinella lutea* Redeuilh & Soop because of the color and form of the basidioma. However, *F. lutea* differs from *Z. citrina* in its yellow to bright yellow stipe without any pink or pinkish red scabrous, white to grayish white context in stipe without color change when injured, white mycelium on the base of stipe, cuticle pileipellis without inflated cells, and larger basidiospores measuring 13–17.5 × 4.5–6.5 µm (Redeuilh and Soop 2006).

24.3 *Zangia erythrocephala* Yan C. Li & Zhu L. Yang, Fungal Divers 49: 134 (2011)

Basidioma small to medium-sized. Pileus 3–8 cm in diam., subhemispherical to plano-convex, or convex to applanate, red to dark red, purple to brownish purple with red tinge; surface viscid when wet, rugose, finely pulverescent; context white to cream, becoming asymmetrically blue when injured. Hymenophore adnate when young, depressed around apex of stipe when mature; surface white to pinkish when young, pinkish to dingy pink when mature, without color change when injured; pores angular to roundish, 0.5–1 mm wide; tubes up to 20 mm long, dingy white to pinkish, without color change when injured. Spore print pinkish to pink or purplish pink (Fig. 3.4 o). Stipe 3–9 × 0.5–1.8 cm, clavate to subcylindrical, enlarged downwards, whitish to yellowish at apical part, yellowish to yellow downwards but chrome-yellow to bright yellow at base; surface covered with red to pink-red scabrous squamules, sometimes lower part with indistinct shallow reticulum; context cream to yellowish at upper part, yellow to bright yellow downwards and chrome-yellow at base, asymmetrically bluish when bruised; mycelium at base golden yellow or chrome-yellow. Taste and odor mild.

Basidia 30–41 × 13–16 μm, clavate, 4-spored, hyaline in KOH. Basidiospores 11.5–16.5 × 5.5–6.5 (7) μm (Q = 1.89–2.62, Q_m = 2.26 ± 0.16), elongated to subcylindrical and inequilateral in side view with slight suprahilar depression, elongated to cylindrical in ventral view, smooth, somewhat thick-walled (up to 1 μm thick), nearly colorless to pale olivaceous in KOH and yellow-brown to brown in Melzer's reagent. Pleuro- and cheilocystidia 26–75 × 5–11 μm, lanceolate, subfusiform to fusiform or ventricose, thin-walled, yellowish to yellowish brown in KOH and yellow-brown to brown in Melzer's reagent. Caulocystidia forming the scabrous squamules over the surface of stipe, morphologically similar to hymenial cystidia

Fig. 24.3 a *Zangia erythrocephala* Yan C. Li & Zhu L. Yang. Photos by J. Qin (KUN-HKAS 83435).

Fig. 24.3 b *Zangia erythrocephala* Yan C. Li & Zhu L. Yang. Photos by C. Yan (KUN-HKAS 85770).

but much shorter. Pileipellis an ixohyphoepithelium: outer layer consisting of 4–6 μm wide gelatinous interwoven hyphae with subcylindrical terminal cells 18–63 × 3–5 μm, inner layer made up of inflated (up to 20 μm wide) concatenated cells arising from 2–5 μm wide radially arranged filamentous hyphae, yellowish to pale brownish in KOH and yellowish brown to brownish in Melzer's reagent. Pileal trama composed of 4–20 μm wide filamentous hyphae, hyaline to yellowish in KOH and yellowish to brownish in Melzer's reagent. Clamp connections absent in all tissues.

Habitat: Solitary to scattered on soil in temperate to subalpine mixed forests dominated by plants of the families Fagaceae and Pinaceae.

Known distribution: Currently known from southwestern China.

Specimens examined: CHINA, SICHUAN PROVINCE: Muli County, 913 Forestry Station, alt. 3250 m, 18 August 1992, P.G. Liu & M.S. Yuan 1413 (KUN-HKAS 25632). YUNNAN PROVINCE: Yulong County, Heibaishui River, 30 July 2001, alt. 2900 m, Z.L. Yang 3121 (KUN-HKAS 38298), the same location, 22 July 2008, alt. 2900 m, Y.C. Li 1325 (KUN-HKAS 56179); Yulong County, Yulong Snow Mountain, 4 August 2005, Z.L. Yang 4532A (KUN-HKAS 48696A), the same location, 8 August 2014, J. Qin 1018 (KUN-HKAS 83435), the same location, 29 August 2014, C. Yan 172 (KUN-HKAS 85770); Shangri-La County, Geza Town, Hongshan Mountain, 9 September 2007, alt. 3400 m, B. Feng 122 (KUN-HKAS 52843) and B. Feng 123 (KUN-HKAS 52844, type); Shangri-La County, Haba Snow Mountain, 12 August. 2008, alt. 3800 m, Y.C. Li 1433 (KUN-HKAS 56273).

Commentary: *Zangia erythrocephala* originally described from southwestern China by Li *et al.* (2011) is characterized by the rugose and red to dark red or brownish red pileus, the white to cream context becoming asymmetrically blue when injured, the white to pinkish or dingy pink hymenophore without color change when injured, the yellowish red or pinkish to pink stipe becoming asymmetrically blue when touched, the ixohyphoepithelium pileipellis, and the temperate to subalpine distribution. This species is phylogenetically related and morphologically similar to *Z. chlorinosma* and *Z. roseola*. For morphological comparisons of these species see the commentary on *Z. chlorinosma*.

24.4 *Zangia olivacea* Yan C. Li & Zhu L. Yang, Fungal Divers 49: 137 (2011)

Basidioma small to medium-sized. Pileus 4–8 cm in diam., subhemispherical to convex or plano-convex to applanate, olive-green to brownish olivaceous or greenish brown; surface dry, but subviscid when wet, rugose; context whitish to pallid, asymmetrically bluish when injured. Hymenophore adnate when young, depressed around apex of stipe when mature; surface initially whitish to pinkish and then pinkish to pink, without color change when touched; pores subangular to roundish, up to 1 mm wide; tubes up to 15 mm long, concolorous or much paler than hymenophoral surface, without color change when injured. Spore print pinkish to pink. Stipe 5–13 × 0.8–2 cm, clavate to subcylindrical, enlarged downwards, white to cream at upper part, but yellowish to yellow downwards and chrome-yellow or golden yellow at base; surface covered with reddish pink to pink scabrous squamules, asymmetrically bluish when injured; context cream to yellowish at upper part, yellowish to yellow downwards and golden yellow at base, asymmetrically bluish when injured; mycelium on the base golden yellow. Taste and odor mild.

Fig. 24.4 *Zangia olivacea* Yan C. Li & Zhu L. Yang. Photos by Y.Y. Cui (KUN-HKAS 79802)

Basidia 20–37×10–15 μm, clavate, 4-spored, hyaline in KOH. Basidiospores 12–17 × 6–7 μm (Q = 1.88–2.6, Q_m = 2.17 ± 0.12), elongated to subcylindrical and inequilateral in side view with slight suprahilar depression, elongated to cylindrical in ventral view, smooth, somewhat thick-walled (up to 0.5 μm thick), hyaline to pale olivaceous in KOH and yellow to yellowish brown or olivaceous brown in Melzer's reagent. Pleuro- and cheilocystidia 32–61 × 6–12 μm, subfusiform to lanceolate or ventricose-mucronate with a long pedicel, thin-walled, yellowish to brownish in KOH and yellow to yellow-brown in Melzer's reagent. Caulocystidia forming the scabrous squamules over the surface of stipe, 20–56 × 6–11 μm, similar to hymenial cystidia. Pileipellis an ixohyphoepithelium: outer layer consisting of 3.5–7 μm wide filamentous hyphae with subcylindrical terminal cell 20–44 × 4–5 μm, inner layer composed of 11–23 μm wide concatenated cells arising from 3–6 μm wide radially arranged filamentous hyphae, hyaline to yellowish or yellowish brown in KOH, yellowish to yellowish brown or yellow-brown in Melzer's reagent. Pileal trama composed of 3–10 μm wide interwoven hyphae. Clamp connections absent in all tissues.

Habitat: Solitary on soil in temperate to subalpine mixed forests dominated by plants of the families Fagaceae and Pinaceae.

Known distribution: Currently known from southwestern China.

Specimens examined: CHINA, YUNNAN PROVINCE: Shangri-La County, Daxue Mountain, alt. 3100 m, 6 July 2004, Z.L. Yang 3960 (KUN-HKAS 45445, type); Shangri-La County, 26 July 2006, alt. 3300 m, Z.W. Ge 1086 (KUN-HKAS 55830); Yulong County, National Astronomical Observatory, 20 July 2008, alt. 3200 m, Y.C. Li 1294 (KUN-HKAS 55831); Yulong County, Yulong Snow Mountain, alt. 3200 m, 12 July 2012, J. Qin 30 (KUN-HKAS 67715); Yulong County, Lijiang Alpine Botanic Garden, alt. 3000 m, 20 August 2013, Y.Y. Cui 132 (KUN-HKAS 79802).

Commentary: *Zangia olivacea* originally described from southwestern China by Li *et al.* (2011) is characterized by the rugose and olive-green to brownish olivaceous or greenish brown pileus, the whitish to pallid context of pileus, the asymmetrically bluish color change when injured, the pink to red scabrous stipe, the ixohyphoepithelium pileipellis, and the distribution in temperate to subalpine forests.

In our multi-locus phylogenetic analysis (Fig. 4.1), *Z. olivacea* is related to *Z. roseola*, *Z. erythrocephala*, and *Z. chlorinosma*. However, *Z. roseola* and *Z. erythrocephala* have red to dull red or purple-red pileus without any olivaceous tinge, while *Z. chlorinosma* has a brownish olivaceous to olivaceous yellow or yellowish brown pileus, relatively large hymenial cystidia measuring 61–75 × 8–12 μm (Li *et al.* 2011).

24.5 *Zangia olivaceobrunnea* Yan C. Li & Zhu L. Yang, Fungal Divers 49: 138 (2011)

Basidioma small to medium-sized. Pileus 4–7 cm in diam., subhemispherical to convex or plano-convex to applanate, reddish olivaceous, purplish brown to reddish brown with olivaceous tinge, or olive-brown to olive-red, much darker in the center; context whitish to pallid, without color change or asymmetrically bluish when injured. Hymenophore adnate when young, deeply depressed around apex of stipe when mature; surface initially white to pinkish and then pinkish to pink or grayish pink when mature, without discoloration when injured; pores angular to roundish, 0.5–1 mm wide; tubes up to 12 mm long, concolorous or a little paler than hymenophoral surface, without discoloration when

Fig. 24.5 a *Zangia olivaceobrunnea* Yan C. Li & Zhu L. Yang. Photos by Y.C. Li (KUN-HKAS 107173).

338 The Boletes of China: *Tylopilus* s.l.

injured. Spore print purplish pink or pale purple. Stipe 6–12 × 0.6–1.6 cm, clavate to subcylindrical, always enlarged downwards, dingy white to cream at upper part, pale yellow to yellow downwards and chrome-yellow at base, bluish when injured; surface densely covered with pink to red or purple scabrous squamules, staining a bluish tinge slowly when injured; context cream to yellowish at apical part, yellowish to yellow downwards and golden yellow at base, staining asymmetric blue slowly when injured; basal mycelium golden yellow or chrome-yellow. Taste and odor mild.

Basidia 20–36 × 10–15 μm, clavate, 4-spored, hyaline in KOH. Basidiospores 12–17 × (4.5) 5–6 (6.5) μm (Q = 2.26–3.08, Q_m = 2.56 ± 0.16), subcylindrical to subfusiform in side view with

Fig. 24.5 b *Zangia olivaceobrunnea* Yan C. Li & Zhu L. Yang. Photos by Z.L. Yang (KUN- HKAS 80032).

slight suprahilar depression, cylindrical to fusiform in ventral view, smooth, somewhat thick-walled (up to 0.5 μm thick), hyaline to pale olivaceous in KOH and yellowish to yellowish brown or olivaceous brown in Melzer's reagent. Pleuro- and cheilocystidia 6–70 × 4–12.5 μm, subfusiform, ventricose mucronate to fusiform, or lanceolate, thin-walled, yellowish to pale brownish in KOH and yellow to yellowish brown in Melzer's reagent. Caulocystidia forming the scabrous squamules over the surface of stipe, clavate to narrowly fusiform. Pileipellis an ixohyphoepithelium: outer layer composed of 4–13 μm wide interwoven hyphae with cylindrical terminal cells 30–63 × 4–6 μm, inner layer made up of 15–21 μm wide concatenated cells arising from 4–7 μm wide radially arranged filamentous hyphae, pale yellowish to brownish yellow in KOH and yellow to brownish in Melzer's reagent. Pileal trama composed of 4–11 μm wide interwoven hyphae, hyaline to yellowish in KOH and yellowish to brownish in Melzer's reagent. Clamp connections absent in all tissues.

Habitat: Solitary to scattered, in subtropical mixed forests dominated by plants of the families Fagaceae and Pinaceae.

Known distribution: Currently known from southwestern China.

Specimens examined: CHINA, YUNNAN PROVINCE: Nanhua County, Wild Mushroom Market, altitude unknown, 3 August 2009, Y.C. Li 1961 (KUN-HKAS 59220) and Y.C. Li 1962 (KUN-HKAS 59221); Binchuan County, Jizu Mountain, alt. 1980 m, 11 August 2011, Y.C. Li 2676 (KUN-HKAS 107173); Kunming, Heilongtan Park, alt. 1980 m, 8 September 2007, Z.L. Yang 4955 (KUN-HKAS 52272), the same location, 9 September 2007, Z.L. Yang 4960 (KUN-HKAS 52275, type), the same location, 16 August 2008, Z.L. Yang 5145 (KUN-HKAS 54442), the same location and date, Y.C. Li 1575 (KUN-HKAS 55511) and Y.C. Li 1576 (KUN-HKAS 55512), the same location, 6 August 2007, Y.C. Li 961 (KUN-HKAS 52648); Kunming, Yeya Lake, alt. 2000 m, 1 September 2013, Z.L. Yang 5754 (KUN-HKAS 80032).

Commentary: *Zangia olivaceobrunnea* was described from southwestern China by Li *et al.* (2011). This species is characterized by the reddish olivaceous pileus, the initially whitish to pallid and then pinkish to grayish pink hymenophoral surface without color change when injured, the dingy white to cream or pale yellow stipe covered with pink to red or purple-red scabrous squamules, the whitish to pallid context of pileus which is asymmetrically bluish when injured, the ixohyphoepithelium pileipellis, and the distribution in subtropical forests. Morphologically, *Z. olivaceobrunnea* is similar to *Z. olivacea* and *Z. chlorinosma* in their somewhat olivaceous pileus. However, *Z. olivacea* has an olive-green to brownish olivaceous or greenish brown pileus without any reddish tinge, broad basidiospores up to 7 μm wide, and distribution in temperate to subalpine forests. *Zangia chlorinosma* has a brownish olivaceous or olivaceous yellow to yellowish brown pileus without any reddish tinge, relatively broad basidiospores up to 7.5 μm wide, and a distribution in temperate to subalpine forests.

In our multi-locus molecular phylogenetic analysis (Fig. 4.1), *Z. olivaceobrunnea* is closely related to *Z. citrina*. However, *Z. citrina* has a distinct lemon-yellow to yellowish or pale yellow pileus and stipe, relatively narrow basidiospores measuring 10–14 × 4–5.5 μm, and a distribution in tropical forests (Li *et al.* 2011).

24.6 *Zangia roseola* (W.F. Chiu) Yan C. Li & Zhu L. Yang, Fungal Divers 49: 140 (2011)

Basionym: *Boletus roseolus* W.F. Chiu, Mycologia 40: 208 (1948).

Synonym: *Tylopilus roseolus* (W.F. Chiu) F.L. Tai, Syll Fung Sinicorum 758 (1979).

Basidioma very small to small. Pileus 2–5 cm in diam., hemispherical subhemispherical or convex, red to purple-red or carnelian when young, purplish red to dull red when mature, pinkish or even white towards margin; surface dry, rugose or roughened, viscid when wet, always covered with white to cream pulverescent squamules easily washed away; context whitish to pallid, asymmetrically bluish when injured. Hymenophore adnate

Fig. 24.6 a *Zangia roseola* (W.F. Chiu) Yan C. Li & Zhu L. Yang. Photos by Z.L. Yang (KUN-HKAS 107179).

when young, adnate to depressed around apex of stipe when mature; surface white to pinkish when young, pinkish to pink when mature; pores angular to roundish, about 0.5–2 mm wide; tubes 5–10 mm long, concolorous or a little paler than hymenophoral surface, without color change when injured. Spore print pink to pale purple. Stipe 4–7 × 0.3–0.8 cm, subcylindrical, sometimes with a bulbar base, whitish to pallid at upper part, cream to yellowish or yellow downwards and chrome-yellow or golden yellow at base; surface covered with red to pinkish red minute scabrous squamules; context whitish to cream at apical part, cream to yellowish downwards but golden yellow or chrome-yellow at base, becoming asymmetric blue slowly when injured; mycelium on base golden yellow to chrome-yellow. Taste and odor indistinct.

Basidia 26–38 × 10–15 μm, clavate, 4-spored, hyaline in KOH. Basidiospores 13–16 (17) × (5.5) 6–7 (8) μm [Q = (2) 2.14–2.54 (2.83), Q_m = 2.31 ± 0.19], subcylindrical to subfusiform in side view with slight suprahilar depression, cylindrical to fusiform in ventral view, smooth, yellowish to light olivaceous in KOH and yellow to yellowish brown in Melzer's reagent. Pleuro- and cheilocystidia 30–83 × 5–18 μm, subfusiform to fusiform or clavate, thin-walled, yellowish to pale brownish in KOH and yellowish brown to brown in Melzer's reagent. Caulocystidia forming the scabrous squamules over the surface of stipe, clavate, lanceolate, ventricose, or subfusiform, concolorous with pleuro- and cheilocystidia. Pileipellis an ixohyphoepithelium: outer layer composed of 4–6 μm wide filamentous hyphae, inner layer made up of 8–20 μm wide concatenated cells arising from 3–6 μm wide radially arranged filamentous hyphae, pale yellowish to brownish yellow in KOH and yellow to brownish in Melzer's reagent. Pileal trama composed of 5–10 μm wide filamentous hyphae, colorless to yellowish in KOH and yellow to yellowish brown in Melzer's reagent. Clamp connections absent in all tissues.

Habitat: Solitary to scattered on soil in subtropical mixed forests dominated by plants of the families Fagaceae and Pinaceae.

Known distribution: Currently known from central and southwestern China.

Specimens examined: CHINA, YUNNAN PROVINCE: Nanhua County, Wild Mushroom Market, altitude unknown, 3 August 2009, Y.C. Li 1958 (KUN-HKAS 59219); Kunming, free market, altitude unknown, 27 July 1938, J.C. Zhou 7889 (HMAS 3889, isotype of *Boletus roseolus* in Chiu 1948); Kunming, Heilongtan Park, alt. 1900 m, 9 September 2007, Z.L. Yang 4962 (KUN-HKAS 52277), the same location, 18 September 2020, Z.L. Yang 6428 (KUN-HKAS 107179); Kunming, Qiongzhu Temple, alt. 2100m, 21 September 2006, Y.C. Li 700 (KUN-HKAS 51137). HENAN PROVINCE: Luanchuan County, Laojun Mountain, alt. 2200 m, 13 August 2015, B. Li 55 (KUN-HKAS 89819). HUBEI PROVINCE: Xingshan County, Muyu Town, alt. 1850 m, 12 July 2012, Q. Cai 740 (KUN-HKAS 75495).

Commentary: *Zangia roseola* originally described as *Boletus roseolus* W.F. Chiu from southwestern China by Chiu (1948) is characterized by the very small to small sized basidioma, the carnelian to purplish red or purple-red to dull red pileus, the initially whitish to pallid and then pinkish to pink hymenophore, the pinkish to purple-pink scabrous stipe, the broad basidiospores which is up to 7 μm wide, the ixohyphoepithelium pileipellis, and the distribution in subtropical forests.

Our multi-locus molecular phylogenetic analysis (Fig. 4.1) indicates *Z. roseola* is closely related to *Z. erythrocephala* and *Z. chlorinosma*. Indeed, they are morphologically similar to each other. However, *Z. erythrocephala* differs from *Z. roseola* in its relatively large basidioma, relatively narrow basidiospores measuring 11.5–16.5 × 5.5–6.5 (7) μm, and distribution in temperate to subalpine forests (Li *et al.* 2011), while *Z. chlorinosma* has a medium-sized basidioma, an initially brownish olivaceous or olivaceous yellow and then yellowish brown pileus, and a distribution in temperate to subalpine forests (Li *et al.* 2011).

Fig. 24.6 b *Zangia roseola* **(W.F. Chiu) Yan C. Li & Zhu L. Yang. Photo by Z.L. Yang (KUN-HKAS 52277).**

Chapter 25

Summary and conclusion

Located in East Asia with an area of 9.6 million km^2 and a span more than 50 degrees in latitude, with great physical diversity, from the "Third Pole" mountains and the Qinghai-Tibet Plateau in the west to low lands in the east, China is one of the diversity hotspots for boletes, the fungi of the Boletaceae family (Zang 2006, 2013; Wu *et al.* 2016a, b). Many boletes were traditionally treated as members of the genus *Tylopilus* s.l. based on the white to pinkish, pink, purplish pink or reddish brown color of the hymenophores or the spore prints. Some of these species are undescribed, their phylogenetic relationships are not known, ecology and distribution patterns are unclear. Such kind of information is crucial for all other biological studies on forest ecosystems including conservation, restoration, and management. In addition, species in this group are also economically important and some species are very popular in China, especially in southwestern China because of their good flavor. Some species are also used in herbal medicines, while other species are toxic. However, without proper names, the development and utilization of this group of boletes is not possible.

Before this study, fewer than 40 species of *Tylopilus* s.l. were reported from China (Li and Song 2003; Fu *et al.* 2006; Li *et al.* 2011; Li *et al.* 2014a, b; Zang 2006, 2013; Wu *et al.* 2014, 2016a, b). With polyphasic approach, our present long-term field investigations and examinations of historical collections from different herbaria in China, reveal that boletes with white to pinkish, pink, purplish pink, or reddish brown colored hymenophores or spore prints are very diverse in morphology, complex in structure and phylogeny, and wide in ecological niches.

Species of *Tylopilus* s.l. from China could be divided into nineteen genera using multi-disciplinary methods. In total, 105 species belonging to seventeen known genera (*Austroboletus*, *Chiua*, *Fistulinella*, *Harrya*, *Hymenoboletus*, *Indoporus*, *Leccinellum*, *Mucilopilus*, *Porphyrellus*, *Pseudoaustroboletus*, *Retiboletus*, *Royoungia*, *Sutorius*, *Tylocinum*, *Tylopilus*, *Veloporphyrellus*, and *Zangia*) and two new genera (*Abtylopilus* and *Anthracoporus*) were recognized (Table 25.1 and Fig. 25.1). Among of them, five genera [*Abtylopilus* (2 species), *Chiua* (4 species), *Hymenoboletus* (4 species), *Tylocinum* (1 species), and *Zangia* (6 species)] are currently known from China. *Tylopilus* s.s. is the most species-rich genus and account for 30% of the overall species of *Tylopilus* s.l. in China. In this book, thirty-four species are newly described, four combinations are newly proposed, and six species are reported from China for the first time. In addition, each species is documented and illustrated based on our first-hand knowledge of field work and detailed observations on herbarium specimens, along with molecular data. For the first time, we provided the images for each species to show the unbelievable high genera and species diversities of this group of boletes from China. The line drawings are all drawn by freehand from the first author, which are helpful for readers to understand the concepts of this kind of boletes. The information of type localities, distribution in China, and endemic species of each genus is listed in Table 25.1. It should be noticed that the diversity of *Tylopilus* s.l. is uneven in China, as indicated in Fig. 25.2, most species (more than 70%) are generally

Table 25.1 Information of species of each genus in *Tylopilus* s.l. from China and their type localities, distributions, and endemics.

Genus	Species	Type locality	Distribution in China	Endemics to China	Endemics to SW China	No. in the world	No. in China	No. newly described from China
Abtylopilus	*Ab. alborubellus*	China	S	Y	N	2	2	2
	Ab. scabrosus	China	SW, S	Y	N			
Anthracoporus	*An. cystidiatus*	China	C, SW	Y	N	3	3	1
	An. holophaeus	Brunei	SW	N	N			
	An. nigropurpureus	Singapore	SE, S, SW	N	N			
Austroboletus	*A. albidus*	China	SE, SW	Y	N	36	7	3
	A. albovirescens	China	SW	Y	Y			
	A. dictyotus	Indonesia	C, S	N	N			
	A. fusisporus	Japan	C, S, SW	N	N			
	A. olivaceobrunneus	China	SW	Y	Y			
	A. olivaceoglutinosus	India	SW	N	N			
	A. subvirens	Japan	SW	N	N			
Chiua	*C. angusticystidiata*	China	C, SE, S, SW	Y	N	4	4	0
	C. olivaceoreticulata	China	NE, C, SW	Y	N			
	C. virens	China	SW	Y	Y			
	C. viridula	China	C, SW	Y	N			
Fistulinella	*F. olivaceoalba*	Vietnam	C, SE, S	N	N	26	2	1
	F. salmonea	China	S	Y	N			
Harrya	*Ha. alpina*	China	SW	Y	Y	6	5	0
	Ha. atrogrisea	China	SW	Y	Y			
	Ha. chromipes	USA	C, NE, SW	N	N			

Chapter 25 Summary and conclusion   345

Continued

Genus	Species	Type locality	Distribution in China	Endemics to China	Endemics to SW China	No. in the world	No. in China	No. newly described from China
Harrya	Ha. moniliformis	China	SW	Y	Y	6	5	0
	Ha. subalpina	China	SW	Y	Y			
Hymenoboletus	Hy. filiformis	China	SW	Y	Y			3
	Hy. griseoviridis	China	C	Y	N	4	4	
	Hy. jiangxiensis	China	SE	Y	N			
	Hy. luteopurpureus	China	SW	Y	Y			
Indoporus	I. squamulosus	China	S, SW	Y	N	2	1	1
Leccinellum	L. castaneum	China	SW	Y	Y			3
	L. citrinum	China	C	Y	N			
	L. cremeum	China	SW	Y	Y	18	6	
	L. griseopileatum	China	SE	Y	N			
	L. onychinum	China	S	Y	N			
	L. sinoaurantiacum	China	SW	Y	Y			
Mucilopilus	M. cinnamomeus	China	SE, SW	Y	N	5	3	3
	M. paracastaneiceps	China	SW	Y	Y			
	M. ruber	China	SW	Y	Y			
Porphyrellus	P. castaneus	China	S, SW	Y	N			3
	P. cyaneotinctus	USA	C	N	N			
	P. griseus	China	SW	Y	Y	18	7	
	P. orientifumosipes	China	SE, SW	Y	N			
	P. porphyrosporus	Sweden	NW, C, SW	N	N			
	P. pseudofumosipes	China	SW	Y	Y			

Continued

Genus	Species	Type locality	Distribution in China	Endemics to China	Endemics to SW China	No. in the world	No. in China	No. newly described from China
Porphyrellus	P. scrobiculatus	China	SE	Y	N	18	7	3
Pseudoaustroboletus	Ps. valens	Singapore	C, S, SE, SW	N	N	1	1	0
Retiboletus	R. ater	China	SW	Y	Y	15	6	1
	R. brunneolus	China	SE, S	Y	N			
	R. fuscus	Japan	C, SW	N	N			
	R. nigrogriseus	China	S, SW	Y	N			
	R. pseudogriseus	China	SE, S	Y	N			
	R. zhangfeii	China	C, E, SE, SW	Y	N			
Royoungia	Ro. coccineinana	Malaysia	SW	N	N	6	4	0
	Ro. grisea	China	C, S, SW	Y	N			
	Ro. reticulata	China	SW	Y	Y			
	Ro. rubina	China	SE	Y	N			
Sutorius	S. alpinus	China	SW	Y	Y	13	6	2
	S. eximius	USA	NE, C	N	N			
	S. microsporus	China	SW	Y	Y			
	S. obscuripellis	Thailand	S, SW	N	N			
	S. pseudotylopilus	Thailand	SE, SW, S	N	N			
	S. subrufus	China	S, SW	Y	N			
Tylocinum	Ty. griseolum	China	SW	Y	Y	1	1	0
Tylopilus	T. albopurpureus	China	S, SW	Y	N	140	32	11
	T. alpinus	China	SW	Y	Y			
	T. argillaceus	Japan	S	N	N			

Chapter 25 Summary and conclusion   347

Continued

Genus	Species	Type locality	Distribution in China	Endemics to China	Endemics to SW China	No. in the world	No. in China	No. newly described from China
Tylopilus	T. atripurpureus	Malaysia	SW	N	N	140	32	11
	T. atroviolaceobrunneus	China	SW	Y	Y			
	T. aurantiacus	China	S, SW	Y	N			
	T. brunneirubens	Singapore	E, S, SE, SW	N	N			
	T. castanoides	Japan	C	N	N			
	T. felleus	France	N, NE, NW	N	N			
	T. fuligineoviolaceus	Japan	NE	N	N			
	T. fuscatus	Singapore	S	N	N			
	T. griseipurpureus	Singapore	S	N	N			
	T. griseiviridus	China	SW	Y	Y			
	T. griseolus	China	C, S	Y	N			
	T. himalayanus	India	C, E, NE,	N	N			
	T. jiangxiensis	China	SE	Y	N			
	T. neofelleus	Japan	C, SW	N	N			
	T. obscureviolaceus	Japan	S	N	N			
	T. olivaceobrunneus	China	E	Y	N			
	T. otsuensis	Japan	C, SW	N	N			
	T. phaeoruber	China	SW	Y	Y			
	T. plumbeoviolaceoides	China	S	Y	N			
	T. pseudoalpinus	China	SW	Y	Y			
	T. pseudoballoui	India	SW	N	N			
	T. purpureorubens	China	C, E, S	Y	N			

Continued

Genus	Species	Type locality	Distribution in China	Endemics to China	Endemics to SW China	No. in the world	No. in China	No. newly described from China
	T. rubrotinctus	China	S	Y	N			
	T. rufobrunneus	China	S	Y	N			
	T. subotsuensis	China	S	N	N			
	T. vinaceipallidus	Singapore	S, SW	N	N	140	32	11
	T. violaceobrunneus	China	E, SW	Y	N			
	T. violaceorubrus	China	SW	Y	Y			
	T. virescens	Japan	S, SW	N	N			
	V. alpinus	China	SE, SW	Y	N			
	V. castaneus	China	C, SW	Y	N			
Veloporphyrellus	V. gracilioides	China	NE, SW	Y	N	8	5	1
	V. pseudovelatus	China	SW	Y	Y			
	V. velatus	Thailand	S	N	N			
	Z. chlorinosma	China	SW	Y	Y			
	Z. citrina	China	C, SE, S, SW	Y	N			
	Z. erythrocephala	China	SW	Y	Y			
Zangia	Z. olivacea	China	SW	Y	Y	6	6	0
	Z. olivaceobrunnea	China	SW	Y	Y			
	Z. roseola	China	C, SW	Y	N			

Information of species from different geographical areas in China are also presented. (C = central China, E = eastern China, N = northern China, NE = northeastern China, NW = northwestern China, S = southern China, SE = southeastern China, SW = southwestern China; Y = yes, N = no; in this study, species in *Leccinellum* only including those that easily misidentified as *Tylopilus* s.l., while species in *Retiboletus* only including those with white or pinkish hymenophores and easily misidentified as *Tylopilus* s.l.).

distributed in southern parts of China including southern, southwestern and southeastern China. Even in the southern parts, taxa are also unevenly distributed. Yunnan, southwestern China, harbors the richest species diversity, with 66 species known from the province. Guangdong, Hainan and Sichuan are the other three top-ranked species diversity provinces and harbor 26, 21 and 17 species, respectively. Southwestern China, especially the Hengduan Mountains is the hotspot distribution area of most genera and harbors the richest species diversity with 35 species endemic to this area. Based on our statistical analyses, there are twelve genera (*Abtylopilus*, *Anthracoporus*, *Chiua*, *Hymenoboletus*, *Pseudoaustroboletus*, *Tylocinum*, *Zangia*, *Harrya*, *Royoungia*, *Veloporphyrellus*, *Mucilopilus*, and *Indoporus*) with the proportion of the native species to the total species in the world (Index Fungorum, accessed on 1 July 2021) up to or more than 50%; five genera (*Abtylopilus*, *Anthracoporus*, *Hymenoboletus*, *Mucilopilus*, and *Indoporus*) with the proportion of the newly described native species to the total species in the world up to or more than 30%; and five genera (*Hymenoboletus*, *Tylocinum*, *Zangia*, *Harrya*, and *Mucilopilus*) with the proportion of the endemic species in southwestern China to the total species in the world up to or more than 30% (Fig. 25.2).

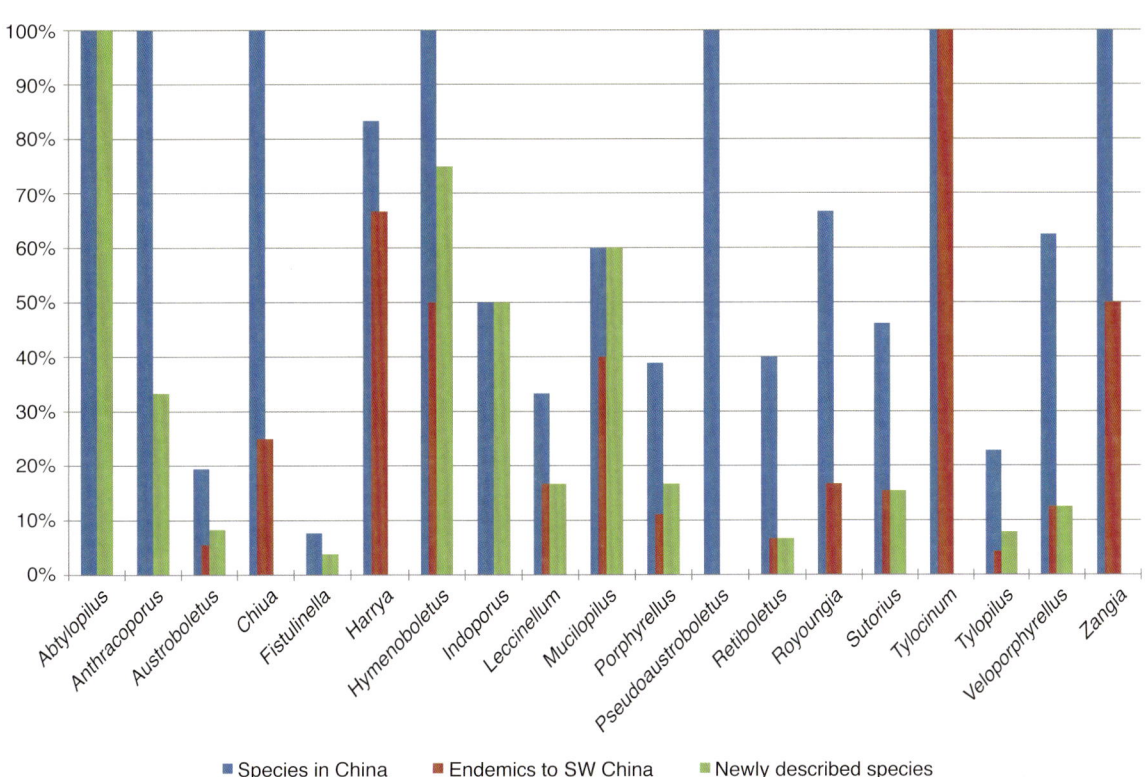

Fig. 25.1 Proportions of native species to the total species of each genus in the world (shown with blue column); proportions of newly described species from China to the total species of each genus in the world (shown with green column); proportions of endemic species of southwestern China to the total species of each genus in the world (shown with red column). Data for the information of each genus are indicated in Table 25.1. The genera of *Tylopilus* s.l. are shown on the horizontal axis and the species percentages are shown on the vertical axis.

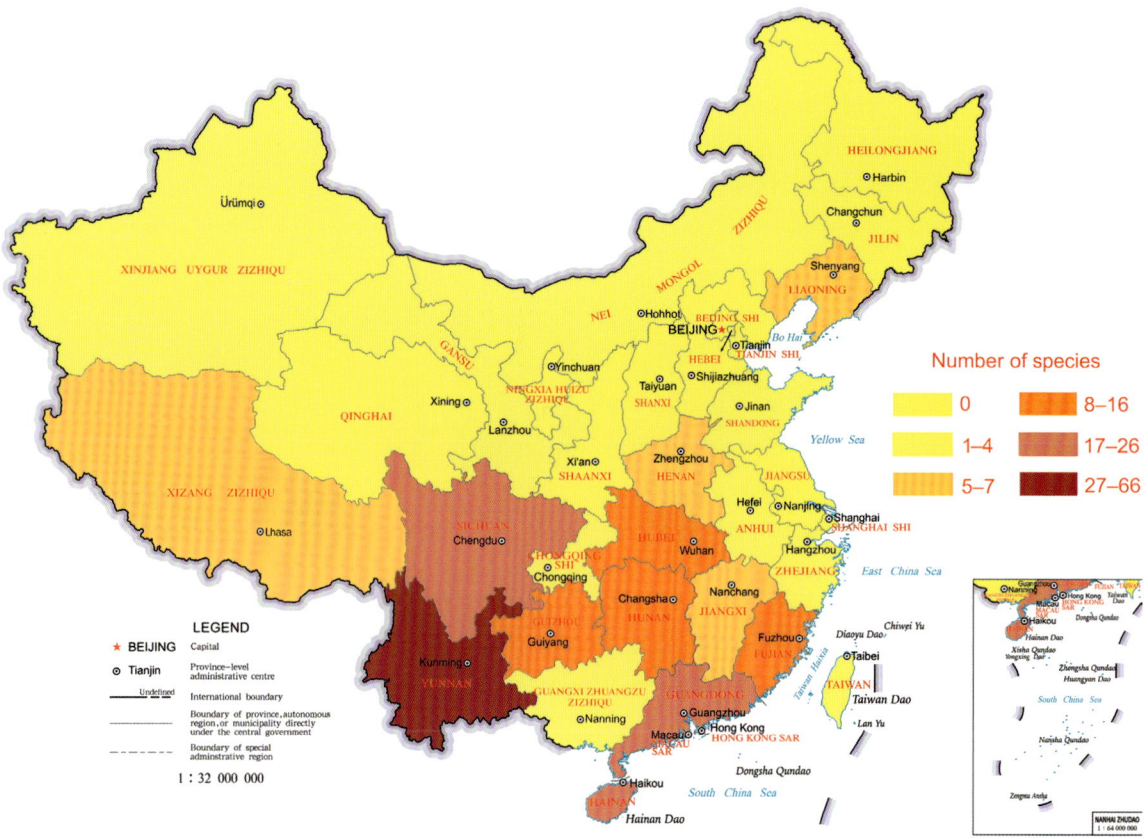

Fig. 25.2 The diversity of species of *Tylopilus* s.l. in China.

The species diversity of most genera in the central and south part of China is higher than that of the north, northeast, east, and northwest parts of China. One major reason is that central, southern, southeastern, and southwestern China has superior natural and geographical conditions. The complex topography and the suitable hydrothermal system are good for the boletes to thrive. In addition, many parts of these areas are refugia for higher plants and fungal species during Earth's recent geological history (Yang 2005; Qiu *et al.* 2011; Feng *et al.* 2012; Matheny *et al.* 2009). Thus, the high species diversity of boletes in southwestern China, especially in the Hengduan Mountains, could have resulted from the comprehensive impacts of the uplifts of mountains, the monsoon formations and the climate fluctuations in the geological history.

References

An YH, Kim YS, Seok SJ et al. (1998) Notes on Korean Strobilomycetaceae (iii) –on *Austroboletus* and the key to genera of the Strobilomycetaceae. Korean J Mycol 26(2): 230–238

Arora D, Frank JL (2014) Clarifying the butter Boletes: a new genus, *Butyriboletus*, is established to accommodate *Boletus* sect. *Appendiculati*, and six new species are described. Mycologia 106: 464–480

Baroni TJ, Both EE (1998) *Tylopilus violantinctus*, a new species of *Tylopilus* for North America, with comments on other violaceous colored *Tylopilus* taxa. Bull Buffalo Soc Nat Sci 36: 261–264

Bas C (1969) Morphology and subdivision of *Amanita* and a monograph of its section *Lepidella*. Persoonia 5(4): 285–579

Benjamin DR (1995) Mushrooms: Poisons and Panaceas—A Handbook for Naturalists, Mycologists and Physicians. WH Freeman and Company, New York

Bessette AE, Roody WC, Bessette AR (2000) North American Boletes: A Color Guide to the Fleshy Pored Mushrooms. Syracuse University Press, New York

Bessette AE, Roody WC, Bessette AR (2016) Boletes of Eastern North America. Syracuse University Press, New York

Bi ZS, Li TH (1990) New taxon and new records of the genus *Suillus* from Guangdong. Acta Microbiol Sin 9: 20–24

Bi ZS, Li TH, Zhang WM et al. (1997) A Preliminary Agaric Flora of Hainan Province. Guangdong Higher Education Press, Guangzhou

Bi ZS, Zheng GY, Li TH (1994) Macrofungi Flora of Guangdong Province. Guangdong Science and Technology Press, Guangzhou

Binder M, Besl H (2000) 28S rDNA sequence data and chemotaxonomical analyses on the generic concept of *Leccinum* (Boletales). *In*: Associazione Micologica Bresadola (ed) Micologia 2000, pp 75–86

Binder M, Bresinsky A (2002a) Derivation of a polymorphic lineage of gasteromycetes from boletoid ancestors. Mycologia 94: 85–98

Binder M, Bresinsky A (2002b) *Retiboletus*, a new genus for a species complex in the Boletaceae producing retipolides. Fedd Repert 113: 30–40

Binder M, Hibbett DS (2006) Molecular systematics and biological diversification of Boletales. Mycologia 98: 971–981

Boedijn KB (1960) The Strobilomycetaceae of Indonesia. Persoonia 1: 315–318

Both EE (1993) The Boletes of North America, A Compendium. Buffalo Museum of Science, New York

Bougher NL, Thiers HD (1991) An indigenous species of *Leccinum* (Boletaceae) from Australia. Mycotaxon 42: 255–262

Bresinsky A, Besl H (2003) Beiträge zu einer Mykoflora Deutschlands: Schlüssel zur Gattungsbestimmung der Blätter-, Leisten- und Röhrenpilze mit Literaturhinweisen zur Artbestimmung. Regensb Mykol Schr 11: 1–236

Bruns TD, Fogel R, White TJ et al. (1989) Accelerated evolution of a false—truffle from a mushroom ancestor. Nature 339: 140–142

Bulliard JBF (1788) Herbier de la France, 8. Chez l'auteur, Paris

Cai Q, Tulloss RE, Tang LP et al. (2014) Multi-locus phylogeny of lethal amanitas: implications for species diversity and historical biogeography. BMC Evol Biol 14: 143. doi: 10.1186/1471-2148-14-143

Castellano MA, Trappe JM, Malajczuk N (1992) Australasian truffle-like fungi. III. *Royoungia* gen. nov. and *Mycoamaranthus* gen. nov. (Basidiomycotina). Aust Syst Bot 5: 613–616

Chai H, Liang ZQ, Xue R et al. (2019) New and noteworthy boletes from subtropical and tropical China. MycoKeys 46: 55–96

Chakraborty D, Vizzini A, Das K (2018) Two new species and one new record of the genus *Tylopilus* (Boletaceae) from Indian Himalaya with morphological details and phylogenetic estimations. MycoKeys 33: 103–124

Chen CM, Huang HW, Yeh KW (1997a) The boletes of Taiwan (VI). Taiwania 42(2): 154–160

Chen CM, Huang HW, Yeh KW (1997b) The boletes of Taiwan (VII). Taiwania 42(3): 174–179

Chen CM, Perng JJ, Yeh KW (1997c) The boletes of Taiwan (VIII). Taiwania 42(4): 316–323

Chen CM, Perng JJ, Yeh KW (1998a) The boletes of Taiwan (IX). Taiwania 43(2): 132–139

Chen CM, Yeh KW (2000) The boletes of Taiwan (XI). Taiwania 45(2): 201–206

Chen CM, Yeh KW, Hsu HK (1998b) The boletes of Taiwan (X). Taiwania 43(2): 140–149

Chiu WF (1948) The boletes of Yunnan. Mycologia 40: 200–231

Chiu WF (1957) Atlas of the Yunnan Boletes. Science Press, Beijing

Corner EJH (1972) *Boletus* in Malaysia. Government Printer, Singapore

Corner EJH (1974) *Boletus* and *Phylloporus* in Malaysia: further notes and descriptions. Gardens' Bull Singapore 27: 1–16

Crous PW, Cowan DA, Maggs-Kölling G et al. (2020) Fungal Planet description sheets: 1112–1181. Persoonia 45: 251–409

Crous PW, Luangsa-Ard JJ, Wingfield MJ et al. (2018) Fungal Planet description sheets: 785–867. Persoonia 41: 238–417

Cui YY, Feng B, Wu G et al. (2016) Porcini mushrooms (*Boletus* sect. *Boletus*) from China. Fungal Divers 81: 189–212

Das K, Dentinger BT (2015) *Austroboletus olivaceoglutinosus*, a new mushroom species from Sikkim, India with a distinctive green, glutinous pileus. Kew Bull 70(1): 1–7

den Bakker HC, Noordeloos ME (2005) A revision of European species of *Leccinum* Gray and notes on extralimital species. Persoonia 18: 511–587

Dentinger BTM, Ammirati JF, Both EE et al. (2010) Molecular phylogenetics of porcini mushrooms (*Boletus* section *Boletus*). Mol Phyl Evol 57: 1276–1292

Desjardin DE, Binder M, Roekring S et al. (2009) *Spongiforma*, a new genus of gasteroid boletes from Thailand. Fungal Divers 37: 1–8

Desjardin DE, Wilson AW, Binder M (2008) *Durianella*, a gasteroid genus of bolete from Malaysia. Mycologia 100: 956–961

Dettman JR, Jacobson DJ, Taylor JW (2003) A multilocus genealogical approach to phylogenetic species recognition in the model eukaryote *Neurospora*. Evolution 57(12): 2703–2720

Dillenius JJ (1719) Catalogus plantarum sponte circa Gissam nascentium. Kessinger legacy reprints. Kessinger Publishing, Whitefish

Doyle JJ, Doyle JL (1987) A rapid DNA isolation procedure for small quantities of fresh leaf tissue. Phytochem Bull 19: 11–15

Eastwood DC, Floudas D, Binder M *et al.* (2011) The plant cell wall-decomposing machinery underlies the functional diversity of forest fungi. Science 333: 762–765

Feng B, Xu J, Wu G *et al.* (2012) DNA sequence analyses reveal abundant diversity, endemism and evidence for Asian origin of the porcini mushrooms. PLoS ONE 7(5): e37567. doi: 10.1371/journal.pone.0037567

Fournier E (1876) Sur Les Graminees mexicaines a sexes separes. Bull Soc Roy Bot Belgique 15 (3): 459–476

Fries EM (1821) Systema Mycologicum I. Gryphiswaldiae, Port Louis

Fries EM, Hök CT (1835) Boleti, Fungorum Generis, Illustratio: 13. Regiae Academiae, Uppsala

Frost CC (1874) Catalogue of boleti of New England, with descriptions of new species. Bull Buffalo Soc Nat Sci 2: 100–105

Fu SZ, Wang QB, Yao YJ (2006) *Tylopilus microsporus*, a new species from southwest China. Mycotaxon 96: 41–46

Fulgenzi TD, Halling RE, Henkel TW (2010) *Fistulinella cinereoalba* sp. nov. and new distribution records for *Austroboletus* from Guyana. Mycologia 102: 224–232

Fulgenzi TD, Henkel TW, Halling RE (2007) *Tylopilus orsonianus* sp. nov. and *Tylopilus eximius* from Guyana. Mycologia 99: 622–627

Gelardi M (2011) A noteworthy British collection of *Xerocomus silwoodensis* and a comparative overview on the European species of *X. subtomentosus* complex. Boll Assoc Micol Ecol Rom 84: 28–38

Gelardi M (2018) Contribution to the knowledge of Chinese boletes. II. *Aureoboletus thibetanus* s.l., *Neoboletus brunneissimus*, *Pulveroboletus macrosporus* and *Retiboletus kauffmanii* (Part I). Rivista Micologica Romana 102(3): 13–30

Gelardi M, Angelini C, Costanzo F *et al.* (2019) *Neoboletus antillanus* sp. nov. (Boletaceae), first report of a red-pored bolete from the Dominican Republic and insights on the genus *Neoboletus*. MycoKeys 49: 73–97

Gelardi M, Simonini G, Ercole E *et al.* (2014) *Alessioporus* and *Pulchroboletus* (Boletaceae, Boletineae), two novel genera for *Xerocomus ichnusanus* and *X. roseoalbidus* from the European Mediterranean basin: molecular and morphological evidence. Mycologia 106: 1168–1187

Gilbert E (1931) Les Livres du Mycologue Tome I-IV, Tom. III: Les Bolets. Paris

Giraud T, Refrégieri G, Le Gac M *et al.* (2008) Speciation in fungi. Fungal Genet Biol 45(6): 791–802

Gómez LD, Singer R (1984) *Veloporphyrellus*, a new genus of Boletaceae from Costa Rica. Brenesia 22: 293–298

Gray SF (1821) A natural arrangement of British plants. According to their relations to each other as pointed out by Jussieu, De Candolle, Brown, &c. including those cultivated for use; with an introduction to botany, in which the terms newly introduced are explained (I). Baldwin, Cradock and Joy, London

Halling RE, Baroni TJ, Binder M (2007) A new genus of Boletaceae from eastern North America. Mycologia 99: 310–316

Halling RE, Fechner N, Nuhn M *et al.* (2015) Evolutionary relationships of *Heimioporus* and *Boletellus* (Boletales), with an emphasis on Australian taxa including new species and new combinations in

Aureoboletus, Hemileccinum and *Xerocomus*. Austr Syst Bot 28: 1–22

Halling RE, Mueller GM (2005) Common mushrooms of the Talamanca Mountains, Costa Rica. Mem New York Bot Gard 90: 1–195

Halling RE, Nuhn M, Fechner NA *et al.* (2012a) *Sutorius*: a new genus for *Boletus eximius*. Mycologia 104: 951–961

Halling RE, Nuhn M, Osmundson T *et al.* (2012b) Affinities of the *Boletus chromapes* group to *Royoungia* and the description of two new genera, *Harrya* and *Australopilus*. Austral Syst Bot 25: 418–431

Han LH, Feng B, Wu G *et al.* (2018) African origin and global distribution patterns: evidence inferred from phylogenetic and biogeographical analyses of ectomycorrhizal fungal genus *Strobilomyces*. J Biogeogr 45: 201–212

Han LH, Wu G, Horak E *et al.* (2020) Phylogeny and species delimitation of *Strobilomyces* (Boletaceae), with an emphasis on the Asian species. Persoonia 44: 113–139

Heim R (1963) Diagnoses latines des espèces de champignons, ou nonda, associés à la folie du komugl tai et dundaal. Rev Mycol 28: 277–283

Heinemann P (1951) Champignons récoltés: Au congo belge par madame M Goossens-Fontana. I. Boletineae. Bull Jardin Bot Bruxelles 21: 223–346

Heinemann P, Rammeloo J (1983) Flore illustrée des champignons d'Afrique centrale, Gyrodontaceae (Boletineae). Bull Jard Bot Nat Belg 10: 173–198

Heleno SA, Barros L, Sousa MJ *et al.* (2011) Targeted metabolites analysis in wild *Boletus* species. LWT- Food Sci Techn 44: 1343–1348

Henkel TW (1999) New taxa and distribution records of *Tylopilus* from *Dicymbe* forests of Guyana. Mycologia 91: 655–665

Henkel TW, Obase K, Husbands DR (2017) New Boletaceae taxa from Guyana: *Binderoboletus segoi* gen. and sp. nov., *Guyanaporus albipodus* gen. and sp. nov., *Singerocomus rubriflavus* gen. and sp. nov., and a new combination for *Xerocomus inundabilis*. Mycologia 108: 157–173

Hennings P (1901) Fungi Camerunenses novi. III. Englers bot. Jahrbuch, 30: 43–44

Hibbett DS, Ohman A, Glotzer D *et al.* (2011) Progress in molecular and morphological taxon discovery in fungi and options for formal classification of environmental sequences. Fungal Biol Rev 25: 38–47

Hongo T (1960) Agaricales of Japan I-(2). Rhodophyllaceae, Paxillaceae, Gomphidiaceae, Boletaceae and Strobilomycetaceae. Acta Phytotax Geobot 18: 97–112

Hongo T (1963) Notes on Japanese Larger fungi (16). J Jap Bot 38: 233–240

Hongo T (1964) Notulae mycologicae (3). Mem Shiga Univ 14: 43–47

Hongo T (1966) Notulae mycologicae (5). Mem Shiga Univ 16: 57–62

Hongo T (1967) Notes on Japanese larger fungi (19). J Jap Bot 42: 151–160

Hongo T (1968) Notulae mycologicae (7). Mem Shiga Univ 18: 47–52

Hongo T (1973) Enumeration of the Hygrophoraceae, Boletaceae and Strobilomycetaceae. Bull Nat Sci Mus, Tokyo 16: 537–557

Hongo T (1974a) Notes on Japanese larger fungi (21). J Jap Bot 49: 294–305

Hongo T (1974b) Notulae mycologicae (13). Mem Shiga Univ 24: 44–51

Hongo T (1984a) Materials for the fungus flora of Japan (35). Trans Mycol Soc Japan 25: 281–285

Hongo T (1984b) On some interesting boletes from the warm-temperate zone of Japan. Mem Shiga Univ

(Nat Sci) 34: 29–32

Hongo T (1985) Notes on Japanese larger fungi. J Jap Bot 60: 370–378

Hongo T, Nagasawa E (1976) Notes on some boleti from Tottori II. Rept Tottori Mycol Inst 14: 85–89

Horak E (2011) Revision of Malaysian species of Boletales s.l. (Basidiomycota) described by E.J.H. Corner (1972, 1974). Malay Fores Rec 51: 1–283

Husbands DR, Henkel TW, Bonito G (2013) New species of *Xerocomus* (Boletales) from the Guiana Shield, with notes on their mycorrhizal status and fruiting occurrence. Mycologia 105: 422–435

Jarosch M (2001) Zur molekularen Systematik der Boletales: Coniophorineae, Paxillineae und Suillineae. Bibl Mycol 191: 1–158

Karsten PA (1881) Enumeratio Boletinearum et Pollyporearum Fennicarum systemate novo dispositarum. Rev Mycol 3: 16

Kernaghan G (2005) Mycorrhizal diversity: Cause and effect? Pedobiologia 49: 511–520

Khmelnitsky O, Davoodian N, Singh P et al. (2019) *Ionosporus*: a new genus for *Boletus longipes* (Boletaceae), with a new species, *I. australis*, from Australia. Mycol Progr 18: 439–451

Kirk PM, Cannon PF, Minter DW et al. (2008) Dictionary of the Fungi, 10th ed. CABI Publishing, Wallingford

Kornerup A, Wanscher JH (1981) Taschenlexikon der Farben. 3. Aufl. Muster-Schmidt Verlag, Göttingen

Kuntze O (1898) Revisio generum plantarum, Vol 3. Felix, Leipzig

Kuo M, Ortiz-Santana B (2020) Revision of leccinoid fungi, with emphasis on North American taxa, based on molecular and morphological data. Mycologia 112: 197–211

Lan M (1436) Diannan Materia Medica, Vol 3. Re-edited and published in 1975. Yunnan People's Publishing House, Kunming

Lannoy G, Estades A (1995) Monographie des *Leccinum* d'Europe. Fédération Mycologique Dauphiné-Savoie, France

Lebel T, Orihara T, Maekawa N (2012) The sequestrate genus *Rosbeeva* T. Lebel & Orihara gen. nov. (Boletaceae) from Australasia and Japan: new species and new combinations. Fungal Divers 22: 49–71

Lee KJ, Koo CD, Kim YS (1982) Four new species of mushrooms collected from a *Pinus rigida* stand in Suweon. Kor J Mycol 10: 125–129

Li F, Zhao K, Deng QL et al. (2016) Three new species of Boletaceae from the Heishiding Nature Reserve in Guangdong Province, China. Mycol Progr 15(12): 1269–1283

Li TH, Song B (2003) Bolete species known from China. Guizhou Sci 21: 78–86

Li TH, Song B, Shen YH (2002) A new species of *Tylopilus* from Guangdong. Mycosystema 21: 3–5

Li TH, Watling R (1999) New taxa and combinations of Australian boletes. Edinburgh Journ Bot 56: 143–148

Li YC, Feng B, Yang ZL (2011) *Zangia*, a new genus of Boletaceae supported by molecular and morphological evidence. Fungal Divers 49: 125–143

Li YC, Li F, Zeng NK et al. (2014b) A new genus *Pseudoaustroboletus* (Boletaceae, Boletales) from Asia as inferred from molecular and morphological data. Mycol Progr 13: 1207–1216

Li YC, Ortiz-Santana B, Zeng NK et al. (2014a) Molecular phylogeny and taxonomy of the genus *Veloporphyrellus*. Mycologia 106: 291–306

Li YC, Yang ZL (2011) Notes on tropical boletes from China. J Fungal Res 9: 204–211

Li YC, Yang ZL (2021) The Boletes of China: *Tylopilus* s.l. Springer, Singapore

Li YC, Yang ZL, Tolgor B (2009) Phylogenetic and biogeographic relationships of *Chroogomphus* species as inferred from molecular and morphological data. Fungal Divers 38: 85–104

Liang ZQ, Su MS, Jiang S et al. (2018) *Tylopilus callainus*, a new species with a sea-green color change of hymenophore and Context from the south of China. Phytotaxa 343: 269–276

Linnaeus C (1753) Species Plantarum. Laurentius Salvius, Stockholm

Liu HY, Li YC, Bau T (2020) New species of *Retiboletus* (Boletales, Boletaceae) from China based on morphological and molecular data. MycoKeys 67: 33–44

Magnago AC, Neves MA, Silveira RMB (2017) *Fistulinella ruschii*, sp. nov., and a new record of *Fistulinella campinaranae* var. *scrobiculata* for the Atlantic Forest, Brazil. Mycologia 109: 1003–1013

Mao XL (2000) The macrofungi in China. Henan Science and Technology Press, Zhengzhou

Massee GE (1909) Fungi exotici. 9. Bull Misc Inf. Kew 1909: 204–209

Matheny PB, Aime MC, Bougher NL et al. (2009) Out of the Palaeotropics? Historical biogeography and diversification of the cosmopolitan ectomycorrhizal mushroom family Inocybaceae. J Biogeogr 36: 577–592

Matsuura M, Yamada M, Saikawa Y et al. (2007) Bolevenine, a toxic protein from the Japanese toadstool *Boletus venenatus*. Phytochemistry 68: 893–898

Mazzer SJ, Smith AH (1967). New and interesting boletes from Michigan. Michigan Botan 6: 57e67

McNabb RFR (1967) The Strobilomycetaceae of New Zealand. New Zealand J Bot 5: 532–547

Murrill WA (1909) The Boletaceae of North America – II. Mycologia 1: 140–158

Myers N, Mittermeier RA, Mittermeier CG et al. (2000) Biodiversity hotspots for conservation priorities. Nature 403: 853–858

Nagasawa E (1997) A preliminary checklist of the Japanese *Agaricales*. I. The Boletineae. Rept Tottori Mycol Inst 35: 39–78

Nelson SF (2010) Bluing components and other pigments of boletes. Fungi 3(4): 11–14

Neves MA, Binder M, Halling RE et al. (2012) The phylogeny of selected *Phylloporus* species, inferred from nuc–LSU and ITS sequences, and descriptions of new species from the Old World. Fungal Divers 55: 109–123

Noordeloos ME, den Bakker HC, van der Linde S (2018) Boletales. In: Noordeloos ME, Kuyper THW, Somhorst I et al. Flora Agaricina Neerlandica, Vol 7. Candusso Editrice, Origgio

Nuhn ME, Binder M, Taylor AF et al. (2013) Phylogenetic overview of the Boletineae. Fungal Biolo 117: 479–511

Orihara T, Lebel T, Ge ZW et al. (2016) Evolutionary history of the sequestrate genus *Rossbeevera* (Boletaceae) reveals a new genus *Turmalinea* and highlights the utility of ITS minisatellite-like insertions for molecular identification. Persoonia 3: 173–198

Orihara T, Sawada F, Ikeda S et al. (2010) Taxonomic reconsideration of a sequestrate fungus, *Octaviania columellifera*, with the proposal of a new genus, *Heliogaster*, and its phylogenetic relationships in the Boletales. Mycologia 102: 108–121

Ortiz-Santana B, Lodge DJ, Baroni TJ et al. (2007) Boletes from Belize and the Dominican Republic. Fungal Divers 27: 247–416

Osmundson TW, Halling RE (2010) *Tylopilus oradivensis* sp nov.: a newly described member of the

Tylopilus balloui complex from Costa Rica. Mycotaxon 113(1):475–483

Parihar A, Hembrom ME, Vizzini A *et al.* (2018) *Indoporus shoreae* gen. et sp. nov. (Boletaceae) from tropical India. Cryptog Mycol 39(4): 447–466

Peck CH (1872) Report of the Botanist. Ann Rep N Y St Mus Nat Hist 24: 41–108

Peck CH (1873) Annual report of the State Botanist. Ann Rep N Y St Mus Nat Hist 25: 57–123

Peck CH (1887) Notes on the boleti of the United States. J Mycol 3: 53–55

Peck CH (1888) Annual Report of the Trustees of the State Museum of Natural History. Albany, New York

Peck CH (1898) Annual Reports of the State Botanist. Albany, New York

Pegler DN, Young TWK (1981) A natural arrangement of the Boletales, with reference to spore morphology. Trans Br Mycol Soc 76: 103–146

Redeuilh G, Soop K (2006) Nomenclature et taxinomie des genres affines à *Fistulinella* (Boletaceae) et *Fistulinella lutea* sp. nov. de Nouvelle-Zélande. Bull Soc mycol Fr 122: 291–304

Rinaldi AC, Comandini O, Kuyper TW (2008) Ectomycorrhizal fungal diversity: separating the wheat from the chaff. Fungal Divers 33: 1–45

Rostrup E (1902) Flora of Koh Chang. Contributions to the knowledge of the vegetation in the Gulf of Siam. Part. VI. Fungi. Bot Tidsskrift 24: 355–367

Saccardo PA, Saccardo D (1905) Supplementum universale. Pars VI. Hymenomycetae-Laboulbeniomycetae. Syll Fung 17: 1–991

Sánchez-Ramírez S, Tulloss RE, Amalfi M *et al.* (2015) Palaeotropical origins, boreotropical distribution and increased rates of diversification in a clade of edible ectomycorrhizal mushrooms (Amanita section Caesareae). J Biogeogr 42: 351–363

Schweinitz LD von (1822) Synopsis fungorum Carolinae superioris secundum observationes Ludovici Davidis de Schweinitz. Schriften Naturf Ges Leipzig 1: 20–131

Singer R (1942) Das System der Agaricales. II. Annales Mycologici 40: 1–132

Singer R (1945) The Boletineae of Florida with notes on extralimital species. I. The Strobilomycetaceae. Farlowia 2: 97–141

Singer R (1947) The Boletoideae of Florida with notes on extralimital species. Amer Midl Nat 37: 89–125

Singer R (1962) The Agaricales in Modern Taxonomy. 2nd ed. J Cramer, Weinheim

Singer R (1973) Notes on bolete taxonomy. Persoonia 7: 313–320

Singer R (1978) Notes on Bolete Taxonomy II. Persoonia 9: 421–438

Singer R (1986) The Agaricales in Modern Taxonomy. 4th ed. Koeltz Scientific Books, Koenigstein

Singer R, García J, Gómez LD (1991) The Boletineae of Mexico and Central America. III. Beih Nov Hedwig 102: 1–99

Smith AH, Thiers HD (1968) Notes on Boletes: 1. Generic position of *Boletus subglabripes* and *Boletus chromapes* 2. A comparison of 4 species of *Tylopilus*. Mycologia 60: 943–954

Smith AH, Thiers HD (1971) The Boletes of Michigan. The University of Michigan Press, Ann Arbor

Smith AH, Thiers HD, Watling R (1966) A preliminary account of the North American species of *Leccinum*, Section *Leccinum*. Michigan Botan 5: 131–178

Smith AH, Thiers HD, Watling R (1967) A preliminary account of the North American species of *Leccinum*, Sections *Luteoscabra* and *Scabra*. Michigan Botan 6: 107–154

Smith AH, Thiers HD, Watling R (1968) Notes on species of *Leccinum*. I. Additions to section *Leccinum*.

Lloydia 31: 252–267

Smith ME, Amses K, Elliott T *et al.* (2015) New sequestrate fungi from Guyana: *Jimtrappea guyanensis* gen. sp. nov., *Castellanea pakaraimophila* gen. sp. nov., and *Costatisporus cyanescens* gen. sp. nov. (Boletaceae, Boletales). IMA Fungus 6(2): 297–317

Snell WH (1942) New proposals relating to the genera of the Boletaceae. Mycologia 34: 403–411

Snell WH, Dick EA (1941) Notes on boletes. VI. Mycologia 33: 23–37

Snell WH, Dick EA (1970) The Boleti of Northeastern North America. Verlag von J Cramer, Lehre

Stamatakis A (2006) RAxML-VI-HPC: maximum likelihood-based phylogenetic analyses with thousands of taxa and mixed models. Bioinformatics 22: 2688–2690

Stevenson G (1962) The Agaricales of New Zealand: I. Boletaceae and Strobilomycetaceae. Kew Bull 15: 381–385

Šutara J (1989) The delimitation of the genus *Leccinum*. Czech Mycol 43: 1–12

Tai FL (1979) Sylloge Fungorum Sinicorum. Science Press, Beijing

Takahashi H (1988) A new species of *Boletus* sect. *Luridi* and a new combination in *Mucilopilus*. Trans Mycol Soc Jap 29: 115–123

Takahashi H (2002) Two new species and one new combination of Agaricales from Japan. Mycoscience 43: 397–403

Takahashi H (2004) Two new species of Agaricales from southwestern islands of Japan. Mycoscience 45: 372–376

Takahashi H (2007) Five new species of the *Boletaceae* from Japan. Mycoscience 48: 90–99

Taylor JW, Jacobson D, Kroken S *et al.* (2000) Phylogenetic species recognition and species concepts in fungi. Fungal Genet Biol 31: 21–32

Teng SQ (1963) Fungi of China. Science Press, Beijing

Terashima Y, Takahashi H, Taneyama Y (2016) The Fungal Flora in Southwestern Japan: Agarics and Boletes. Tokai University Press, Tokyo

Thiers B (2018) Index Herbariorum: a global directory of public herbaria and associated staff. New York Botanical Garden's Virtual Herbarium. http://sweetgum.nybg.org/science/ih/ (continuously updated)

Turland NJ, Wiersema JH, Barrie FR *et al.* (2017) International Code of Nomenclature for algae, fungi, and plants (Shenzhen Code) adopted by the Nineteenth International Botanical Congress Shenzhen, China, July 2017. Koeltz Botanical Books, Glashütten

Vadthanarat S, Halling RE, Amalfi M *et al.* (2021) An unexpectedly high number of new *Sutorius* (Boletaceae) species from northern and northeastern Thailand. Front Microbiol 12(no. 643505): 1–27

van der Heijden MGA, Klironomos JN, Ursic M *et al.* (1998) Mycorrhizal fungal diversity determines plant biodiversity, ecosystem variability and productivity. Nature 396: 69–72

Vasco-Palacios AM, López-Quintero C, Franco-Molano AE (2014) *Austroboletus amazonicus* sp. nov. and *Fistulinella campinaranae* var. *scrobiculata*, two commonly occurring boletes from a forest dominated by *Pseudomonotes tropenbosii* (Dipterocarpaceae) in Colombian Amazonia. Mycologia 106: 1004–1014

Vellinga EC, Kuyper TW, Ammirati J *et al.* (2015) Six simple guidelines for introducing new genera of fungi. IMA Fungus 6(2): 65–68

Vialle A, Feau N, Frey P *et al.* (2013) Phylogenetic species recognition reveals host-specific lineages among poplar rust fungi. Mol Phylogenet Evol 66(3): 628–644

Vizzini A (2014a) Nomenclatural novelties. Index Fung, 146. http://www.indexfungorum.org/names/IndexFungorumPublicationsListing.asp

Vizzini A (2014b) Nomenclatural novelties. Index Fung, 147. http://www.indexfungorum.org/names/IndexFungorumPublicationsListing.asp

Vizzini A (2014c) Nomenclatural novelties. Index Fung, 176. http://www.indexfungorum.org/names/IndexFungorumPublicationsListing.asp

Vizzini A (2014d) Nomenclatural novelties. Index Fung, 183. http://www.indexfungorum.org/names/IndexFungorumPublicationsListing.asp

Vizzini A (2014e) Nomenclatural novelties. Index Fung, 192. http://www.indexfungorum.org/names/IndexFungorumPublicationsListing.asp

Wang B, Qiu YL (2006) Phylogenetic distribution and evolution of mycorrhizas in land plants. Mycorrhiza 16: 299–363

Wang QB, Yao YJ (2005) *Boletus reticuloceps*, a new combination for *Aureoboletus reticuloceps*. Sydowia 57: 131–136

Watling R (1970) Boletaceae: Gomphidiaceae: Paxillaceae. Royal Botanical Garden Edinburgh, Edinburgh

Watling R, Li TH (1999) Australian *Boletus*, A Preliminary Survey. Royal Botanic Garden Edinburgh, Edinburgh

Watling R, Turnbull E (1994) Boletes from South & East Central Africa - II. Edinb J Bot 51: 331–353

Wolfe CB (1979a) *Austroboletus* and *Tylopilus* subg. *Porphyrellus*, with emphasis on North American taxa. Biblthca Mycol 69: 1–148

Wolfe CB (1979b) *Mucilopilus*, a new genus of the Boletaceae, with emphasis on North American taxa. Mycotaxon 10: 116–132

Wolfe CB (1981) Nomenclature and taxonomy of the tribe *Ixechineae* (Boletaceae). Taxon 30: 36–38

Wolfe CB (1982) A taxonomic evaluation of the generic status of *Ixechinus* and *Mucilopilus* (Ixechineae, Boletaceae). Mycologia 74: 36–43

Wolfe CB, Bougher NL (1993) Systematics, mycogeography, and evolutionary history of *Tylopilus* subg. *Roseoscabra* in Australia elucidated by comparison with Asian and American species. Aust Syst Bot 6: 187–213

Wolfe CB, Petersen RH (1978) The type of *Boletus fumosipes*. Mycologia 70: 676–679

Wu G, Feng B, Xu JP et al. (2014) Molecular phylogenetic analyses redefine seven major clades and reveal 22 new generic clades in the fungal family Boletaceae. Fungal Divers 69: 93–115

Wu G, Li YC, Zhu XT et al. (2016a) One hundred noteworthy boletes from China. Fungal Divers 81: 25–188

Wu G, Zhao K, Li YC et al. (2016b) Four new genera of the fungal family Boletaceae. Fungal Divers 81: 1–24

Yang ZL (2005) Diversity and biogeography of higher fungi in China. *In*: Xu J (ed). Evolutionary Genetics of Fungi. pp. 35–62. Horizon Bioscience, Norfolk (UK)

Yang ZL, Trappe JM, Binder M et al. (2006) The sequestrate genus *Rhodactina* (Basidiomycota, Boletales) in northern Thailand. Mycotaxon 96: 133–140

Yang ZL, Wu G, Li YC et al. (2021). Common Edible and Poisonous Mushrooms of Southwestern China. Science Press, Beijing

Yeh KW, Chen ZC (1980) The boletes of Taiwan (I). Taiwania 25: 166–184

Yeh KW, Chen ZC (1981) The boletes of Taiwan (II). Taiwania 26: 100–115

Yeh KW, Chen ZC (1982) The boletes of Taiwan (III). Taiwania 27: 52–63

Yeh KW, Chen ZC (1983) Boletes of Taiwan (IV). Taiwania 28: 122–127

Yeh KW, Chen ZC (1985) Boletes of Taiwan (V)—two new species. Trans Mycol Soc Repub China 1: 71–76

Ying JZ, Zang M (1994) Economic Macrofungi from Southwestern China. Science Press, Beijing

Zang M (1980) Some new species of Basidiomycetes from the Xizang autonomous region of China. Acta Microbiol Sinica 20: 29–34

Zang M (1985) Notes on the Boletales from eastern Himalayas and adjacent of China. Acta Bot Yunnanica 7: 383–401

Zang M (1986) Notes on the Boletales from eastern Himalayas and adjacent of China (2). Acta Bot Yunnanica 8: 1–22

Zang M (1996) A contribution to the taxonomy and distribution of the genus *Xerocomus* from China. Fungal Sci 11: 1–15

Zang M (1999) An annotated checklist of the genus *Boletus* and its sections from China. Fungal Sci 14: 79–87

Zang M (2006) Flora Fungorum Sinicorum. Vol 22, Boletaceae (I). Science Press, Beijing

Zang M (2013) Flora Fungorum Sinicorum. Vol 44, Boletaceae (II). Science Press, Beijing

Zang M, Li TH, Petersen RH (2001) Five new species of Boletaceae from China. Mycotaxon 80: 481–488

Zeng NK, Chai H, Jiang S *et al.* (2018) *Retiboletus nigrogriseus* and *Tengioboletus fujianensis*, two new boletes from the south of China. Phytotaxa 367: 045–054

Zeng NK, Liang ZQ, Wu G *et al.* (2016) The genus *Retiboletus* in China. Mycologia 108: 363–380

Zeng NK, Tang LP, Li YC *et al.* (2013) The genus *Phylloporus* (Boletaceae, Boletales) from China: morphological and multilocus DNA sequence analyses. Fungal Divers 58: 73–101

Zeng NK, Wu G, Li YC *et al.* (2014) *Crocinoboletus*, a new genus of Boletaceae (Boletales) with unusual polyene pigments boletocrocins. Phytotaxa 175: 133–140

Zhao K, Wu G, Yang ZL (2014) A new genus, *Rubroboletus*, to accommodate *Boletus sinicus* and its allies. Phytotaxa 188: 61–77

Zhao K, Zhang FM, Zeng QQ *et al.* (2020) *Tylopilus jiangxiensis*, a new species of *Tylopilus* s. str. from China. Phytotaxa 434 (3): 281–291

Index of scientific names

Species are listed in alphabetical order. Regular number indicate the page(s) where a taxon is cited within the text. Numbers in boldface mark the description of a taxon.

A

Abtylopilus	8, 29, 33, **36**, 43, 121, 343, 344, 349
Abtylopilus alborubellus	10, **37**, 39, 42, 344
Abtylopilus scabrosus	10, 26, 36, 39, **40**, 42, 344
Anthracoporus	8, 29, 33, **43**, 121, 343, 344, 349
Anthracoporus cystidiatus	10, 44, **45**, 47, 50, 344
Anthracoporus holophaeus	10, 43, 44, **48**, 49, 53, 344
Anthracoporus nigropurpureus	10, 25, 44, **51**, 52, 53, 344
Aureoboletus catenarius	97
Austroboletoideae	29, 89
Austroboletus	4, 8, 27, 29, 33, **54**, 343, 344
Austroboletus albidus	10, 55, 56, **57**, 58, 344
Austroboletus albovirescens	10, 25, 55, 56, **60**, 62, 72, 344
Austroboletus dictyotus	10, 55, 56, **64**, 65, 75, 344
Austroboletus fusisporus	10, 25, 26, 55, 56, **66**, 67, 344
Austroboletus longipes var. *albus*	58
Austroboletus mucosus	67
Austroboletus olivaceobrunneus	11, 55, 56, **68**, 69, 75, 344
Austroboletus olivaceoglutinosus	11, 25, 55, 56, **71**, 72, 344
Austroboletus subvirens	11, 55, 56, 70, **73**, 75, 344

B

Boletoideae	29, 36, 121
Boletus	2, 4, 8, 140
Boletus albellus	174, 175
Boletus atripurpureus	234, 235
Boletus balloui var. *fuscatus*	250, 251
Boletus brunneirubens	241
Boletus castanopsidis	170
Boletus chromipes	94, 101, 103
Boletus coccineinanus	193
Boletus dictyotus	64
Boletus edulis	1, 11, 140
Boletus eximius	201, 206
Boletus felleus	4, 223, 246, 247
Boletus griseipurpureus	253
Boletus griseus var. *fuscus*	183, 184
Boletus holophaeus	43, 48, 49
Boletus nigropurpureus	51, 52, 53
Boletus olivaceirubens	275
Boletus ornatipes	176
Boletus porphyrosporus	4, 153, 165
Boletus reticuloceps	11, 170
Boletus roseolus	340, 341
Boletus sinoaurantiacus	139, 140
Boletus subgenus *Austroboletus*	4
Boletus subgenus *Tylopilus*	4
Boletus valens	173, 174, 175
Boletus velatus	325
Boletus vinaceipallidus	301
Boletus violaceofuscus	305
Boletus virens	76, 82, 83
Boletus virescens	310, 312
Butyriboletus roseoflavus	1

C

Castellanea	9
Ceriomyces	103
Ceriomyces chromipes	101
Chiua	4, 8, 29, 34, **76**, 83, 94, 126, 343, 344, 349
Chiua angusticystidiata	11, 77, **78**, 81, 83, 85, 344
Chiua olivaceoreticulata	11, 77, 78, 79, **80**, 81, 83,

	85, 344	*Indoporus shoreae*	13, 121
Chiua virens	11, 26, 77, 79, **82**, 83, 196, 344	*Indoporus squamulosus*	13, 25, **122**, 123, 345
Chiua viridula	12, 77, 79, 83, **84**, 85, 344	*Ionosporus*	9
Costatisporus	9		

J

Jimtrappea 9

F

Fistulinella	4, 8, 29, 34, **86**, 89, 142, 343, 344
Fistulinella campinaranae	92
Fistulinella cinereoalba	89
Fistulinella lutea	131, 332
Fistulinella olivaceoalba	12, **87**, 89, 344
Fistulinella prunicolor	12, 89
Fistulinella salmonea	12, 89, **91**, 92, 344
Fistulinella staudtii	4, 86
Fistulinella viscida	12, 89

K

Krombholzia	103
Krombholzia chromipes	101

L

Leccinellum	8, 29, 35, **125**, 343, 345
Leccinellum castaneum	13, 25, 26, 126, **127**, 129, 345
Leccinellum citrinum	13, 126, **130**, 345
Leccinellum cremeum	13, 126, **133**, 345
Leccinellum griseopileatum	14, 126, **134**, 136, 137, 345
Leccinellum nigrescens	125
Leccinellum onychinum	14, 126, **137**, 345
Leccinellum sinoaurantiacum	14, 126, **139**, 140, 141, 345
Leccinum	125, 126
Leccinum cartagoense	101
Leccinum chromipes	101
Leccinum eximium	206
Leccinum section *Luteoscabra*	125

G

Gyroporus	250, 251
Gyroporus fuscatus	250, 251

H

Harrya	4, 8, 29, 34, 76, **94**, 126, 343, 344, 349
Harrya alpina	12, 95, **96**, 97, 100, 107, 344
Harrya atriceps	12, 94, 103
Harrya atrogrisea	12, 95, **99**, 100, 105, 107, 344
Harrya chromipes	12, 94, 95, **101**, 103, 344
Harrya moniliformis	12, 26, 95, 100, **104**, 105, 107, 345
Harrya subalpina	12, 95, 97, **106**, 107, 345
Hymenoboletus	4, 8, 29, 34, 76, 94, **108**, 126, 343, 345, 349
Hymenoboletus filiformis	13, 109, **110**, 112, 114, 120, 345
Hymenoboletus griseoviridis	13, 109, 110, 112, **113**, 114, 118, 345
Hymenoboletus jiangxiensis	13, 109, 114, **116**, 118, 345
Hymenoboletus luteopurpureus	13, 108, 109, 112, **119**, 120, 345

M

Mucilopilus	4, 8, 27, 29, 34, 86, **142**, 343, 345, 349
Mucilopilus castaneiceps	144, 149
Mucilopilus cinnamomeus	14, 25, **143**, 144, 345
Mucilopilus paracastaneiceps	14, 25, 26, **147**, 149, 345
Mucilopilus ruber	14, **150**, 151, 152, 345

P

Porphyrellus	4, 8, 29, 35, **153**, 343, 345
Porphyrellus castaneus	15, 25, 26, 129, 154, **155**, 156, 164, 169, 345
Porphyrellus cyaneotinctus	15, 154, 156, **157**,

I

Indoporus 4, 8, 9, 29, 33, **121**, 343, 345, 349

Porphyrellus dictyotus 158, 164, 169, 345
Porphyrellus dictyotus 54, 64
Porphyrellus fumosipes 160, 168
Porphyrellus fusisporus 66
Porphyrellus griseus 15, 154, 156, 158, **160**, 164, 169, 345
Porphyrellus holophaeus 48
Porphyrellus nigropurpureus 51, 53
Porphyrellus orientifumosipes 15, 154, 156, 160, **163**, 164, 168, 169, 312, 345
Porphyrellus porphyrosporus 15, 154, **165**, 166, 345
Porphyrellus pseudofumosipes 15, 154, 156, 164, **167**, 168, 345
Porphyrellus scrobiculatus 15, 154, **170**, 171, 346
Porphyrellus section *Graciles* 4
Porphyrellus subvirens 73
Porphyrellus viscidus 142
Pseudoaustroboletus 4, 8, 29, 34, **173**, 343, 346, 349
Pseudoaustroboletus valens 15, 26, **174**, 175, 346

R
Retiboletus 8, 29, 34, **176**, 343, 346
Retiboletus ater 177, **178**, 179, 182, 346
Retiboletus brunneolus 15, 176, 177, **180**, 182, 346
Retiboletus fuscus 15, 177, 179, 182, **183**, 184, 188, 346
Retiboletus griseus 15, 179, 184
Retiboletus nigerrimus 190
Retiboletus nigrogriseus 15, 177, **185**, 346
Retiboletus pseudogriseus 16, 177, 179, 184, **187**, 188, 346
Retiboletus zhangfeii 16, 177, 186, **189**, 190, 346
Royoungia 4, 8, 29, 34, 76, 94, 126, **191**, 343, 346, 349
Royoungia boletoides 16, 191
Royoungia coccineinana 16, 192, **193**, 346
Royoungia grisea 16, 192, **195**, 196, 346
Royoungia reticulata 16, 26, 192, **197**, 198, 346
Royoungia rubina 16, 192, 193, **199**, 346
Rubinoboletus balloui var. *fuscatus* 250

S
Strobilomyces 36, 123
Suillus eximius 206
Suillus velatus 325
Sutorius 4, 8, 29, 35, 126, **201**, 343, 346
Sutorius alpinus 16, 25, 26, 202, **203**, 205, 211, 346
Sutorius australiensis 17, 211
Sutorius eximius 17, 202, 205, **206**, 207, 346
Sutorius microsporus 17, 202, **209**, 210, 346
Sutorius obscuripellis 17, 202, 205, **212**, 214, 346
Sutorius pseudotylopilus 17, 202, 205, 214, **215**, 217, 346
Sutorius subrufus 17, 202, 205, 214, **218**, 219, 346

T
Turmalinea 9
Tylocinum 4, 8, 29, 35, **220**, 343, 346, 349
Tylocinum griseolum 17, 26, 220, **221**, 222, 346
Tylopilus 4, 8, 29, 35, **223**, 343, 346
Tylopilus alboater 17, 47
Tylopilus albopurpureus 17, 225, **227**, 229, 249, 346
Tylopilus alpinus 17, 224, **230**, 231, 263, 272, 284, 294, 346
Tylopilus argillaceus 17, 225, **232**, 269, 290, 346
Tylopilus atripurpureus 18, 225, 229, **234**, 235, 249, 281, 347
Tylopilus atroviolaceobrunneus 18, 224, 229, **236**, 237, 249, 270, 309, 347
Tylopilus aurantiacus 18, 226, **238**, 239, 240, 286, 292, 347
Tylopilus balloui 18, 260, 286
Tylopilus brunneirubens 18, 224, 231, **241**, 243, 263, 272, 274, 284, 294, 296, 347
Tylopilus callainus 310
Tylopilus cartagoensis 101
Tylopilus castanoides 18, 225, **244**, 245, 274, 279, 347
Tylopilus chlorinosmus 329, 330
Tylopilus chromipes 101
Tylopilus chromoreticulatus 82

Tylopilus coccineinanus 193
Tylopilus cyaneotinctus 157, 158
Tylopilus eximius 206
Tylopilus felleus 4, 18, 83, 224, **246**, 247, 266, 268, 294, 296, 305, 347
Tylopilus fuligineoviolaceus 18, 225, 229, **248**, 249, 347
Tylopilus fuscatus 18, 226, **250**, 251, 347
Tylopilus griseipurpureus 19, 225, 229, 249, **253**, 347
Tylopilus griseiviridus 19, 226, **255**, 257, 276, 347
Tylopilus griseolus 19, 226, 251, **258**, 259, 260, 347
Tylopilus himalayanus 19, 226, **261**, 263, 272, 274, 347
Tylopilus holophaeus 48
Tylopilus hongoi 101
Tylopilus jiangxiensis 19, 226, **264**, 265, 347
Tylopilus microsporus 267, 268
Tylopilus neofelleus 19, 26, 225, 229, 247, 249, **267**, 268, 347
Tylopilus nigropurpureus 51, 53
Tylopilus obscureviolaceus 19, 225, 229, 249, **269**, 270, 347
Tylopilus olivaceirubens 275
Tylopilus olivaceobrunneus 19, 226, 245, 257, 263, **272**, 274, 347
Tylopilus otsuensis 19, 25, 226, 257, 263, 272, 274, **275**, 276, 300, 347
Tylopilus phaeoruber 19, 225, **277**, 279, 303, 347
Tylopilus pinophilus 82
Tylopilus plumbeoviolaceoides 19, 224, 229, 249, **280**, 281, 347
Tylopilus plumbeoviolaceus 19, 235, 281
Tylopilus porphyrosporus 165
Tylopilus pseudoalpinus 20, 224, 243, **282**, 347
Tylopilus pseudoballoui 1, 20, 25, 226, 240, **285**, 286, 292, 347
Tylopilus purpureoniger 51, 53
Tylopilus purpureorubens 20, 225, 229, 249, 270, **287**, 289, 347
Tylopilus roseolus 340

Tylopilus rubrobrunneus 20
Tylopilus rubrotinctus 20, 226, 240, 286, **291**, 292, 348
Tylopilus rufobrunneus 20, 224, **294**
Tylopilus subotsuensis 20, 226, **298**, 299, 348
Tylopilus valens 174
Tylopilus velatus 325
Tylopilus vinaceipallidus 20, 225, 232, **301**, 303, 348
Tylopilus violaceobrunneus 20, 224, 229, 247, 249, 294, 296, **304**, 305, 348
Tylopilus violaceorubrus 20, 224, 229, **306**, 309, 348
Tylopilus virens 82
Tylopilus virescens 21, 225, 260, **310**, 312, 348

V

Veloporphyrellus 4, 8, 27, 29, 33, **313**, 325, 343, 348, 349
Veloporphyrellus alpinus 314, **315**, 316, 324, 348
Veloporphyrellus castaneus 314, **317**, 319, 321, 348
Veloporphyrellus gracilioides 314, 319, **320**, 321, 348
Veloporphyrellus pantoleucus 313
Veloporphyrellus pseudovelatus 314, 316, **322**, 324, 348
Veloporphyrellus velatus 314, **325**, 348

Z

Zangia 4, 8, 29, 34, 76, 94, 126, **327**, 343, 348, 349
Zangia chlorinosma 328, **329**, 330, 334, 339, 342, 348
Zangia citrina 328, **331**, 332, 339, 348
Zangia erythrocephala 328, 330, **333**, 334, 336, 341, 348
Zangia olivacea 85, 105, 328, 330, **335**, 336, 339, 348
Zangia olivaceobrunnea 328, **337**, 339, 348
Zangia roseola 327, 328, 334, 336, **340**, 341, 348
Zangioideae 29